Bioinorganic Chemistry of Nickel

Bioinorganic Chemistry of Nickel

Special Issue Editors
Michael J. Maroney
Stefano Ciurli

MDPI • Basel • Beijing • Wuhan • Barcelona • Belgrade

Special Issue Editors
Michael J. Maroney
University of Massachusetts Amherst
USA

Stefano Ciurli
University of Bologna
Italy

Editorial Office
MDPI
St. Alban-Anlage 66
4052 Basel, Switzerland

This is a reprint of articles from the Special Issue published online in the open access journal *Inorganics* (ISSN 2304-6740) 2019 (available at: https://www.mdpi.com/journal/inorganics/special_issues/bioinorganic_chemistry_nickel).

For citation purposes, cite each article independently as indicated on the article page online and as indicated below:

LastName, A.A.; LastName, B.B.; LastName, C.C. Article Title. *Journal Name* **Year**, *Article Number*, Page Range.

ISBN 978-3-03928-066-7 (Pbk)
ISBN 978-3-03928-067-4 (PDF)

© 2020 by the authors. Articles in this book are Open Access and distributed under the Creative Commons Attribution (CC BY) license, which allows users to download, copy and build upon published articles, as long as the author and publisher are properly credited, which ensures maximum dissemination and a wider impact of our publications.

The book as a whole is distributed by MDPI under the terms and conditions of the Creative Commons license CC BY-NC-ND.

Contents

About the Special Issue Editors vii

Michael J. Maroney and Stefano Ciurli
Bioinorganic Chemistry of Nickel
Reprinted from: *Inorganics* 2019, 7, 131, doi:10.3390/inorganics7110131 1

Samuel Buxton, Emily Garman, Katherine E. Heim, Tara Lyons-Darden, Christian E. Schlekat, Michael D. Taylor and Adriana R. Oller
Concise Review of Nickel Human Health Toxicology and Ecotoxicology
Reprinted from: *Inorganics* 2019, 7, 89, doi:10.3390/inorganics7070089 3

Robert J. Maier and Stéphane L. Benoit
Role of Nickel in Microbial Pathogenesis
Reprinted from: *Inorganics* 2019, 7, 80, doi:10.3390/inorganics7070080 41

Xinyue Liu and Thomas C. Pochapsky
Human Acireductone Dioxygenase (HsARD), Cancer and Human Health: Black Hat, White Hat or Gray?
Reprinted from: *Inorganics* 2019, 7, 101, doi:10.3390/inorganics7080101 72

Yusha Zhu, Qiao Yi Chen, Alex Heng Li and Max Costa
The Role of Non-Coding RNAs Involved in Nickel-Induced Lung Carcinogenic Mechanisms
Reprinted from: *Inorganics* 2019, 7, 81, doi:10.3390/inorganics7070081 85

Per E. M. Siegbahn, Shi-Lu Chen and Rong-Zhen Liao
Theoretical Studies of Nickel-Dependent Enzymes
Reprinted from: *Inorganics* 2019, 7, 95, doi:10.3390/inorganics7080095 98

Uthaiwan Suttisansanee and John F. Honek
Preliminary Characterization of a Ni^{2+}-Activated and Mycothiol-Dependent Glyoxalase I Enzyme from *Streptomyces coelicolor*
Reprinted from: *Inorganics* 2019, 7, 99, doi:10.3390/inorganics7080099 127

Brenna C. Keegan, Daniel Ocampo and Jason Shearer
pH Dependent Reversible Formation of a Binuclear Ni_2 Metal-Center within a Peptide Scaffold
Reprinted from: *Inorganics* 2019, 7, 90, doi:10.3390/inorganics7070090 144

Khadine Higgins
Nickel Metalloregulators and Chaperones
Reprinted from: *Inorganics* 2019, 7, 104, doi:10.3390/inorganics7080104 162

Yap Shing Nim and Kam-Bo Wong
The Maturation Pathway of Nickel Urease
Reprinted from: *Inorganics* 2019, 7, 85, doi:10.3390/inorganics7070085 194

Elia Barchi and Francesco Musiani
Molecular Modelling of the Ni(II)-Responsive *Synechocystis* PCC 6803 Transcriptional Regulator InrS in the Metal Bound Form
Reprinted from: *Inorganics* 2019, 7, 76, doi:10.3390/inorganics7060076 210

Marila Alfano, Julien Pérard and Christine Cavazza
Nickel-Induced Oligomerization of the Histidine-Rich Metallochaperone CooJ from *Rhodospirillum Rubrum*
Reprinted from: *Inorganics* **2019**, *7*, 84, doi:10.3390/inorganics7070084 . . . . **221**

About the Special Issue Editors

Michael J. Maroney (Professor Emeritus) was born in Ames (Iowa, USA) and received a B.S. in chemistry from Iowa State University in 1977. He received his Ph.D. from the University of Washington Seattle in 1981. Following a short stint at Chevron Research Co., in Point Richmond, CA, he did postdoctoral work at Northwestern University and at the University of Minnesota Minneapolis before joining the faculty at the University of Massachusetts Amherst in 1985.

Stefano Ciurli was born in Rosignano Marittimo (Tuscany, Italy) and received a Laurea in chemistry from the University of Pisa (Italy) in 1986, with a thesis carried out at the Department of Chemistry of Columbia University (New York, USA). He received his Ph.D. from Harvard University (Cambridge, MA, USA) in 1990. After two years of postdoctoral studies at the University of Bologna (Italy), he joined the faculty of the University of Bologna in 1992 as an associate professor before becoming a full professor of general and inorganic chemistry in 2001.

Editorial

Bioinorganic Chemistry of Nickel

Michael J. Maroney [1,*] and Stefano Ciurli [2,*]

1. Department of Chemistry and Program in Molecular and Cellular Biology, University of Massachusetts Amherst, 240 Thatcher Rd. Life Sciences, Laboratory Rm N373, Amherst, MA 01003, USA
2. Laboratory of Bioinorganic Chemistry, Department of Pharmacy and Biotechnology, University of Bologna, Viale G. Fanin 40, I-40127 Bologna, Italy
* Correspondence: mmaroney@chem.umass.edu (M.J.M.); stefano.ciurli@unibo.it (S.C.)

Received: 11 October 2019; Accepted: 11 October 2019; Published: 30 October 2019

Following the discovery of the first specific and essential role of nickel in biology in 1975 (the dinuclear active site of the enzyme urease) [1], nickel has become a major player in bioinorganic chemistry, particularly in microorganisms, having impacts on both environmental settings and human pathologies. At least nine classes of enzymes are now known to require nickel in their active sites, including catalysis of redox [(Ni,Fe) hydrogenases, carbon monoxide dehydrogenase, methyl coenzyme M reductase, acetyl coenzyme A synthase, superoxide dismutase] and nonredox (glyoxalase I, acireductone dioxygenase, lactate isomerase, urease) chemistries. In addition, the dark side of nickel has been illuminated in regard to its participation in microbial pathogenesis, cancer, and immune responses. Knowledge gleaned from the investigations of inorganic chemists into the coordination and redox chemistry of this element have boosted the understanding of these biological roles of nickel in each context. In this issue, eleven contributions, including four original research articles and seven critical reviews, will update the reader on the broad spectrum of the role of nickel in biology.

The understanding of the biological role of nickel from the inorganic chemistry side is reviewed on a theoretical basis by Siegbahn et al. [2], who discuss the enzyme mechanisms, including the canonical mechanism of urease, in view of the recently reported crystal structure of the enzyme-substrate complex [3]. This chemistry is further elucidated by original contributions on the pH dependence of binuclear nickel peptide complexes by Keegan et al. [4].

The knowledge of proteins involved in cellular nickel trafficking (metalloregulators and metallochaperones) is summarized by Higgins in a review [5], which is complemented by a second monographic article by Nim and Wong [6], that focuses more specifically on the maturation of the nickel enzyme urease as a paradigmatic example of how cells balance nickel essentiality and toxicity. These two reviews are augmented by two original research papers on this aspect of the nickel bioinorganic chemistry field: the paper by Alfano et al. [7] is focused on CooJ, an accessory protein necessary for the maturation of the nickel-dependent enzyme carbon monoxide dehydrogenase, while the paper by Barchi and Musiani [8] describes the structure-function relationships in InrS, a nickel-dependent transcription factor from cyanobacteria.

Other reviews in this issue focus on aspects of nickel in human health, with the goal of making this literature more accessible to the bioinorganic community. The general aspects of the field are surveyed by Buxton et al. [9], while a more focused review by Maier and Benoit [10] discusses the role of nickel in microbial pathogenesis. The role of noncoding RNA in nickel-induced human cancer is discussed in a review by Zhu et al. [11], while the role of human acireductone dioxygenase in human health and its metal-dependent function are discussed in the monograph by Liu and Pochapsky [12].

The range of nickel containing systems is still expanding, as demonstrated by the original research paper by Suttisansanee and Honek, which reports a preliminary characterization of a nickel activated and mycothiol-dependent glyoxalase I from fungi [13].

In conclusion, we hope that these open-access contributions will serve as guiding lights for future research into the biological role of nickel. We thank the authors for their original contributions for the special issue, and we thank the reviewers for their insightful comments on each article.

References

1. Dixon, N.E.; Gazzola, C.; Blakeley, R.; Zerner, B. Jack bean urease (EC 3.5.1.5). A metalloenzyme. A simple biological role for nickel? *J. Am. Chem. Soc.* **1975**, *97*, 4131–4132. [CrossRef] [PubMed]
2. Siegbahn, P.E.M.; Chen, S.-L.; Liao, R.-Z. Theoretical Studies of Nickel-Dependent Enzymes. *Inorganics* **2019**, *7*, 95. [CrossRef]
3. Mazzei, L.; Cianci, M.; Benini, S.; Ciurli, S. The structure of the elusive urease-urea complex unveils a paradigmatic case of metallo-enzyme catalysis. *Angew. Chem. Int. Ed.* **2019**, *131*, 7493–7497. [CrossRef]
4. Keegan, B.C.; Ocampo, D.; Shearer, J. pH Dependent Reversible Formation of a Binuclear Ni2 Metal-Center within a Peptide Scaffold. *Inorganics* **2019**, *7*, 90. [CrossRef]
5. Higgins, K. Nickel Metalloregulators and Chaperones. *Inorganics* **2019**, *7*, 104. [CrossRef]
6. Nim, Y.S.; Wong, K.-B. The Maturation Pathway of Nickel Urease. *Inorganics* **2019**, *7*, 85. [CrossRef]
7. Alfano, M.; Pérard, J.; Cavazza, C. Nickel-Induced Oligomerization of the Histidine-Rich Metallochaperone CooJ from Rhodospirillum Rubrum. *Inorganics* **2019**, *7*, 84. [CrossRef]
8. Barchi, E.; Musiani, F. Molecular Modelling of the Ni(II)-Responsive Synechocystis PCC 6803 Transcriptional Regulator InrS in the Metal Bound Form. *Inorganics* **2019**, *7*, 76. [CrossRef]
9. Buxton, S.; Garman, E.; Heim, K.E.; Lyons-Darden, T.; Schlekat, C.E.; Taylor, M.D.; Oller, A.R. Concise Review of Nickel Human Health Toxicology and Ecotoxicology. *Inorganics* **2019**, *7*, 89. [CrossRef]
10. Maier, R.J.; Benoit, S.L. Role of Nickel in Microbial Pathogenesis. *Inorganics* **2019**, *7*, 80. [CrossRef]
11. Zhu, Y.; Chen, Q.Y.; Li, A.H.; Costa, M. The Role of Non-Coding RNAs Involved in Nickel-Induced Lung Carcinogenic Mechanisms. *Inorganics* **2019**, *7*, 81. [CrossRef]
12. Liu, X.; Pochapsky, T.C. Human Acireductone Dioxygenase (HsARD), Cancer and Human Health: Black Hat, White Hat or Gray? *Inorganics* **2019**, *7*, 101. [CrossRef]
13. Suttisansanee, U.; Honek, J.F. Preliminary Characterization of a Ni2+-Activated and Mycothiol-Dependent Glyoxalase I Enzyme from Streptomyces coelicolor. *Inorganics* **2019**, *7*, 99. [CrossRef]

© 2019 by the authors. Licensee MDPI, Basel, Switzerland. This article is an open access article distributed under the terms and conditions of the Creative Commons Attribution (CC BY) license (http://creativecommons.org/licenses/by/4.0/).

Review

Concise Review of Nickel Human Health Toxicology and Ecotoxicology

Samuel Buxton *, Emily Garman, Katherine E. Heim, Tara Lyons-Darden, Christian E. Schlekat, Michael D. Taylor and Adriana R. Oller *

NiPERA Inc., 2525 Meridian Pkwy Ste 240, Durham, NC 27713, USA
* Correspondence: sbuxton@nipera.org (S.B.); aoller@nipera.org (A.R.O.)

Received: 24 May 2019; Accepted: 4 July 2019; Published: 12 July 2019

Abstract: Nickel (Ni) metal and Ni compounds are widely used in applications like stainless steel, alloys, and batteries. Nickel is a naturally occurring element in water, soil, air, and living organisms, and is essential to microorganisms and plants. Thus, human and environmental nickel exposures are ubiquitous. Production and use of nickel and its compounds can, however, result in additional exposures to humans and the environment. Notable human health toxicity effects identified from human and/or animal studies include respiratory cancer, non-cancer toxicity effects following inhalation, dermatitis, and reproductive effects. These effects have thresholds, with indirect genotoxic and epigenetic events underlying the threshold mode of action for nickel carcinogenicity. Differences in human toxicity potencies/potentials of different nickel chemical forms are correlated with the bioavailability of the Ni^{2+} ion at target sites. Likewise, Ni^{2+} has been demonstrated to be the toxic chemical species in the environment, and models have been developed that account for the influence of abiotic factors on the bioavailability and toxicity of Ni^{2+} in different habitats. Emerging issues regarding the toxicity of nickel nanoforms and metal mixtures are briefly discussed. This review is unique in its covering of both human and environmental nickel toxicity data.

Keywords: nickel; bioavailability; carcinogenicity; genotoxicity; allergy; reproductive; asthma; nanoparticles; ecotoxicity; environment

1. Nickel Occurrence and Uses

Nickel (Ni) is a naturally occurring element and is found in abundance in the earth's crust and core. Nickel occurs in air, water, sediments, and soil from various natural sources and anthropogenic processes. Nickel is introduced into the environment and is circulated through the system by chemical and physical processes and through biological transport mechanisms of living organisms [1]. Nickel is essential for the normal growth of many species of microorganisms and plants [1].

Nickel exists in nature mainly in the form of sulfide, oxide, and silicate minerals, and is an important commercial element in industrialized societies. Thus, human and environmental Ni exposures are ubiquitous. Anthropogenic nickel releases to the environment occurs locally from emissions of metal mining, smelting, and refining operations; from industrial activities, such as nickel plating and alloy manufacturing; from land disposal of sludges, solids, and slags; and from disposal as effluents. Other diffuse sources may arise from combustion of fossil fuels, waste incineration, and wood combustion.

Nickel compounds can be water-insoluble, like oxidic (such as black nickel oxide) and sulfidic (such as nickel subsulfide), with the latter being sparingly soluble in some media. A third group of nickel compounds are water-soluble (such as nickel sulfate). Metallic nickel (nickel metal and alloys) are nickel substances with very low or no water solubility.

The important uses of nickel substances in transportation products, aerospace equipment, paints, ceramics, medical applications, electronics, food and beverage production, batteries, chemicals, and many other uses indicate that potential exposure to nickel metal, nickel compounds, and nickel-containing alloys is wide ranging. Although many sources and types of exposure exist, potential toxicity is dependent on the physico-chemical characteristics of the nickel substance, as well as the amount, duration, and route of exposure.

2. Nickel Exposure to Humans and Toxicokinetics

2.1. Nickel Exposures

2.1.1. Occupational

Occupational exposure to Ni is primarily associated with workers in the nickel-producing and nickel-using industry sectors. Industries associated with nickel production include mines, mills, refineries, and smelters, whereas, nickel using/processing industries include alloy and stainless-steel production, catalysts, pigments, batteries, and electroplating. In addition, many industrial processes can generate Ni exposures such as welding and grinding of Ni-containing alloys [2–4]. Workplace exposure is primarily to airborne nickel; inhalation is therefore the major route of exposure of toxicological importance in occupational settings. To a lesser extent, skin contact can occur during certain processes or physical handling of nickel and/or nickel containing-products [3,4]. Oral exposure also occurs as a consequence of swallowing inhaled coarse particles or through hand to mouth contact.

2.1.2. General Public

Nickel is naturally present in the air, soil, water, plants, and various foods. Nickel in the air can result from forest fires, volcanoes, and anthropogenic activities [4,5]. Anthropogenic activities can also contribute to nickel levels in the soil, water and plants; their Ni levels are generally higher near industrial sites involved with the mining and processing of nickel. The primary source of nickel exposure for the general public is via dietary intake from foods like chocolate, coffee, teas, legumes, and nuts that tend to have naturally higher Ni levels and to a lesser extent from drinking water [2,5]. Other nickel exposure sources for the general public include commonly used products like cooking pots and pans, jewelry, and medical devices like dental appliances and joint prostheses [4,6]. Exposure of the public to nickel and its compounds is generally low [7].

2.2. Toxicokinetics and Bioavailability of Nickel

The absorption, distribution and elimination of nickel is affected by factors like route of exposure, physical form of the material (massive or powder), metal release and in the case of dusts or powders, the aerodynamic size of the nickel particles. While most of the historically available information relates to micron-size particles of nickel-containing substances, recent studies have looked at the toxicokinetics of nickel nanoparticles to characterize how they differ from those of the corresponding micron size. This is further discussed under Section 5.2.

2.2.1. Gastrointestinal

Gastrointestinal absorption of nickel comes from nickel present in ingested beverages, drinking water and foods. Nickel is naturally present in foodstuff as it is essential to plants. Ingestion of soil is also possible, particularly in small children. For the general public, oral ingestion of nickel is the most relevant exposure pathway for systemic absorption and toxicity. In occupational settings, mucociliary clearance of inhaled nickel dust that is swallowed can contribute an appreciable amount to absorption via the gastrointestinal tract [2]. The gastrointestinal absorption and bioavailability of ingested nickel is affected by the type of matrix (food, soil) ingested and the prior presence of food in the stomach. For example, the absorption of nickel in fasted subjects given water soluble nickel sulfate in drinking water was higher (up to 27%) than in subjects given nickel sulfate with food (0.7–5%) [8,9].

Other factors like the chemical form of nickel also affect absorption. Generally, water soluble nickel compounds have a greater oral absorption than poorly soluble nickel substances.

2.2.2. Respiratory

In toxicological terms, inhalation is the most important exposure route for nickel particles in occupational settings. Absorption of nickel particles deposited in the nasopharyngeal, tracheobronchial, or alveolar regions of the respiratory tract is dictated by several factors, such as the particles aerodynamic diameter (d_{ae}), solubility, surface area, amount deposited, ventilation rate, clearance, and retention rates [10–12]. Only particles sufficiently small (<100 μm d_{ae}) can enter the respiratory tract and be inhaled. An important first step in inhalation absorption is particle deposition. Particle size dictates both the depth of deposition along the respiratory tract and the subsequent absorption. Particles deposited in the lower regions of the respiratory tract (alveolar) are predominantly ≤4 μm d_{ae}; particles deposited in the tracheobronchial region are 4–10 μm d_{ae} and particles deposited in the nasopharyngeal region are between 10 μm and 100 μm. Less than or equal to 10% of inhaled respirable size aerosol are deposited in the pulmonary region of the human respiratory tract [13]; the fraction is even lower for workplace aerosols. While this small percentage is expected to be cleared via dissolution, macrophages, or lymph nodes, the mucociliary clearance of the majority of deposited undissolved particles leads to their expectoration or swallowing, contributing to the gastrointestinal absorption. Similar to gastrointestinal absorption, the water-soluble nickel compounds are more readily absorbed in the respiratory tract than the poorly soluble compounds [5]. Animal studies with respirable size aerosols have shown that the poorly soluble nickel particles have long lung retention times and slow clearance, and thus accumulate over time.

2.2.3. Dermal

A very small fraction of nickel that is dermally exposed is absorbed. Following dermal exposure, nickel ions (Ni^{2+}) and particles can penetrate the skin, especially at sweat ducts and hair follicles. Here too, particle size is a limiting factor for absorption; smaller sized particles are absorbed more readily than larger sized particles. Additionally, dermal absorption of nickel is affected by solubilizing agents like detergents, solvents, and the presence of barriers to the skin, such as clothes or gloves [14–17]. During exposure to metallic nickel in massive forms (e.g., jewelry) corrosion and Ni^{2+} ion release must occur prior to absorption. Research has shown that approximately <2% of soluble compounds [18,19] and <0.2% of metallic and insoluble nickel [20] is absorbed.

2.3. Distribution, Metabolism and Excretion of Nickel

The distribution and elimination of nickel is influenced by the route of administration and binding to proteins. Nickel in the bloodstream is bound to albumin and metalloproteins, which modulates their tissue distribution and elimination. Postmortem analysis of nickel in human tissues shows that the highest amounts of absorbed nickel is distributed to the lungs, thyroid glands, and adrenal glands, with lesser amounts to brain, kidneys, heart, liver, spleen, and pancreas [2,21]. Inhaled nickel is predominantly distributed in the respiratory tract (lungs, nasal sinus), followed by the kidneys [22]. Inhaled soluble nickel is eliminated primarily in the urine, while mucociliary clearance leads to a fraction of the inhaled poorly soluble nickel particles being eliminated in the feces. Orally absorbed nickel is distributed to the kidneys, followed by the liver, brain, and heart [23]. Nickel absorbed via the gastrointestinal tract is excreted predominantly in urine; unabsorbed nickel is eliminated with the feces. Hair is another distribution and elimination tissue for absorbed nickel. Nickel can also be eliminated via sweat and human breast milk [24–26]. The majority of dermally exposed nickel is not absorbed and thus not available for distribution.

3. Toxicity of Nickel

3.1. Toxicity and Nickel Ion

As with other metals, the toxicity of nickel-containing substances is considered to be related to the bioavailability of the metal ion (Ni^{2+}) at systemic or local target sites [10]. The main human health effects of concern associated with Ni exposure include nickel allergic contact dermatitis, respiratory carcinogenicity, reproductive toxicity and non-cancer respiratory effects.

3.2. Nickel Allergic Contact Dermatitis (NACD)

3.2.1. Prevalence in General and Clinical Populations

Nickel is one of the most common causes of allergic contact dermatitis (ACD). An estimated 12–19% of females and 3–6% of males in the general population are allergic to nickel (i.e., nickel-sensitized) [27]. Higher percentages are recorded in dermaotology clinics [28]. The reason for the relatively high prevalence of nickel sensitization is due to the use of nickel-releasing consumer items that come in direct and continuous prolonged contact with the skin. Although exposure may occur in some occupational settings (generally associated with soluble nickel salts), the marked prevalence of nickel sensitization in the general population is primarily due to consumer dermal exposure to nickel released from articles (e.g., in jewelry, watches, eyeglasses) that are made of nickel-plated materials or high nickel-releasing alloys.

3.2.2. Induction vs. Elicitation

Many chemical agents, including nickel, can cause allergic contact dermatitis (ACD) which results in inflammation of areas of the skin in sensitized (i.e., allergic) individuals. While nickel ACD can cause pain, inflammation and discomfort, it is not life threatening because it causes a delayed-type allergy (type 4), which cannot trigger anaphylactic shock, contrary to some other types of allergies (type 1, 2, or 3). The development of nickel ACD requires that an individual become immunologically sensitized to nickel. This is termed the induction phase or sensitization phase and the length of this phase varies between individuals. It can range from 1–3 weeks to develop, following days to weeks of prolonged intimate contact in a piercing or on the skin with a nickel-containing article that has released a sufficient amount of solubilized Ni^{2+} onto the skin. The quantity of Ni^{2+} that is sufficient to induce sensitivity varies with the individual. If the skin is already damaged, sensitization may be induced more quickly and by lower amounts of the solubilized Ni^{2+}. Temperature, the presence of other allergic conditions, gender, and age may also be determining factors for (1) susceptibility, (2) the amount of Ni^{2+} required for a reaction, and (3) the time to develop sensitization to nickel. Induction of nickel sensitization most commonly originates from body piercing but is also more likely if skin exposure to Ni is combined with irritants and/or moisture that could also compromise the skin barrier.

A nickel-sensitized individual, when re-exposed to Ni^{2+} on the skin in sufficient amounts, may have an allergic response within several hours. This is termed the elicitation phase, which often occurs at a lower concentration of Ni^{2+} than required for inducing sensitization in the first place. The elicitation of nickel ACD usually only occurs at the site of exposure but it can also occur in skin remote from the current site of contact with nickel, (e.g., at the location where previous nickel sensitization reactions have occurred) [29].

While oral systemic elicitation of ACD in individuals previously sensitized by direct and continuous prolonged skin contact is well documented to occur in a small proportion of nickel-sensitized individuals (e.g., hypersensitive people), there exists some controversy about the ability to sensitize individuals when nickel exposure is oral, intravenous, or inhaled [30].

3.2.3. Mechanisms of Nickel ACD

Nickel ions released from nickel compounds, nickel metal, and various alloys may trigger skin reactions when they are absorbed into the skin. These Ni^{2+} ions can then bind to and activate epithelial cells such as Langerhans or dendritic cells in the basal layer of the epidermis (see Figure 1). These cells produce cytokines or chemokines, triggering complex immune reactions that activate antigen-presenting cells and T cells [31–33]. As part of this process, migration of activated antigen-presenting cells to the draining lymph nodes occurs, where the bound nickel, as a hapten, is presented to the naive CD4-positive T cells [34]. Nickel differs from classical haptens by its ability to form coordinative bonds with proteins and to directly activate human innate immune cells via the toll-like receptor (TLR) 4 [35].

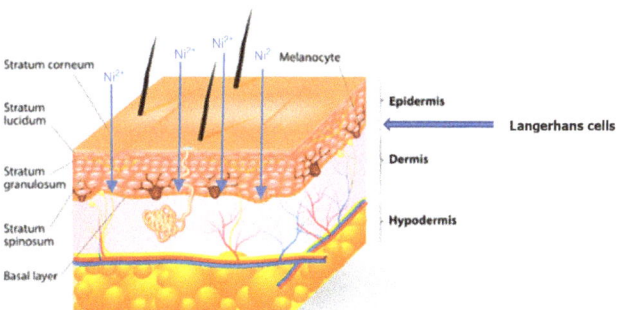

Figure 1. Nickel ions from exposure to solubilized nickel compounds or released from corrosion of nickel metal or alloys must cross the skin barrier and reach the basal layer of the epidermis to cause an allergic reaction. Modified with permission from original image: Designua/Shutterstock.com.

Future exposure to nickel in sufficiently high amounts (above threshold) would lead to the activation of the nickel-specific T-cells. Migration of these cells into the bloodstream triggers visible signs of allergic reactivity after hours of Ni^{2+} exposure [36]. The exact sequence of events and interactions between antigen presenting and immune cells involved in nickel allergy are still being elucidated.

3.2.4. Sources of Exposure: Nickel Release versus Content

Nickel ACD was first noticed in occupational settings where soluble forms of nickel came into contact with worker's skin [37]. Individuals working in electroplating shops, in battery manufacturing, and with nickel catalysts were the most susceptible to nickel ACD. However, workplace-related nickel dermatitis is now relatively rare due to improved production processes and occupational hygiene measures that limit exposure.

Non-occupational nickel sensitization is well documented. It was first observed in individuals who had skin contact with suspenders in the 1950s–1960s, followed by jean buttons and zippers, then nickel-releasing ear-piercing items and nickel-plated jewelry [38]. The significant differences in prevalence between females and males is sometimes correlated with the much higher prevalence of ear-piercing among women, but other factors such as hormone differences and the tendency for young women to wear more and/or low-quality jewelry than males may also play a role [39].

The release of Ni^{2+} is necessary for causing nickel sensitization and nickel ACD, which are threshold effects (requiring release of ions above a specific amount to cause a reaction). Alloys such as many stainless steels contain nickel but do not release a sufficient amount of Ni^{2+} to cause an individual to become nickel sensitized or elicit a nickel ACD reaction if they are nickel-sensitized.

To have nickel release from metallic nickel or nickel alloys, the nickel metal must be corroded and the corrosion product dissolved into Ni^{2+}. For this reason, sweat or other wet conditions can increase the release rate compared to dry conditions.

The risk of nickel sensitization or elicitation of nickel ACD can be managed and minimized through reduced exposure to nickel-releasing items. In the workplace, exposure reduction includes personal protective equipment and other risk management measures. For consumers, exposure can be reduced through avoidance of direct and continuous prolonged exposure to items releasing nickel in amounts greater than the threshold for nickel ACD, and switching to items made from surgical stainless steel (AISI 316L) and other low nickel-releasing alloys, or non-nickel-containing materials.

Accordingly, the European Union (EU) nickel restriction (REACH, Entry 27 of Annex XVII) [40] is based on nickel release, rather than nickel content. Articles intended to come into direct and continuous prolonged contact with the skin must release less than 0.5 µg Ni/cm^2 surface area of the item per week using the standardized methodology to assess conformity and compliance with this specific regulation (EN1811:2011+A1) [41]. Items known to be associated with nickel ACD that are included under the EU nickel restriction include necklaces, bracelets and chains, anklets, finger rings, wrist-watch cases, watch straps, rivet buttons, tighteners, rivets, earrings, and zippers.

3.2.5. Susceptible Populations

Nickel sensitization is not an inherited condition. It is related to direct and continuous prolonged skin contact (i.e., exposure) to materials releasing an amount of nickel sufficiently high to cause sensitization reactions, be that nickel metal, nickel alloys, or soluble nickel salts.

A common cause of nickel sensitization and nickel ACD is body piercing, which involves inserting high nickel-releasing studs into the wound to prevent closure during healing and bypassing the skin barrier. Once healed, with the stud removed, additional contact with nickel in the pierced area may occur by wearing jewelry or posts in piercings that release a significant amount of Ni^{2+}.

Individuals who have reactions to other allergens, have overly sensitive skin or other skin diseases, and individuals who sweat excessively have been considered to be more susceptible to nickel allergy. This susceptibility is not unique to nickel but is rather a function of increased immunological reactivity, decreased skin barrier function, or increased corrosion due to sweating. These individuals would be more likely to attend dermatology clinics for treatment of nickel allergy and other skin problems, and would have to avoid not only high nickel-releasing items but also other skin allergens and irritants.

A very small part of the nickel-allergic population is hypersensitive to nickel. These individuals react to lower concentrations of nickel on the skin than most nickel-sensitive individuals. A small fraction of these people also react to oral nickel exposure. Prevention of elicitation in these individuals is important and is done on a case-by-case basis. Regulation and prevention of nickel sensitization and nickel ACD of the general population is not intended to protect these hypersensitive individuals [42], as they are a small subset of the general population and may need more specific medical advice. While a low-nickel diet is helpful for some of these individuals who react to ingestion of nickel [43], oral hyposensitization, using gradually increasing low doses of nickel has also been shown to increase the amount of nickel needed to cause a nickel allergic reaction [44].

3.3. Nickel Carcinogenicity

Nickel carcinogenicity is an occupational concern due to the required inhalation route of exposure and high exposure levels. The evidence for carcinogenicity, or lack thereof, of nickel metal and nickel compounds come from epidemiological and animal (rats and mice) studies. These studies indicate that the inhalation route is the exposure route of concern and the respiratory system (lungs and nasal sinus) the target organ for carcinogenicity of nickel compounds. The human and animal evidence supports the respiratory carcinogenicity of nickel compounds but do not identify nickel metal as a respiratory carcinogen. The hazard classifications reflect this difference in carcinogenicity between nickel metal and nickel compounds.

Under the European Union Classification, Labeling and Packaging (CLP) legislation, many soluble and insoluble nickel compounds are classified as *Carc 1A*, stating that these compounds are *known to have carcinogenic potential for humans*, based largely on human evidence. This classification specifies inhalation as the only route of concern [45]. Nickel metal is classified as *Carc 2, suspected human carcinogen based on* insufficient evidence from human studies with suggestive evidence from animal studies via non-relevant routes of exposure. Likewise, the International Agency for Research on Cancer (IARC) classified soluble and insoluble nickel compounds under *Group 1, carcinogenic to humans*, and nickel metal and alloys under *Group 2B, possibly carcinogenic to humans* [46].

3.3.1. Human and Animal Evidence for Nickel Carcinogenicity

High exposures to mixtures of water-soluble and complex-insoluble nickel compounds in workers involved with mining, refining, and processing of sulfidic nickel ores have been associated with excess respiratory cancer risks. No excess respiratory cancer risks in workers at lateritic ore refineries, alloy manufacturing, or electroplating have been observed. A seminal comprehensive study by the International Committee on Nickel Carcinogenesis in Man (ICNCM) examining cancer risks in 10 cohorts of about 80,000 nickel processing and nickel alloy production workers reported an association between exposure to certain sulfidic, oxidic and water-soluble nickel compounds, and respiratory cancer of the lungs and nasal sinus; no association with exposure to metallic nickel was identified [47]. Among the nickel compounds, different chemical forms appear to have different carcinogenic potentials and potencies in the human studies.

Animal studies are useful in elucidating mechanisms of carcinogenesis and determining the source of the carcinogenicity observed in humans (mixed exposures) to specific nickel substances. There are eight relevant lifetime inhalation and oral carcinogenicity studies in rats and mice [48–52]. The animal studies support the conclusions from human studies that the inhalation route and the respiratory tract are the relevant exposure route and target organ, respectively, for nickel compounds carcinogenicity. No carcinogenicity is associated with the oral exposure route. A recent review of the human and animal evidence for the respiratory carcinogenicity of nickel metal and nickel compounds is provided in the European Chemicals Agency (ECHA) background document in support of occupational exposure limit values [2].

In interpreting nickel carcinogenicity studies, it is important to realize that exposures in animal studies are to a "pure" nickel compound (a single nickel compound), while exposures in human epidemiological studies are to mixtures of nickel compounds (plus other inorganic compounds). Any potential co-carcinogenic or promoting effect of the different nickel compounds and other inorganic compounds (e.g., arsenic, acid mists) in the human studies will not exist in the single exposure animal studies.

There is generally a good correlation between the human occupational exposure studies and animal studies on the carcinogenicity of nickel and nickel compounds. The evidence from both human and animal studies point to the absence of carcinogenic effects of nickel metal but the presence of carcinogenic effects for sulfidic and oxidic nickel compounds. The only inconsistency between the human and animal evidence relates to the carcinogenicity of soluble nickel compounds [53]. The animal studies have failed to show carcinogenic effect of pure soluble nickel compounds following inhalation and oral exposures. In the human studies, an association between inhalation exposure to soluble nickel (with additional exposures to insoluble nickel compounds) and/or smoking and lung cancer was observed in some groups of workers.

3.3.2. Inhalation Exposure Route

The 1990 ICNCM report [47] concluded that inhalation exposure to mixtures of water-soluble nickel compounds (e.g., nickel sulfate, nickel chloride) and water insoluble nickel compounds (e.g., nickel subsulfide, nickel oxide, complex Ni-Cu oxides) were associated with excess respiratory cancer risk in workers. Much of the excess respiratory cancer risk was associated with exposure to high concentrations

($\geq$1 mg Ni/m^3) of soluble compounds or ($\geq$10 mg Ni/m^3) of a mixture of sulfidic and oxidic nickel compounds. Excess nasal and/or lung cancer risks have been observed in different cohorts of workers. More recently, analyses of dose-responses for the main chemical forms of nickel (soluble, oxidic and sulfidic compounds) that included 13 cohorts of nickel workers (~100,000 workers), indicated that no excess cancer risk were observed in these studies when exposures to nickel in the inhalable aerosol fraction were kept $\leq$0.1 mg Ni/m^3 [13].

The ICNCM report and subsequent studies found no association between metallic nickel and excess risks of lung or nasal cancer [54]. In two reports where hints of possible correlations between excess cancer risk and nickel metal exposure have been indicated, failure of cross-validation of the test model and non-significant odds ratio after adjusting for confounding exposures suggest that the risks for metallic nickel may have been overestimated [55,56]. No association has been found between increased respiratory cancer risk and inhalation exposure to metallic nickel outside the nickel refineries, when local populations were used as controls [57,58]. More recent cohort studies support the lack of association between exposure to metallic nickel and excess respiratory cancer [59–63].

In rats and mice, seven lifetime carcinogenicity studies via the inhalation route have been conducted in which the maximum tolerated doses were reached [48–50,52,64]. A 30-month inhalation study with nickel metal powder in Wistar rats did not increase the incidence of lung tumors [52] consistent with an earlier study in rats and mice, that, although compromised by high mortality, also suggested that metallic nickel did not cause cancer [65]. While positive results have been found in one intratracheal instillation study of nickel metal in rats [66], the doses in that study were shown to not be achievable via the normal inhalation route; intratracheal instillation is a non-physiological and non-relevant exposure route for workers and the public. Other inhalation studies in guinea pigs and hamsters have buttressed the negative carcinogenicity of metallic nickel [65,67]. Thus, there has been a consistent lack of increased respiratory cancer risk associated with metallic nickel exposures in animals and humans.

The ability of nickel substances to induce respiratory tumors after inhalation may be related to the bioavailability of the Ni^{2+} ions at target sites within epithelial cells (Figure 2). The bioavailability of Ni^{2+} ions in the nucleus of target respiratory epithelial cells is not dictated by just the water solubility of the nickel particle but by the interplay of factors like respiratory toxicity, extracellular and intracellular dissolution, and lung clearance [10]. According to the bioavailability model, it is the delivery of sufficient (above threshold) Ni^{2+} ions to the nucleus of respiratory epithelial cells, which is influenced by other factors, that governs the respiratory carcinogenicity of nickel.

Nickel subsulfide exposure induced lung tumors in male and female rats at levels $\geq$0.1 mg Ni/m^3 respirable aerosol, but not in mice. At the lowest concentration level at which increased tumors were detected, the incidence and severity of chronic lung inflammation in rats was similar between nickel sulfate and subsulfide, even though the tumor outcome was different. According to the Ni^{2+} bioavailability model, nickel subsulfide has a high carcinogenic potential due to its low extracellular, but high intracellular dissolution, resulting in the highest dose of bioavailable Ni^{2+} in the nucleus of respiratory epithelial cells.

Green nickel oxide induced some tumors in rats, but only at higher exposures ($\geq$1.0 mg Ni/m^3 respirable aerosol) and with no clear dose-response [49]. The low extracellular but medium intracellular dissolution of nickel oxide results in the delivery of above-threshold doses Ni^{2+} ions to the nucleus of respiratory epithelial cells, causing the tumors observed in rats. Nickel oxide had equivocal evidence of tumors in female mice. Perhaps mice are not very susceptible to the carcinogenic effects of nickel even tough other metals, such as cobalt, have been able to induce excess lung tumors in mice [68,69]. Another possibility is that mice have higher thresholds for the nickel cancer-causing mechanisms.

Nickel sulfate inhalation exposures in rats have not induced tumors at exposure levels up to 0.11 mg Ni/m^3 respirable aerosol, and 0.22 mg Ni/m^3 respirable aerosol in mice. A plausible explanation for this is that nickel sulfate is not carcinogenic by itself at the exposure levels that can be tolerated by rats without overt toxicity; in human studies exposures are mixed and soluble Ni can enhance the lung carcinogenicity of more insoluble Ni compounds through inflammatory and proliferative effects. However, the lack of tumors in rats is likely due to the high extracellular but low intracellular dissolution of nickel sulfate that results in the delivery of below threshold or insufficient dose of Ni^{2+} ions to the nuclei of respiratory epithelial cells.

Likewise, metallic nickel may have failed to induce lung tumors in rats following inhalation exposure due to its high extracellular but low intracellular dissolution, resulting in low bioavailability of Ni^{2+} ions in the nucleus of target respiratory epithelial cells.

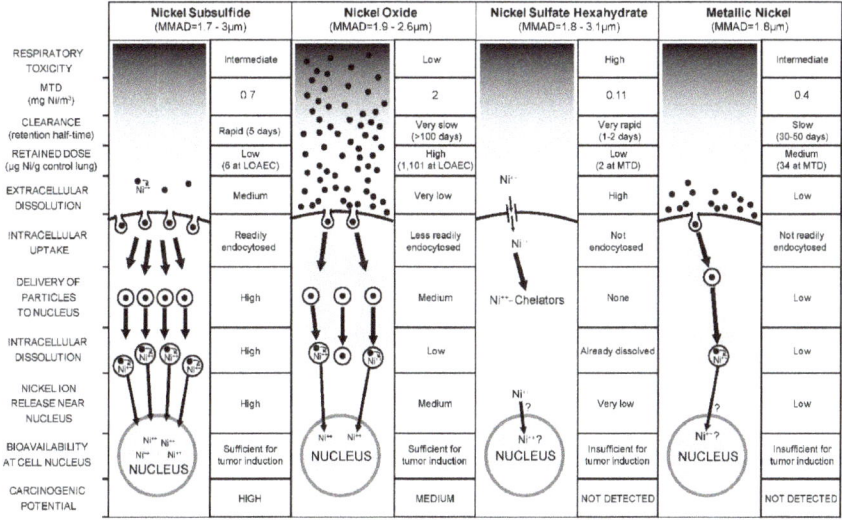

Figure 2. Summary of the bioavailability model for nickel carcinogenicity potential in humans and animals. According to this model, the interplay of factors like respiratory toxicity, maximum tolerated dose in animals, lung clearance, and delivery of sufficient nickel ions to the nuclei of respiratory epithelial cells influence the carcinogenicity observed with nickel in animal inhalation studies. DMT—Divalent Metal Transporter. Reproduced with permission from [10], Taylor & Francis Ltd. www.tandfonline.com.

3.3.3. Oral Exposure Route

Besides respiratory cancer, there are currently no consistent and reliable epidemiological data to suggest that nickel causes excess cancer risks at other organ sites. Although excess cancer of the buccal cavity and pharynx, or stomach, or prostate have been reported in some workers exposed to nickel, these findings have been rare and have not been consistently reproduced [46,70].

Exposure to soluble nickel compounds, by themselves, via the oral route has not produced tumors in rats [51] or mice [71]. No robust animal studies exist for oral administration of metallic or insoluble nickel compounds. However, the negative results with the most bioavailable of the nickel compounds via the oral route are also relevant for these less bioavailable substances since absorption is needed for any systemic effects.

3.3.4. Dermal Exposure Route

There is presently no human epidemiological study linking dermal exposure to metallic nickel and/or nickel compounds with excess local or systemic cancer risks. For the dermal route, no skin cancers have been reported with exposure to nickel compounds or alloys, and the systemic absorption of nickel through the skin is very low ($\leq$2% for soluble Ni substances and 0.2% for nickel metal). Thus, the epidemiological data do not suggest an association between cancer and dermal nickel exposure.

No cancer studies in animals using dermal administration of soluble, insoluble nickel compounds, or metallic nickel have been conducted. However, based on the low dermal bioavailability of Ni^{2+} and other physiological considerations, it is not expected that dermal exposure to metallic nickel and/or nickel compounds will be associated with cancer.

3.3.5. Other Exposure Routes

No consistent epidemiological data currently exists linking nickel exposure via other routes (e.g., implants) to cancer. Given the specificity of nickel cancer on the respiratory system, it is not expected that nickel cancer will be associated with routes other than inhalation.

There are various animal studies assessing the carcinogenicity of the different nickel chemical forms following parenteral, intratracheal instillation, intraperitoneal and injection administrations [2]. However, these routes of exposure are not appropriate nor physiologically relevant for metallic nickel and nickel compounds. For example, while local sarcomas at sites of injection of nickel metal powder were found, the relevance of these findings for the assessment of nickel metal respiratory carcinogenicity or other carcinogenicity in humans is highly questionable [54].

3.3.6. Nickel Compounds' Genotoxic and Carcinogenic Mode of Action

Generally, carcinogenic substances can be direct-acting genotoxicants, indirect-acting genotoxicants, or substances that cause cancer via non-genotoxic modes of action. For direct-acting genotoxicants, all exposure levels are assumed to be associated with some degree of excess cancer risk. Direct-acting genotoxic agents are positive in bacterial, germ cell, and mammalian cell mutagenicity tests and interact directly with DNA. Nickel compounds have been consistently negative or showed weak effects in bacterial, in vitro and in vivo mutagenicity tests. Nickel ions have a weak interaction with DNA but a preferentially stronger interaction with proteins. For example, Ni^{2+} has binding constants of 6.7×10^{-1} M^{-1} and 4.37×10^9 M^{-1} for adenosine (nucleic acid) and cysteine (amino acid), respectively [72,73]. Nickel and nickel compounds are therefore not considered to have a direct genotoxic mode of action (MOA) [2].

For indirect-acting genotoxic agents and non-genotoxic agents, threshold exposure levels can be identified below which cancer risks are expected to be negligible. Indirect genotoxicants can damage DNA via secondary mechanisms like generation of reactive oxygen species (ROS) or inhibition of repair. Table 1 summarizes the genotoxicity evidence for nickel and provides examples of relevant references. In vitro, nickel compounds have been shown to cause DNA damage (such as fragmentation, single-strand breaks) indirectly through increased formation of oxidative radicals. Nickel compounds enhanced the induction of sister chromatid exchange, chromosomal aberrations and micronucleus formation in vitro. Nickel compounds have also been shown to increase formation of DNA-Protein crosslinks.

Additionally, nickel compounds are known to inhibit DNA repair enzymes in vitro; Ni^{2+} ions competitively inhibit the repair enzyme ABH2 by binding to the same site as Fe^{2+} [74]. DNA repair inhibition may also occur via inhibition of DNA ligation and post replication repair. Non-genotoxic modes of action include the induction of epigenetic effects that can affect gene expression. Nickel compounds can increase histone phosphorylation (H3S10), methylation (H3K4), ubiquitination (H2B and H2A) and decrease histone acetylation (H4) through decreased histone acetyltransferase activity (Table 1). Nickel compounds can induce selective fragmentation or decondensation of heterochromatic long arms of X-chromosomes (Table 1). Recent reports also suggest that non-coding RNAs play a role in nickel respiratory carcinogenesis [75–77]; see the review "The role of Non-coding RNAs Involved in Nickel Induced Lung Carcinogenic Mechanisms" by Zhu et al. in this issue.

The in vivo genotoxic effects of nickel compounds in the lung are observed following inflammation and macrophage activation that results in indirect oxidative DNA damage. *Preventing inflammation would therefore prevent the indirect genotoxic effects and prevent tumor formation.* The indirect genotoxic and non-genotoxic effects of nickel compounds have thresholds below which these effects are not observed.

This furthers support the bioavailability model wherein nickel carcinogenicity depends on the delivery of sufficient amounts of Ni^{2+} ions to the nucleus of respiratory epithelial cells [10]. Taken together, the totality of the evidence supports an indirect genotoxic and/or non-genotoxic MOA with thresholds for nickel carcinogenesis.

Table 1. Summary of genotoxicity evidence of nickel and nickel compounds in vitro, in vivo and in humans.

Genotoxicity/Carcinogenicity	In Vitro Studies	In Vivo Animal Studies	Human Studies
DNA damage	Positive, indirect damage; e.g., single strand breaks, DNA fragmentation [78–80]	Positive [81–83]	Positive in welders but unassignable to Ni [84,85]; positive in mucosa of orthodontic patients but unassignable to Ni [86–88]
Impair DNA repair	Positive [74,89–91]		
Gene Mutations	Low-to-none [92–95]	Low mutations [23,92,96]	
Chromosomal alterations	ᵃ Mixed results [23,96–98]	ᵃ Mixed results [99–101]	ᵃ Mixed results [102–107]
HIF-1α-dependent changes	Stabilization of HIF-1α, increased HIF-1α-dependent gene expressions, e.g., Cap43 [108–111]	Increased HIF-1α-dependent Cap43 expression [108], hematocrit and/or hemoglobin [51]	
Apoptosis/Autophagy	Increase [109,112–115]		Increase via increased expression of caspase 8 in workers [116]
Histone modifications			
*Hypermethylation	Positive at H3K9 [117,118]		In workers at H3K4 [119–121]
*Hyperphosphorylation	Positive at H3S10 [122]		
*Hyperubiquitination	Positive, inhibiting deubiquitinating enzyme activity [123]		
*Hypo/hyperacetylation	Hypoacetylation via inhibiting histone acetyltransferase activity [124,125]		Increased H3K9 acetylation in steel workers [121]
Oxidative stress, ROS production	Positive [78,124,126,127]	Positive in rats and mice [80,128,129]	Oxidative DNA lesions linked to nickel levels [130]

ᵃ Mixed results indicates that both positive and negative effects were observed.

3.4. Reproductive and Developmental Toxicity

3.4.1. Human Epidemiological Studies

In 1994, Chaschschin et al. [131] published a preliminary qualitative report of an apparent increase in spontaneous abortions and structural malformations in newborn babies whose mothers were employed in the Russian nickel refinery at Monchegorsk. These workers were exposed via inhalation to a mixture of water-soluble and water-insoluble nickel compounds, as well as metallic nickel. Due to this concern, a large epidemiological study of the Russian cohort was launched in 1995 to determine whether the observed effects were due to their workplace nickel exposures or to other factors in the workplace and/or ambient environment. Researchers set up a birth registry for all births occurring in the region during the period of the study, which included information on 22,836 newborns and 2793 pregnancy outcomes surveyed for spontaneous abortions [132,133]. Exposures were reconstructed (using air and urinary nickel measurements) for the female workers at the refineries to be able to link specific pregnancy outcomes with occupational exposures via inhalation and systemically bioavailable doses of nickel (i.e., urinary nickel levels). A series of reports by Vaktskjold et al. [132–135] based on results from this study indicated that exposure of female refinery workers to soluble nickel was not associated with adverse pregnancy outcomes for (1) male newborns with genital malformations, (2) spontaneous abortions, (3) small-for-gestational-age newborns, or (4) musculoskeletal effects in newborns.

Background urinary nickel levels in the female Monchegorsk population had a geometric mean of 5.9 µg/L, the low-exposure refinery workers had urinary levels up to 70 µg/L (~12-fold increase in urinary levels) and the high-exposure workers had urinary levels between 70 and 179 µg/L (up to 30-fold increase in urinary levels). Urinary nickel levels are better indicators of fetal exposure, as they account for systemically absorbed Ni^{2+} from occupational and non-occupational sources (e.g., diet) by all routes of exposure. The data from this study showed no correlation between nickel exposures (urinary levels as high as 30-fold over background) and observed reproductive impairment. Further evidence that nickel exposure was not adversely affecting the reproduction of these women was provided by the lack of a "small-for-gestational-age" finding and the lack of an association of male genital malformations with nickel exposure. Both endpoints are considered "sentinel" effects (i.e., sensitive endpoints) for developmental toxicity in humans.

Other epidemiological studies have been carried out to examine potential associations between nickel exposure and reproductive and developmental outcomes in both occupational and environmental exposure. While there are population-based studies that examined potential associations between nickel exposure and reproductive and developmental outcomes; these studies suffer from limitations derived from their reliance on single pollutant models to assess risks in multi-pollutant studies. These studies are useful to generate hypotheses; yet, they are not robust enough to establish reliable evidence of causality. For example, a comparison of the low birth weight results from Ebisu and Bell [136] to those in the refinery workers' study of Vaktskjold et al. [132] demonstrates that the risks predicted from the population study were not realized in workers with daily air nickel intake levels 60 to 376-fold higher than the minimal needed to detect these effects with sufficient statistical power.

3.4.2. Studies in Animals

Potential fertility impairment due to exposure to nickel compounds (including the most bioavailable forms) has been extensively studied, and no effects on fertility have been found. There are several reliable 13 week and one- and two-generation studies utilizing inhalation or oral administration of water-soluble nickel compounds in rats that have not indicated adverse effects on fertility, estrous cycling, sperm parameters, vaginal cytology, copulation and fertility indices, precoital intervals, gestation lengths, gross necropsy findings and histopathology [137–142]. Because reproductive toxicity effects are related to the bioavailable Ni^{2+} at target sites, the lack of fertility effects following exposure to water-soluble nickel compounds (which have the highest bioavailability) is also relevant to water-insoluble nickel compounds and nickel metal, which have lower bioavailability [143]. Therefore, based on reliable studies, neither water-soluble nickel compounds, nickel metal, nor insoluble nickel compounds carry harmonized classifications for fertility effects in the EU (see Part 3 of Annex VI; CLP Regulation) [144].

While no fertility effects were reported in these studies, adverse developmental effects have been consistently observed following exposure to water-soluble nickel compounds such as nickel sulfate and nickel chloride in rats and mice. The most sensitive effect observed in rat studies was perinatal mortality, or an increased death rate of the offspring around parturition [138,141,142,145]. While no other developmental effects, including malformations (i.e., teratogenesis), were identified in a rat prenatal developmental toxicity study with water-soluble nickel chloride at the maximum tolerated dose of 42 mg Ni/kg bw/day [139,140], nickel chloride was shown to cause malformations (e.g., microphthalmia) in a prenatal developmental toxicity study in mice at 46 mg/kg bw/day and other teratogenic effects at higher doses [146]. Based on these studies, soluble nickel compounds are classified as Repro 1B for developmental effects, with the most sensitive effect being perinatal mortality in rats. For comparison, toxicokinetic data indicates that oral exposure to insoluble nickel compounds or metallic nickel (micron-size powder or massive form) would not allow enough absorption of Ni^{2+} to exceed the threshold for developmental effects in rodents with soluble nickel compounds (i.e., there is 100-fold lower oral absorption of nickel from nickel metal compared to soluble nickel compounds in rats [143]).

3.4.3. Conclusions on Reproductive Toxicity

While developmental toxicity was observed in high-dose oral animal studies with water-soluble nickel compounds, an epidemiological study of female workers exposed to nickel metal and nickel compounds at exposure levels higher than present in current operations [132–135] has not indicated adverse developmental effects in humans. Thus, the threshold-mediated developmental effects observed in rodents (1) may not be relevant to humans or (2) may not have been observed in the exposed human population because even the highest measured for female workers' exposure (e.g., ~179 µg Ni/L in urine) was lower than the levels achieved in rats at the most sensitive LOAEL for reproductive effects (2300 µg Ni/L in urine). In either case, the relevance of the positive results in rats with soluble nickel compounds (at what seem to be concentrations not relevant to human exposures) needs to be considered together with the negative results for the highest exposed human population in a weight of evidence approach [145]. At the very least, the data indicates that humans do not appear to be more sensitive to developmental toxicity effects of Ni^{2+} than rodents.

3.5. Non-Cancer Lung Effects

Inhalation of nickel-containing aerosols can result in acute as well as chronic effects in the respiratory tract. Acute effects, such as upper respiratory irritation, pneumonitis and even death have been described in workers exposed to very high levels (≥0.5 mg Ni/m^3) of very fine nickel-containing particles (e.g., welding fumes <0.5 µm particle diameter and spraying of nanosize powders <25 nm particle diameter) [147,148]. Some cases of acute exposures to water-soluble nickel substance aerosols have been associated with asthma attacks in sensitive individuals. Potential non-cancer effects associated with repeated exposure to nickel-containing aerosols include chronic bronchitis and lung fibrosis [5,149]. While inhalation animal studies with Ni sulfate, Ni subsulfide, Ni oxide and Ni metal have reported the presence of inflammation at the histopathological level after chronic exposure, fibrosis has also been noted after repeated exposure to Ni sulfate and Ni subsulfide.

Studies of respiratory disease in nickel-exposed workers are limited [150,151] but have not indicated that nickel exposed workers experience pneumoconiosis to any significant extent. The overall incidence of irregular opacities (ILO ≥ 1/0) in X-rays taken at a nickel refinery (4.5%) was not significantly different from the incidence among "normal" X-rays from a hospital (4.2%) and was lower than for X-rays from quarry workers (13.6%) [151]. It should be noted that an X-ray finding does not necessarily correspond with a functional diagnosis of lung fibrosis e.g., [152]. More information on respiratory disease can be obtained from mortality studies. Studies of tens of thousands of workers (many of whom would have experienced very high nickel exposures have not indicated increased mortality from non-cancer respiratory disease) [57,153–156]. A comparison of the animal (histopathology) and workers (X-ray) findings suggest that humans are not more sensitive to lung toxicity effects than rats.

Regarding occupational asthma, even though there are tens of thousands of workers exposed to metallic nickel and water-insoluble nickel compounds in primary and secondary nickel production facilities, grinding applications, catalyst manufacturing etc., only a few anecdotal reports of nickel-related asthma exist e.g., [157,158]. Although nickel is a weak to moderate skin sensitizer, it is not necessarily a respiratory sensitization, since the type of immunological reaction is different for the skin and respiratory sensitizations [158–160]. While soluble nickel compounds are classified as respiratory sensitizers in the EU, the evidence linking soluble nickel exposures and hypersensitivity reactions of the respiratory tract indicate low sensitization potential. Some case reports have shown positive responses of subjects to inhalation challenges; yet, the presence of IgE (indicative of Type

I immune response), the appearance of early or late responses (indicative of Type I or IV immune responses, respectively), the correlation with positive skin tests and other signs of immunological reactions have been quite inconsistent among studies [160].

4. Nickel Exposure in the Environment

4.1. Exposure Sources in the Environment

As a naturally occurring element, nickel substances are present in all compartments of the environment at background concentrations. Chemical and physical degradation of rocks, and soils and atmospheric deposition of nickel-containing particulates release nickel into ambient waters [1,161]. Natural sources of airborne nickel include soil dust, sea salt, volcanoes, forest fires, and vegetation exudates, accounting for about 16% of the atmospheric nickel burden [1,162].

Human activities that contribute to nickel loading in aquatic and terrestrial ecosystems include mining, smelting, refining, alloy processing, scrap metal reprocessing, fossil fuel combustion, and waste incineration. The primary human sources of nickel to soils are emissions from smelting and refining operations and disposal of sewage sludge or application of sludge as a fertilizer. Secondary sources include automobile emissions and emissions from electric power utilities [161]. Weathering and erosion of geological materials release nickel into the environment [162], and acid rain may leach nickel from plants into soils as well [1].

Emissions of nickel can be influenced by many factors, including specific production methods, fuel sources, and pollution control measures. The variability of these factors spatially and over time make it difficult to quantify emissions with accuracy.

4.2. Interactions between Nickel and Natural Chemical Parameters and the Concept of Bioavailability

Nickel substances present in the environment interacts with ecosystems and the organisms that live there, with the type and degree of reaction depending on the speciation of the nickel substance. When nickel in the environment reaches sufficiently high concentrations, it can be toxic to plants and animals. It has been well documented that the amount of nickel that will cause toxicity is highly dependent on a range of factors, including the intrinsic sensitivity of the organism, the geochemical conditions of the environmental media, the presence of other stressors in the environment, and even the route of exposure [163–165]. This concept, known as "metal bioavailability", is a measure that reflects the exposures that organisms actually "experience" and is driven by a combination of the physico-chemical factors governing metal behavior and the specific physiology of the organisms. A substance is bioavailable if it can be adsorbed or absorbed by an organism with the potential for distribution, metabolism, elimination and/or bioaccumulation [166].

Research has demonstrated that the free Ni^{2+} ion is the most bioavailable and toxic form of nickel in the environment [167–171]. Soluble compounds, like soluble nickel salts, are associated with high bioavailability, while sparingly soluble compounds, like some nickel oxides, are less bioavailable. Understanding the chemical characteristics of the media in which exposure occurs is as important as characteristics of the organism (e.g., physiology) when assessing potential bioavailability of nickel. These geochemical characteristics influence the chemical species of nickel in the media, as well as the distribution of nickel between different environmental components, like complexes with organic matter or other inorganic species.

4.2.1. Water

Research has demonstrated that the most important water chemistry parameters affecting bioavailability and toxicity of nickel to freshwater organisms are hardness, pH, and the amount of dissolved organic carbon (DOC) [167–169,171–173]. These parameters drive the two fundamental chemical processes that occur in natural waters when nickel is present. The first is complexation, where DOC forms complexes with dissolved free ionic Ni^{2+}, thereby reducing the quantity of Ni^{2+}

that is available to bind to the site of biological activity on the organism, sometimes referred to as the biotic ligand. The term 'biotic ligand' is used to conceptually describe the fact that binding sites within the organism are subject to the same chemical processes as abiotic ligands (e.g., HCO_3^-, Cl^-, etc.). From this conceptual perspective, the amount of metal that is bound to the biotic ligand is proportional to adverse effects. The second process affecting bioavailability and toxicity is competition, which describes the interaction between similarly charged ions, such as calcium (Ca^{2+}), magnesium (Mg^{2+}), and protons (H^+, expressed as pH). These cations will compete with nickel for binding sites on the biotic ligand. For nickel, it was observed that these interactions result in the following trends: as pH increases, toxicity increases; as hardness increases, toxicity decreases; as DOC increases, toxicity decreases [167–169,171–173].

This means that nickel bioavailability and toxicity can vary considerably among different freshwater systems with different pH, hardness, and DOC conditions. Also, it means that toxicity tests with the same aquatic species that are performed under different water quality conditions can result in different toxicity conclusions. To remove the influence of chemical conditions on the outcome of toxicity tests, tests should be conducted in the same type of water or should be normalized to similar water quality conditions.

One way to normalize a dataset is to "correct" the data to a common water quality condition using a bioavailability-based toxicity model, like the Biotic Ligand Model (BLM) [164,174]. The BLM is a mechanistically based model that was developed to describe metal bioavailability and toxicity to freshwater organisms [175]. For nickel, chronic BLMs have been developed and validated for the invertebrates *Ceriodaphnia dubia* and *Daphnia magna*, the fish *Oncorhynchus mykiss*, and the green alga *Pseudokirchneriella subcapitata*) [167–169]. Acute BLMs are available for invertebrates and fish [176]. For more information on data normalization, see Section 4.3.1.

Currently there are no bioavailability models for assessing the impacts of nickel on marine organisms although recent research suggests that nickel toxicity to marine organisms is highly dependent on quantity and chemical composition of DOC in the ecosystem [177,178]. Toxicity tests with shrimp (*Americamysis bahia*), sea urchin (*Strongylocentrotus purpuratus, Evechinus chloroticus*) and mussels (*Mytilus edulis*) have been used to investigate nickel bioavailability relationships in marine waters [177,178]. Research demonstrated that toxicity is controlled by organic carbon content, but DOC quantity alone may not always be the best predictor of toxicity [178,179]. Efforts are currently underway to develop marine BLMs using the data developed in these experiments.

In recent years, the BLM has increasingly been used in the regulatory setting, for instance to develop new water quality standards that ensure the protection of aquatic plants and animals [180–182]. The Environmental Quality Standard for nickel under the European Water Framework Directive accounts for bioavailability [183,184].

4.2.2. Sediment

Like the aquatic compartment, nickel toxicity from contaminated sediments can rarely be predicted from total metal concentrations because of the influence of the sediment geochemistry and the biochemistry, physiology, and behavior of benthic organisms [185]. Sediment parameters known to affect metal bioavailability include total organic carbon (TOC), iron (Fe) and manganese (Mn) oxides, and the relationship between acid volatile sulfides (AVS) and simultaneously extracted metals (SEM) [186,187]. AVS and SEM can be used to predict bioavailability in anoxic sediments containing sulfides that react to form insoluble metal complexes, while Fe and Mn oxides control bioavailability in oxic sediments. In anoxic sediments, AVS is the amount of amorphous iron sulfide in sediments available for binding to the metals, while SEM is the amount of metals in sediment that could be available to biotic receptors. When the amount of SEM exceeds AVS content, the metal in the sediment may be bioavailable [188,189]. Di Toro et al. [190] demonstrated that normalizing the molar difference between SEM and AVS by the fraction of sediment organic carbon is an even more precise estimate of metal bioavailability.

Nickel sediment bioavailability was investigated in a research program examining chronic toxicity of several nickel-spiked sediments containing a wide range of AVS concentrations on ten test species. Based on the results from these studies, bioavailability models were developed [186,187,191]. For all species tested, the sediment parameter showing the strongest linear relationship with toxicity was AVS, indicating that as AVS concentration increased, nickel toxicity decreased [186,187,192]. The empirical relationships developed between sediment toxicity endpoints and AVS concentration allow nickel ecotoxicity data to be normalized to different sediment conditions. Although the inverse relationship between toxicity and AVS was consistently observed for all species, the magnitude of the effect was not similar among species, and these differences appear to be linked with organism behavior. The strongest mitigating effects of AVS are observed for those species with an epibenthic lifestyle, such as *H. azteca*, *S. corneum*, and *G. pseudolimnaeus* [187].

4.2.3. Soil

Research has demonstrated that when considering the bioavailability of nickel in soils, the most important factors in determining the ecotoxicity to soil organisms are metal form (chemical species), ageing, and soil characteristics [193,194]. Nickel can enter the soil environment in different forms, such as soluble or sparingly soluble compounds. Additionally, laboratory soils spiked with nickel (or other metals) often show greater toxicity than field contaminated soils with the same nickel concentration. The greater toxicity of nickel in spiked soils compared to corresponding field contaminated soils is associated with the time between the addition of nickel to soils and the measurement of toxicity [194]. These concerns can be addressed by leaching and ageing the soil after spiking. It has been demonstrated that the bioavailability and toxicity of nickel in spiked soils tend to decrease with time after spiking and is dependent on soil pH [194].

As with the aquatic and sediment compartments, the toxicity of nickel in soil is highly dependent on the characteristics of the environment. Specifically, nickel toxicity to plants, invertebrates, and microbial processes decreases as the effective cation exchange capacity (eCEC) of the soil increases [194,195]. The eCEC is a measure for the sum of exchangeable cations plus extractable acidity held on or near the surface of negatively charged material, such as clay or organic matter, at native soil pH (measured at the native pH of the soil). Chronic regression bioavailability models for nickel have been developed using laboratory data. These studies show that accounting for ageing and leaching (via a leaching-ageing factor) and accounting for differences in soil properties significantly explains variation in nickel toxicity for all endpoints tested. It was observed that chronic Ni toxicity was best correlated with the eCEC of the soils and the same trends were observed for all the species tested: as soil eCEC increased, chronic toxicity decreased [194]. The application of bioavailability models to normalize nickel soil toxicity data can be found in Section 4.3.4.

4.3. Ecosystem-Specific Nickel Ecotoxicity

4.3.1. Identification, Screening and Aggregation of Nickel Ecotoxicity Data

Ecotoxicity data for nickel are widely available. The data described in this paper were obtained from existing critical reviews of the literature, which are available for temperate [196] and tropical [197,198] ecosystems. These reviews followed data screening approaches consistent with those reported previously [181,196,199,200].

Abundant acute ecotoxicity data are available for nickel for a wide range of organisms; however, given the current preference for chronic ecotoxicity data in global regulatory processes, acute data will not be discussed [181,196]. The current global regulatory focus is toward ecosystem-level protection, with environmental guidelines and standards being based on chronic, rather than acute toxicity endpoints. To this end, chronic laboratory ecotoxicity data based on ecologically relevant endpoints (e.g., growth, reproduction, mortality) are the focus, where "chronic" refers to adverse effects caused by exposure to nickel for a substantial (>10%) portion of the lifespan of the test organism, or effects experienced during the most sensitive life stage [201].

To be classified as reliable chronic nickel ecotoxicity data, several data quality criteria need to be satisfied, including the use of soluble nickel salts (e.g., $NiCl_2$, $NiSO_4$, etc.), full reporting of methods employed during toxicity tests, satisfying minimum performance in control treatments, and full reporting of analytical chemistry data (including the quantification of nickel exposure during the test). Toxicity threshold values calculated as L(E)C10 (the concentration that causes 10% effect during a specified time interval) values are preferred; however, NOEC values (No Observed Effect Concentration) are seen as equivalent.

Nickel ecotoxicity varies as a function of the chemistry of the relevant environmental matrix (i.e., water, sediment, soil). Modern ecological risk assessment of metals relies on the ability to normalize ecotoxicity data to a common set of chemistry parameters to remove the influence of inter-test chemistry differences on test organisms. The ecotoxicity data described in this paper have been normalized for bioavailability using established approaches for freshwater [172], soil [194], and sediments [202]. To perform bioavailability normalizations, ecotoxicity data must be accompanied by the appropriate environmental chemistry data, i.e., pH, dissolved organic carbon (DOC), and hardness (Ca^{2+} and Mg^{2+}) for freshwater data, effective cation exchange capacity (eCEC) and pH for soils, and acid volatile sulfides (AVS) for sediments.

Statistical extrapolation using Species Sensitivity Distributions (SSD) are increasingly used in ecological risk assessments when large ecotoxicity datasets are available. SSDs are appropriate when the ecotoxicity datasets are representative of the ecosystem to which they are applied. For example, for freshwater systems, the ecotoxicity dataset should include data for algae, invertebrates, and fish that include different life history strategies and feeding behaviors. SSDs are constructed by applying an appropriate curve fitting distribution (usually a log-normal distribution) to the normalized high-quality aggregated chronic toxicity data [203]. The concentration associated with the 5th percentile cumulative probability is calculated from the SSD. This value, referred to as the HC5 (hazard concentration at the 5th percentile) represents the concentration that is protective of 95% of organisms in the SSD, and corresponds to protection goals expressed by many regulatory jurisdictions.

4.3.2. Freshwater

High-quality chronic ecotoxicity data are available for a wide range of freshwater aquatic organisms, including unicellular algae, vascular plants, invertebrates, fish, and amphibians. Nys et al. [172] normalized Ni ecotoxicity data (n = 31 species) for a high bioavailability scenario (pH = 8.1, hardness = 165 mg $CaCO_3$/L, and DOC = 3.8 mg/L), and calculated an HC5 of 8.1 µg Ni/L. When normalized to this condition, the most sensitive species include invertebrates, unicellular algae, and vascular plants. Among invertebrates, the most sensitive species is the snail *Lymnaea stagnalis*, which has consistently exhibited sensitivity to Ni exposure in different studies [171,172,204]. Cladocera have also consistently demonstrated sensitivity, with the most sensitive species within this group being *Ceriodaphnia dubia*. Evidence supporting cladoceran sensitivity is broad, with studies from Europe [172], North America [171] and Australia [165] all demonstrating that *C. dubia* is among the most sensitive species in normalized SSDs.

The majority of nickel ecotoxicity data have been generated for temperate species. Application of these data to tropical ecosystems, which are increasingly important in terms of global nickel production, carries uncertainty because of differences in taxonomic groups (e.g., corals live only in tropical systems) and the distribution of chemistry parameters known to influence nickel toxicity. Binet et al. [197] critically reviewed the literature for available nickel ecotoxicity data for tropical species, and only identified high-quality data for four species, including one unicellular alga, two vascular plants, and one invertebrate. The low number of available tropical data suggests that additional testing of key species (crustaceans, gastropods, and fish) may be necessary to develop a robust tropical dataset. On the other hand, Peters et al. [205] demonstrated that normalized sensitivities of tropical and temperate species to nickel overlap, despite differences in species, temperature, and geochemistry. This means that tropical ecotoxicity data, which are relatively few in number, can be pooled with the broader temperate ecotoxicity data to develop a comprehensive and robust database that can be used in ecological nickel risk assessments regardless of location.

While the freshwater ecotoxicity dataset for nickel is among the most extensive for any chemical substance, uncertainty exists when applying data from single species laboratory toxicity tests to field conditions. Mesocosm and field studies can address this uncertainty. Hommen et al. [206] performed a four-month exposure to a freshwater community that included species known to be sensitive to nickel, including *L. stagnalis* and several species of cladocerans. The water used in this study (median pH, DOC, and hardness = 8.6, 3.9 mg/L, and 107 mg $CaCO_3$/L, respectively) represented a high nickel bioavailability condition. The most sensitive species was *L. stagnalis*, which is consistent with the ranking of sensitivities for single species tests. A study-specific No Effects Concentration of 12 µg Ni/L was reported, which is greater than the HC5 from an SSD comprised of laboratory ecotoxicity endpoints normalized to the water chemistry conditions of the mesocosm. Peters et al. [207] determined bioavailable nickel concentrations above which benthic macroinvertebrate communities were affected. They showed that the most sensitive benthic organisms were snails, and that the bioavailability-normalized HC5 based on laboratory results was protective of (i.e., lower than) the effects based on field data. Together, results of Hommen et al. [206] and Peters et al. [207] show consistency in terms of the sensitivity of snails to nickel exposure, and they demonstrate that the bioavailability-normalized HC5 can be used confidently in ecological risk assessments of Ni without the application of additional uncertainty factors.

Several regulatory agencies have used nickel ecotoxicity data to generate threshold concentrations that are meant to be protective of aquatic life. In 2013, the European Union announced an Environmental Quality Standard (EQS) based on chronic ecotoxicity data for 31 species [208]. The EQS is bioavailability-based and uses a reference value of 4 µg Ni/L that represents conditions of high bioavailability in Europe. The EQS is implemented using a tiered approach. In cases where dissolved Ni concentrations exceed the reference value of 4 µg Ni/L, local or regional water chemistry data (pH, DOC, and hardness) are used to calculate a bioavailability-based EQS for that area [164]. If dissolved Ni concentrations are below the site-specific bioavailability-based EQS, then a conclusion of no risk is assessed.

4.3.3. Marine

Chronic nickel sensitivities among marine organisms vary greatly. DeForest and Schlekat [209] compiled available high-quality chronic ecotoxicity data for temperate marine organisms and demonstrated a 7×10^3 difference in EC10s among 17 species. The most sensitive species in this review was the early life stage of the sea urchin *Diadema antillarum*, with an EC10 based on developmental success of 2.9 µg Ni/L. On the other hand, fish are among the least sensitive organisms, with an EC10 for the sheepshead minnow, *Cyprinodon variegatus*, of 2.1×10^4 µg Ni/L. Wide ranges of sensitivities are observed within specific marine phyla. For example, DeForest and Schlekat [209] report additional EC10s for the sea urchins *Paracentrotus lividus* and *Strongylocentrotus purpuratus* ranging from 89 to 335 µg Ni/L. More recently, Blewett et al. [179] reported a developmental EC50 for the sea urchin *Evechinus chloroticus* of 14 µg Ni/L, which is equivalent to an EC10 of approximately 2.8 µg Ni/L, again illustrating the wide range of sensitivities within a given taxonomic groups.

Coastal marine and open ocean environments are highly consistent in terms of pH and ionic composition (e.g., Ca and Mg concentrations). The one water chemistry parameter that does vary is DOC, and while Ni toxicity has been demonstrated to vary as a function of DOC quality [179], the relationship is non-linear, and shows a less than two-fold difference among natural waters varying in DOC concentration from 1 to 5 mg/L [177]. Therefore, the differences observed among closely related species cannot be attributed to differences in water chemistry, and instead reflect true differences in intrinsic sensitivity.

No bioavailability normalization approaches are available for Ni in marine systems, and as a consequence, all of the data reviewed here are expressed as dissolved concentrations. A marine EQS under the EU WFD of 8.6 µg Ni/L was established in 2013 [208], -based largely on the chronic ecotoxicity data reported by DeForest and Schlekat [209]. The US EPA ambient water quality criterion for dissolved Ni in saltwater is 8.2 µg Ni/L and was developed using a completely different ecotoxicity dataset and derivation approach.

Recently, ecotoxicity data have been generated on tropical marine species to address questions about the relative sensitivity of nickel in temperate and tropical ecosystems, and on the sensitivity of obligate tropical species like corals. Gissi et al. [210] tested tropical marine species including a gastropod, a copepod, and a barnacle. While the gastropod (EC10 = 64 µg Ni/L) and barnacle (EC10 = 67 µg Ni/L) were within the range of similar temperate species, the marine copepod was highly sensitive to Ni exposure, with an EC10 of 5.5 µg Ni/L. A range of endpoints have been measured for coral, including fertilization (*Acropora aspera*, NOEC < 280 µg Ni/L; *Acropora digitifera*, EC10 = 2000 µg Ni/L; *Platygyra daedalea*, EC10 > 4610 µg Ni/L; [211]), and the integrity of the coral microbiome associated with *A. muricata* (NOEC > 9050 µg Ni/L; [212]). These studies showed that corals are not especially sensitive to Ni exposure. High-quality chronic ecotoxicity data aggregated in an SSD yielded an HC5 value of 8.2 µg/L Ni, which is in the range of existing threshold concentrations for temperate species.

4.3.4. Sediment

Sediments represent a sink for many contaminants entering aquatic ecosystems, and this includes metals like nickel. The EU REACH legislation includes an assessment of effects to sediment organisms [213], and chronic nickel ecotoxicity data were recently generated to satisfy this obligation. Fewer standardized sediment toxicity test methods are available for sediment organisms compared with water column organisms, so non-standard tests were also used. The generated dataset met the requirements for using refined risk assessment approaches like SSDs and bioavailability normalization.

An important methodological step when testing the toxicity of metals like Ni in sediment phases is controlling the diffusion of Ni to overlying water during testing. Earlier testing efforts showed that Ni concentrations in overlying water reached toxic levels, e.g., [214] creating difficulty in attributing adverse effects to nickel in sediment phases. Brumbaugh et al. [215] developed a methodology that limits diffusion of nickel from sediment phases to the overlying water. The approach includes neutralizing sediment pH to relevant levels (adding soluble metal salts to sediments decreases sediment pH) and exchanging overlying water to maintain non-toxic overlying nickel concentrations. The resulting distribution of nickel among pore water and sediment phases is similar to that of field-contaminated sediment and is therefore relevant for estimating Ni toxicity in real world situations.

High-quality chronic ecotoxicity data exist for eight species of freshwater benthic invertebrates, including amphipods, oligochaetes, insects, and molluscs. In low-bioavailability sediments, the amphipod *Hyalella Azteca* is the most sensitive organism, with an EC10 based on biomass of 149 mg Ni/kg [186]. The least sensitive species is the oligochaete *Tubifex*, with a biomass EC10 of 1100 mg Ni/kg [202]. Using empirical bioavailability relationships to predict nickel toxicity based on sediment chemistry parameters, Schlekat et al. [202] reported ranges of HC5 values from 136 to 437 mg Ni/kg for sediments ranging in AVS concentrations from 0.77 to 38.4 µM/g dry weight (dw), which represents the 10th to 90th percentile distribution of European freshwater sediments.

Few data are available for marine or estuarine sediment organisms. Chandler et al. [216] showed that the benthic copepod *Amphiascus tenuiremis* was more sensitive to nickel exposure than the amphipod *Leptocheirus plumulosus* in water-only exposures. In sediment exposures, however, no adverse effects were observed to *A. tenuiremis* in sediment nickel concentrations as high as 676 mg Ni/kg, whereas *L. plumulosus* experienced toxicity when sediment nickel concentrations exceeded 334 mg Ni/kg. This demonstrates the need to collect sediment toxicity data from appropriate testing as opposed to extrapolating water-only data using approaches like equilibrium partitioning.

Extensive field information is available on effects of nickel to sediment ecosystems [192,217–219]. Costello et al. [192] demonstrated that nickel toxicity decreases over time, and that no effects associated with nickel exposure are detectable after 58 days, despite sediment nickel concentrations exceeding 4500 mg Ni/kg. Costello et al. [220] demonstrated that this decrease in toxicity can be explained by a transition in partitioning and mineralogy from more labile sediment phases like sulfides to more stable phases like iron oxides. Mendonca et al. [221] demonstrated that nickel bioavailability is controlled by iron oxides in a field contaminated site. Together, this information suggests that the bioavailability-based HC5 using laboratory ecotoxicity data is highly protective of effects that are observed in the field.

4.3.5. Soil

Ecotoxicological data are available for a range of soil organisms, including plants and soil invertebrates. Microbial processes and enzymatic activity are also considered relevant for ecological risk assessments of chemical substances on soil ecosystems. When adjusted for leaching and ageing and after normalization to soil eCEC, the most sensitive endpoints-organisms have been shown to be enzymatic activity like dehydrogenase, growth of microbes such as *Aspergillus clavatus*, and growth of plants like *Lolium perenne* [194,222]. The variability within these groups is high, and examples of insensitive endpoints can be found for each of these groups of soil organisms. Invertebrates like earthworms and collembola are the least sensitive groups.

When aggregated in an SSD, a European ecotoxicity soil dataset showed HC5 values ranging from 8.6 to 194.3 mg Ni/kg, for soils ranging in eCEC from 10.4 to 36.0 cmol/kg (Figure 3). This range represents the 10th to 90th percentile of eCEC for European soils.

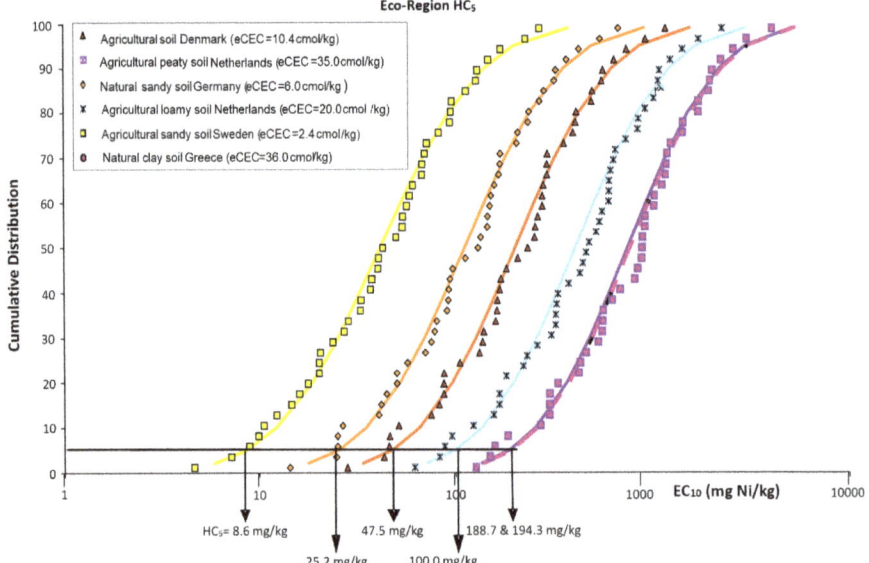

Figure 3. Species sensitivity distributions (SSDs) of soil organisms mean EC10 values ($n = 42$) for soils with different effective cation exchange capacities (eCEC). The 5th percentile of the hazard concentration (HC5) was calculated for each SSD yielding a range of HC5s from 8.6 to 194.3 mg Ni/kg. The "Agricultural peaty soil Netherlands" overlaps with the "Natural clay soil Greece" in the graph, with similar eCEC values of 35.0 cmol/kg and 36.0 cmol/kg, respectively.

Hale et al. [222] compared soil laboratory toxicity data to endpoints measured under field conditions. The field data included a study in which soluble $NiCl_2$ had been added to Chinese soils varying in eCEC (15.2 to 37.1 cmol/kg) and pH (5.3 to 8.9). Laboratory data normalized to the field soil chemistry were protective of the same endpoints measured in field studies with the exception of a soil with low pH (5.2) and relatively low eCEC (15.2 cmol/kg). This suggests that situations of high bioavailability, i.e., low soil pH (<6.5) and low eCEC (<15 cmol/kg), require special attention in terms of risk assessment.

4.4. Mechanisms of Toxicity

Our understanding of the mechanisms of Ni toxicity to aquatic organisms is limited, despite many studies that have explored potential mechanisms of action for nickel. Evidence exists in the literature to support several possible mechanisms, including disruption of trace element and ion homeostasis (e.g., Ca, Mg, and Fe), allergic reactions of the respiratory epithelia, disruption of energy metabolism, and oxidative stress [223].

A negative relationship between increasing water hardness and nickel toxicity to aquatic organisms has been well documented [173]. Studies have shown that increasing Ca^{2+} reduces nickel toxicity to a variety of aquatic and terrestrial animals [173]. In contrast, studies indicate Ca^{2+} has no effect on nickel toxicity to aquatic or terrestrial plants, suggesting nickel uptake occurs via different mechanisms in plants [223]. Additionally, increasing Mg^{2+} concentration reduces Ni accumulation and toxicity in a variety of organisms (daphnids, oligochaetes, rainbow trout, and frog embryos) [167,173,224,225]. Increasing Mg^{2+} also reduces nickel toxicity in terrestrial [170] but not aquatic plants [226]. Competitive interactions between nickel and iron observed in several biological systems in both aquatic and terrestrial environments suggest that disruption of iron homeostasis by nickel may be an important mechanism of action [223].

Nickel is a well-known contact allergen in humans; however, while T cell activation by nickel in fish gills is a potentially viable mechanism for an allergic response, it is not considered a likely mechanism of action for nickel toxicity in fish or invertebrates [223].

Alternatively, formation of reactive oxygen species (ROS) is a potential mechanism for nickel toxicity in all organisms. While there have been few studies investigating the effects of nickel on excessive ROS generation in aquatic animals, there is extensive literature in mammalian and terrestrial plant systems, showing that at elevated levels, ROS will damage various cell structures including lipids, membranes, proteins and nucleic acids (oxidative stress) [227]. Overall, it appears that nickel-induced oxidative stress may be a viable mechanism of action in aquatic and terrestrial systems, although it is not clear if oxidative stress occurs at environmentally relevant concentrations and if so, whether this stress at the cellular level translates into significant effects at higher levels of biological organization for multi-cellular organisms [223].

4.5. Bioaccumulation and Trophic Transfer Potential

The bioaccumulation of chemical substances is addressed in risk assessment because of concerns related to the direct toxicity of a bioaccumulated substance and the potential for trophic transfer of the substance via biomagnification. In general, inorganic metals are subject to homeostatic control, and do not exhibit biomagnification [228]. Bioconcentration factors (BCF), which represent the ratio between tissue concentrations and dissolved concentrations, are relatively low for nickel. For example, a median nickel BCF for bivalves of log 2.43 was reported by the European Commission [196]; this value is below the threshold of log 3.0 that triggers evaluation of secondary poisoning effects. Furthermore, the median soil nickel bioaccumulation factor for earthworms was 0.3, indicating low risk of exposure to small mammals and birds that feed on earthworms [229]. An exception to the general conclusion of low bioaccumulation of Ni by soil organisms occurs in serpentine soils, where natural nickel concentrations can exceed 10,000 mg Ni/kg. Plant species that have developed specific adaptions to tolerate these conditions are referred to as hyperaccumulators, and tissue Ni concentrations in these plants can exceed 24,000 mg/kg [230]. Generic assessments of nickel bioaccumulation intentionally exclude data from hyperaccumulators. In general, the conclusion of the European Commission [196] was that nickel exhibits biodilution, i.e., a decrease in tissue concentration within increasing trophic level. Therefore, effects associated with the trophic transfer of nickel are of low concern.

5. Emerging Issues

5.1. Mixtures

The fate and effects of nickel in the environment are well understood; however, in the environment, metals do not often occur singly, but as mixtures with other metals and contaminants. When metals occur together, they can interact with each other or with other ligands (e.g., DOC) and potentially affect the bioavailability and toxicity of each individual component [231]. To address these complexities, metals mixture models are under varying stages of development [232–234].

Two general modelling approaches can be applied to understanding the toxicity of chemical mixtures to the environment: concentration addition (CA) and independent action (IA) [231]. In CA, substances have a similar mode of action, and in IA (also known as response addition), substances have different modes of action. Nickel in the aquatic compartment, when combined in binary or ternary mixtures with other metals, does not necessarily exhibit a strictly response-additive toxicity in invertebrates or fish [235–237] and responses depend on the other metals present. For instance, Traudt et al. [236] found that acute Ni + Cd mixtures resulted in less-than-additive toxicity to *D. magna* neonates, while Ni + Cu mixtures resulted in greater-than-additive toxicity and Ni + Zn mixtures resulted in response-additive toxicity. Results from chronic studies with *L. stagnalis* similarly showed that trends from binary metal mixtures (including nickel) were highly dependent on the metals combinations [238]. Cremazy et al. [238] found that for chronic binary studies with *L. stagnalis*, the IA model was the most accurate and the CA model was the most conservative. They concluded that for binary combinations, the simple CA model may provide adequate protection from the chronic metal toxicity of metal mixtures to aquatic invertebrates [239]. There have been significant advances in the last ten years on metal mixture toxicity in the environment and though some bioavailability models have begun the effort to accurately predict the resulting toxicity of multi-metal mixtures, more questions remain to be answered before metal–metal interactions are fully understood and metals mixtures can be considered in regulatory frameworks.

Similarly, human health effects in epidemiological studies are associated with exposure to mixtures of substances, making determination of the toxicity relationship (e.g., additivity) of nickel substances with each other and with other substances difficult. Recent rat intraperitoneal toxicity studies and mathematical modeling of binary and/or ternary combinations of manganese, chromium (VI) and nickel demonstrated additivity and/or subadditivity of combined subchronic toxicity effects [240,241]; these effects were influenced by dose, effect level, and type of effect assessed. In another study, with oral gavage administration of soluble forms of zinc, copper, manganese, chromium, cadmium, lead, mercury and nickel to rats, systemic toxicity, mortality and effects on neurobehavioral function were observed dose-dependently at the two highest mixture exposure levels of 464 and 1000 mg/kg bw [242]. Here, the interactions (such as antagonism, synergism, additivity) were not accounted for.

Welding fumes provide a good example of how integrating human and animal toxicology can be used to understand the toxicity of mixtures of nano materials. Exposure to welding fumes has been demonstrated to increase lung cancer risk among welders of stainless or mild steel [243]. During welding, particles in the submicron, nanometer range are generated [244]. These particles are made of oxides of iron and manganese (when welding mild steel) or iron, manganese, chromium and nickel (when welding stainless steel); these particles are generated from the electrode or flux material [243]. In A/J mice studies using a two-stage model for lung carcinogenicity, oropharyngeal aspiration of freshly generated stainless steel and mild steel welding fumes could act as a lung tumor promoter [245,246]. When individual metal oxides were tested in the two-stage mice model it was surprising to find that inhalation of iron oxide (Fe_2O_3) significantly promoted lung tumors, while neither mixed chromium oxides nor nickel oxide did [246]. Combining human studies with exposures to complex mixtures and animal studies with exposures to individual components can help understand risks associated with metal mixtures [247].

The above studies demonstrate the necessity of, but also the unique challenges with, conducting studies on metal mixtures to appropriately mimic human exposures occupationally and in the environment. However, emerging technologies and alternative animal models (such as organotypic cultures or organ-on-a-chip) makes the testing of metal mixtures easier. A case-by-case approach for risk assessment of metal mixtures may be more prudent.

5.2. Nanoparticles

The small size and unique properties of nanoparticles have increased demand for their use in many consumer products today. Manufactured and engineered nickel nanoparticles have primarily commercial and industrial uses, including conductive coatings, lithium ion batteries, fuel cells, electrodes, catalysts, ceramics, additives, etc. [248]. Nanoparticles are generally tightly bound or well-integrated into most end-products for consumer use [249], thus greatly reducing the potential for inhalation or dermal exposure on the part of the general public.

Occupational exposure in manufacturing and research laboratories could be a concern for workers during production and handling of nickel nanoparticles. Additionally, hot processes such as welding can generate exposures in the nano size range. In vitro and in vivo studies suggest that these nickel nanoparticles are primarily associated with lung toxicity, inflammation, oxidative stress, and apoptosis, e.g., [250–257]. Recent inhalation studies indicate possible differences in toxicokinetics between NiO nanoparticle and larger micron particles [48,252,253,256,257].

Although there are many recent studies evaluating both the health and environmental effects of nickel nanoparticles [254,255] there is still uncertainty with regards to the toxicity of nanoparticles in general. For instance, some intratracheal instillation or pharyngeal aspiration studies suggest that nickel oxide nanoparticles are more toxic (e.g., lung inflammation, allergic response) than larger nickel oxide micron particles [254,256]. However, an in vitro study with nickel metal nanoparticles indicated that the nanoparticles were not more toxic than larger nickel metal micron particles [257]. Additionally, reliable studies are needed to increase our knowledge of the reproductive and carcinogenic effects of nickel-containing nano-powders. For example, there is a study indicating that oral exposure to nickel metal nanoparticles at relatively high levels (5–45 mg/kg bw) caused perinatal mortality in rats [258]. However, because of some statistical and methodological limitations of the study, further research is needed to evaluate the accuracy of this report.

In the environment, nickel nanoparticles appear to pose a low ecological risk to aquatic organisms [258–262]. In the Ispas et al. [255] study, zebrafish were treated with three different nickel nanoparticle sizes resulting in less toxicity than the larger soluble nickel. Another aquatic study showed that soluble micron nickel appeared to result in more toxicity compared to nickel nanoparticles in marine calanoid copepods [261]. Unfortunately, very little reliable toxicity and exposure information is available regarding terrestrial organisms. With continued research and increasing knowledge in this growing field, it will be easier to clarify the health and environmental toxicological profiles of nickel nanoparticles.

As discussed in Section 5.1, above, the combination of nickel nanoparticles with other metal nanoparticles can have unexpected outcomes, such as reduced or additive effects, depending on the specific toxicity endpoint, the metal nanoparticles involved, and the percentage of each component in the mixture; this is shown in intraperitoneal injection and in vitro studies comparing and combining nickel oxide and manganese oxide or cobalt oxide nanoparticles [263–266]. The results from these studies suggest that the toxicokinetics and toxicity profile for combined exposures to metal nanoparticles (e.g., occupational exposure) may differ from what could be predicted from each of the individual nanoparticle components alone.

6. Conclusions

The main human and environmental health effects associated with exposure to nickel and its compounds have been summarized in the previous sections. The bioavailable Ni^{2+} ion dictates the human health toxicity and environmental toxicity of metallic nickel and nickel compounds. Thresholds for the toxicity of Ni^{2+} have been identified, below which the toxicity associated with exposure to metallic nickel and nickel compounds may not be observed. Table 2 provides a summary of key threshold values for the main human and environmental toxicity effects of concern.

This review is the first to integrate the human health toxicology and ecotoxicology of nickel, emphasizing the presence of threshold effects and the central role of the bioavailable Ni^{2+} ion in driving toxicity.

Table 2. Thresholds for nickel toxicity corresponding to the main routes and health effects of concern to human health and for the main environmental compartments.

Human Health Effects		
Endpoint (Underlying Studies and Data)	Threshold Values [a]	Reference
NACD (human)	Dermal-NOAEL = 0.44 µg Ni/cm^2 skin/day [a]	[267]
	Oral-$BMDL_{0.1}$ = 4.3 µg Ni/kg bw/day adult (in addition to diet)	[145]
Respiratory cancer (human)	Inhalation—practical threshold = 0.1 mg Ni/m^3 inhalable aerosol	[13]
	Inhalation—practical threshold = 0.5 µg Ni/m^3 ng/m^3 PM10 aerosol	[268]
Respiratory non cancer (animal)	Inhalation-NOAEC = 0.03 mg/m^3 (Ni sulfate) respirable aerosol	[13]
	Inhalation-NOAEC = 9.4 $\mu g/m^3$ (Ni sulfate) PM10 aerosol	[268]
Reproductive (animal)	Oral-$BMDL_{0.05}$ = 1.8 mg Ni/kg bw/day adult	[145]
Environmental Health Effects		
Compartments	Threshold Values [a]	Reference
Freshwater	Bioavailability-based EQS = 4 µg Ni/L	[208]
	HC5 = 7.1 to 43.6 µg Ni/L for EU waters representing the 10th to 90th percentile of bioavailability conditions	[196]
Marine water	Temperate HC5 = 8.6 µg Ni/L	[209]
	Tropical aggregated HC5 = 8.2 µg Ni/L	[212]
Sediment	HC5 = 136–437 mg Ni/kg (AVS: 0.77–38.4 µM/g dw)	[202]
Soil	HC5 = 8.6–194.3 mg Ni/kg (eCEC: 10.4–36.0 cmol/kg)	[194]

[a] Values are based on, or derived from, the most bioavailable chemical form of nickel (i.e., water soluble nickel compounds) but are generally applied to "total nickel".

Funding: This research received no external funding.

Acknowledgments: The authors acknowledge the support of NiPERA Inc. colleague, Connie Lawson, with formatting and reviewing the references in this review article. The authors also thank Mike Dutton and Violaine Verougstraete for their expert review of this article.

Conflicts of Interest: All the authors are Toxicologists at NiPERA Inc., the science division of the Nickel Institute, a global association of leading nickel producers.

References

1. World Health Organization (WHO). *Environmental Health Criteria 108. Nickel*; Prepared as a Part of WHO's International Programme on Chemical Safety; World Health Organization: Geneva, Switzerland, 1991; p. 383.
2. European Chemicals Agency (ECHA). *Annex 1–Background Document in Support of the Committee for Risk Assessment (RAC) for Evaluation of Limit Values for Nickel and Its Compounds in the Workplace*; ECHA/RAC/A77-0-0000001412-86-189/F; European Chemicals Agency: Helsinki, Finland, 2018; pp. 1–211.
3. Nickel Institute. *Safe Use of Nickel in the Workplace: A Guide for Health Maintenance of Workers Exposed to Nickel, Its Compounds and Alloys*, 3rd ed.; Nickel Institute: Brussels, Belgium, 2008.

4. International Agency for Research on Cancer (IARC). *IARC Monographs on the Evaluation of Carcinogenic Risks to Humans: Nickel and Nickel Compounds Monograph*; WHO Press: Geneva, Switzerland, 2017; Volume 100C, pp. 169–218.
5. Agency for Toxic Substances and Disease Registry (ATSDR). *Toxicological Profile for Nickel*; Department of Health and Human Services, Public Health Service: Atlanta, GA, USA, 2005; pp. 1–397.
6. National Toxicology Program (NTP). *Final Report on Carcinogens Background Document for Metallic Nickel and Certain Nickel Alloys, Proceedings of theMeeting of the NTP Board of Scientific Counselors Report on Carcinogens Subcommittee, Durham, NC, USA, 13–14 December 2000*; Technology Planning and Management Corporation: Durham, NC, USA, 2000; pp. 1–102.
7. De Brouwere, K.; Buekers, J.; Cornelis, C.; Schlekat, C.E.; Oller, A.R. Assessment of indirect human exposure to environmental sources of nickel: Oral exposure and risk characterization for systemic effects. *Sci. Total Environ.* **2012**, *419*, 25–36. [CrossRef] [PubMed]
8. Sunderman, F.W., Jr.; Hopfer, S.M.; Sweeney, K.R.; Marcus, A.H.; Most, B.M.; Creason, J. Nickel absorption and kinetics in human volunteers. *Proc. Soc. Exp. Biol. Med.* **1989**, *191*, 5–11. [CrossRef] [PubMed]
9. Nielsen, G.D.; Søderberg, U.; Jørgensen, P.J.; Templeton, D.M.; Rasmussen, S.N.; Andersen, K.E.; Grandjean, P. Absorption and retention of nickel from drinking water in relation to food intake and nickel sensitivity. *Toxicol. Appl. Pharmacol.* **1999**, *154*, 67–75. [CrossRef] [PubMed]
10. Goodman, J.E.; Prueitt, R.L.; Thakali, S.; Oller, A.R. The nickel ion bioavailability model of the carcinogenic potential of nickel-containing substances in the lung. *Crit. Rev. Toxicol.* **2011**, *41*, 142–174. [CrossRef] [PubMed]
11. Oller, A.R.; Oberdörster, G. Incorporation of particle size differences between animal studies and human workplace aerosols for deriving exposure limit values. *Regul. Toxicol. Pharmacol.* **2010**, *57*, 181–194. [CrossRef] [PubMed]
12. Oller, A.R.; Oberdörster, G. Incorporation of dosimetry in the derivation of reference concentrations for ambient or workplace air: A conceptual approach. *J. Aerosol. Sci.* **2016**, *99*, 40–45. [CrossRef] [PubMed]
13. Oller, A.R.; Oberdörster, G.; Seilkop, S.K. Derivation of PM_{10} size-selected human equivalent concentrations of inhaled nickel based on cancer and non-cancer effects on the respiratory tract. *Inhal. Toxicol.* **2014**, *26*, 559–578. [CrossRef]
14. Fullerton, A.; Menné, T. In vitro and in vivo evaluation of the effect of barrier gels in nickel contact allergy. *Contact Dermat.* **1995**, *32*, 100–106. [CrossRef]
15. Fullerton, A.; Andersen, J.R.; Hoelgaard, A. Permeation of nickel through human skin in vitro—Effect of vehicles. *Br. J. Dermatol.* **1988**, *118*, 509–516. [CrossRef]
16. Fischer, T. Occupational nickel dermatitis. In *Nickel and the Skin: Immunology and Toxicology*, 1st ed.; Maibach, H.I., Menné, T., Eds.; CRC Press: Baton Rouge, FL, USA, 1989; pp. 117–132.
17. Wilkinson, D.S.; Wilkinson, J.D. Nickel allergy and hand eczema. In *Nickel and the Skin: Immunology and Toxicology*, 1st ed.; Maibach, H.I., Menné, T., Eds.; CRC Press: Baton Rouge, FL, USA, 1989; pp. 133–163.
18. Fullerton, A.; Andersen, J.R.; Hoelgaard, A.; Menné, T. Permeation of nickel salts through human skin in vitro. *Contact Dermat.* **1986**, *15*, 173–177. [CrossRef]
19. Tanojo, H.; Hostýnek, J.J.; Mountford, H.S.; Maibach, H.I. In vitro permeation of nickel salts through human stratum corneum. *Acta Derm. Venereol.* **2001**, *212*, 19–23. [CrossRef] [PubMed]
20. Hostynek, J.J.; Dreher, F.; Pelosi, A.; Anigbogu, A.; Maibach, H.I. Human Stratum corneum penetration by nickel: *In vivo* study of depth distribution after occlusive application of the metal as powder. *Acta Derm. Venereol.* **2001**, *212*, 5–10. [CrossRef] [PubMed]
21. Rezuke, W.N.; Knight, J.A.; Sunderman, F.W., Jr. Reference values for nickel concentrations in human tissues and bile. *Am. J. Ind. Med.* **1987**, *11*, 419–426. [CrossRef] [PubMed]
22. Dunnick, J.K.; Elwell, M.R.; Benson, J.M.; Hobbs, C.H.; Hahn, F.F.; Haly, P.J.; Cheng, Y.S.; Eidson, A.F. Lung toxicity after 13-week inhalation exposure to nickel oxide, nickel subsulfide, or nickel sulfate hexahydrate in F344/N rats and B6C3F1 mice. *Fundam. Appl. Toxicol.* **1989**, *12*, 584–594. [CrossRef]
23. EFSA (European Food Safety Agency). Scientific opinion on the risks to public health related to the presence of nickel in food and drinking water, EFSA Panel on Contaminants in the Food Chain (CONTAM). *EFSA J.* **2015**, *13*, 4002.
24. Christensen, O.B.; Möller, H.; Andrasko, L.; Lagesson, V. Nickel concentration of blood, urine and sweat after oral administration. *Contact Dermat.* **1979**, *5*, 312–316. [CrossRef]

25. Sunderman, F.W., Jr.; Aitio, A.; Morgan, L.G.; Norseth, T. Biological monitoring of nickel. *Toxicol. Ind. Health* **1986**, *2*, 17–78. [CrossRef] [PubMed]
26. Grandjean, P. Human exposure to nickel. *IARC Sci. Publ.* **1984**, *53*, 469–485.
27. Alinaghi, F.; Bennike, N.H.; Egeberg, A.; Thyssen, J.P.; Johansen, J.D. Prevalence of contact allergy in the general population: A systematic review and meta-analysis. *Contact Dermat.* **2019**, *80*, 77–85. [CrossRef]
28. Warshaw, E.M.; Aschenbeck, K.A.; DeKoven, J.G.; Maibach, H.I.; Taylor, J.S.; Sasseville, D.; Belsito, D.V.; Fowler, J.F., Jr.; Zug, K.A.; Zirwas, M.J.; et al. Epidemiology of pediatric nickel sensitivity: Retrospective review of North American Contact Dermatitis Group (NACDG) data 1994–2014. *J. Am. Acad. Dermatol.* **2018**, *79*, 664–671. [CrossRef]
29. Zenz, C.; Dickerson, O.B.; Horvath, E.P. *Occupational Medicine*, 3rd ed.; Mosby: St. Louis, MO, USA, 1994.
30. Menné, T.; Veien, N.; Sjolin, K.-E.; Maibach, H.I. Systemic contact dermatitis. *Dermatitis* **1994**, *5*, 1–12. [CrossRef]
31. Curtis, A.; Morton, J.; Balafa, C.; MacNeil, S.; Gawkrodger, D.J.; Warren, N.D.; Evans, G.S. The effects of nickel and chromium on human keratinocytes: Differences in viability, cell associated metal and IL-1α release. *Toxicol. Vitr.* **2007**, *21*, 809–819. [CrossRef] [PubMed]
32. Larsen, J.M.; Bonefeld, C.M.; Poulsen, S.S.; Geisler, C.; Skov, L. IL-23 and T(H)17-mediated inflammation in human allergic contact dermatitis. *J. Allergy Clin. Immunol.* **2009**, *123*, 486–492. [CrossRef] [PubMed]
33. Sebastiani, S.; Albanesi, C.; Nasorri, F.; Girolomoni, G.; Cavani, A. Nickel-specific CD4(+) and CD8(+) T cells display distinct migratory responses to chemokines produced during allergic contact dermatitis. *J. Investig. Dermatol.* **2002**, *118*, 1052–1058. [CrossRef] [PubMed]
34. Saito, M.; Arakaki, R.; Yamada, A.; Tsunematsu, T.; Kudo, Y.; Ishimaru, N. Molecular Mechanisms of Nickel Allergy. *Int. J. Mol. Sci.* **2016**, *17*, 202. [CrossRef] [PubMed]
35. Schmidt, M.; Raghavan, B.; Müller, V.; Vogl, T.; Fejer, G.; Tchaptchet, S.; Keck, S.; Kalis, C.; Nielsen, P.J.; Galanos, C.; et al. Crucial role for human Toll-like receptor 4 in the development of contact allergy to nickel. *Nat. Immunol.* **2010**, *11*, 814–819. [CrossRef]
36. Steinman, R.M.; Pack, M.; Inaba, K. Dendritic cells in the T-cell areas of lymphoid organs. *Immunol. Rev.* **1997**, *156*, 25–37. [CrossRef]
37. Thyssen, J.P.; Johansen, J.D.; Menné, T. Contact allergy epidemics and their controls. *Contact Dermat.* **2007**, *56*, 185–195. [CrossRef]
38. Thyssen, J.P.; Menné, T. Metal allergy—A review on exposures, penetration, genetics, prevalence, and clinical implications. *Chem. Res. Toxicol.* **2010**, *23*, 309–318. [CrossRef]
39. Johansen, J.D.; Menné, T.; Christophersen, J.; Kaaber, K.; Veien, N. Changes in the pattern of sensitization to common contact allergens in Denmark between 1985–1986 and 1997–1998, with a special view to the effect of preventive strategies. *Br. J. Dermatol.* **2000**, *142*, 490–495. [CrossRef]
40. European Commission (EC). *EC Regulation No 552/2009 of 22 June 2009, Entry 27 in Annex XVII*; European Union: Brussels, Belgium, 2009.
41. European Committee for Standardisation (CEN). *Reference Test Method for Release of Nickel from All Post Assemblies Which Are Inserted into Pierced Parts of the Human Body and Articles Intended to Come into Direct and Prolonged Contact with the Skin*; EN 1811:2011+A1; European Union: Brussels, Belgium, 2015.
42. Menné, T.; Rasmussen, K. Regulation of nickel exposure in Denmark. *Contact Dermat.* **1990**, *23*, 57–58. [CrossRef]
43. Antico, A.; Soana, R. Chronic allergic-like dermatopathies in nickel sensitive patients. Results of dietary restrictions and challenge with nickel salts. *Allergy Asthma Proc.* **1999**, *20*, 235–242. [CrossRef] [PubMed]
44. Di Gioacchino, M.; Ricciardi, L.; De Pità, O.; Minelli, M.; Patella, V.; Voltolini, S.; Di Rienzo, V.; Braga, M.; Ballone, E.; Mangifesta, R.; et al. Nickel oral hyposensitization in patients with systemic nickel allergy syndrome. *Ann. Med.* **2014**, *46*, 31. [CrossRef] [PubMed]
45. European Chemicals Agency (ECHA). *Classification, Labelling and Packaging. Table of Harmonized Entries in Annex VI to CLP*; European Chemicals Agency: Helsinki, Finland, 2018.
46. International Agency for Research on Cancer (IARC). Nickel and nickel compounds. *IARC Monogr. Eval. Carcinog. Risks Hum.* **2012**, *100C*, 169–218.
47. International Committee on Nickel Carcinogenesis in Man (ICNCM). Report of the International Committee on Nickel Carcinogenesis in Man. *Scand. J. Work Environ. Health* **1990**, *16*, 1–82. [CrossRef]

48. National Toxicology Program (NTP). NTP toxicological and carcinogenesis studies of nickel oxide (CAS No. 1313-99-1) in F344/N rats and B6C3F1 mice (inhalation studies). *Natl. Toxicol. Progr. Tech. Rep. Ser.* **1996**, *451*, 1–381.
49. National Toxicology Program (NTP). NTP toxicological and carcinogenesis studies of nickel subsulfide (CAS No. 12035-72-2) in F344/N rats and B6C3F1 mice (inhalation studies). *Natl. Toxicol. Progr. Tech. Rep. Ser.* **1996**, *453*, 1–365.
50. NTP (National Toxicology Program). NTP toxicological and carcinogenesis studies of nickel sulphate hexahydrate (CAS No. 10101-97-0) in F344/N rats and B6C3F1 mice (inhalation studies). *Natl. Toxicol. Progr. Tech. Rep. Ser.* **1996**, *454*, 1–380.
51. Heim, K.E.; Bates, H.K.; Rush, R.E.; Oller, A.R. Oral carcinogenicity study with nickel sulfate hexahydrate in Fischer 344 rats. *Toxicol. Appl. Pharmacol.* **2007**, *224*, 126–137. [CrossRef]
52. Oller, A.R.; Kirkpatrick, D.T.; Radovsky, A.; Bates, H.K. Inhalation carcinogenicity study with nickel metal powder in Wistar rats. *Toxicol. Appl. Pharmacol.* **2008**, *233*, 262–275. [CrossRef]
53. Oller, A.R. Respiratory carcinogenicity assessment of soluble nickel compounds. *Environ. Health Perspect.* **2002**, *110*, 841–844. [CrossRef]
54. Sivulka, D.J. Assessment of respiratory carcinogenicity associated with exposure to metallic nickel: A review. *Regul. Toxicol. Pharmacol.* **2005**, *43*, 117–133. [CrossRef] [PubMed]
55. Easton, D.F.; Peto, J.; Morgan, L.G.; Metcalfe, L.P.; Usher, V.; Doll, R. Respiratory cancer mortality in Welsh nickel refiners: Which nickel compounds are responsible. In *Nickel and Human Health: Current Perspectives*; Nieboer, E., Nriagu, J.O., Eds.; John Wiley & Sons, Inc.: New York, NY, USA, 1992; pp. 603–619.
56. Grimsrud, T.K.; Berge, S.R.; Haldorsen, T.; Andersen, A. Exposure to different forms of nickel and risk of lung cancer. *Am. J. Epidemiol.* **2002**, *156*, 1123–1132. [CrossRef] [PubMed]
57. Arena, V.C.; Sussman, N.B.; Redmond, C.K.; Costantino, J.P.; Trauth, J.M. Using alternative comparison populations to assess occupation-related mortality risk. Results for the high nickel alloys workers cohort. *J. Occup. Environ. Med.* **1998**, *40*, 907–916. [CrossRef] [PubMed]
58. Arena, V.C.; Costantino, J.P.; Sussman, N.B.; Redmond, C.K. Issues and findings in the evaluation of occupational risk among women high nickel alloys workers. *Am. J. Ind. Med.* **1999**, *36*, 114–121. [CrossRef]
59. Morfeld, P.; Groß, J.V.; Erren, T.C.; Noll, B.; Yong, M.; Kennedy, K.J.; Esmen, N.A.; Zimmerman, S.D.; Buchanich, J.M.; Marsh, G.M. Mortality among hardmetal production workers: German historical cohort study. *J. Occup. Environ. Med.* **2017**, *59*, e288–e296. [CrossRef] [PubMed]
60. Marsh, G.M.; Buchanich, J.M.; Zimmerman, S.; Liu, Y.; Balmert, L.C.; Graves, J.; Kennedy, K.J.; Esmen, N.A.; Moshammer, H.; Morfeld, P.; et al. Mortality among hardmetal production workers: Pooled analysis of cohort data from an international investigation. *J. Occup. Environ. Med.* **2017**, *59*, e342–e364. [CrossRef] [PubMed]
61. Marsh, G.M.; Buchanich, J.M.; Zimmerman, S.; Liu, Y.; Balmert, L.C.; Esmen, N.A.; Kennedy, K.J. Mortality among hardmetal production workers: US cohort and nested case-control studies. *J. Occup. Environ. Med.* **2017**, *59*, e306–e326. [CrossRef]
62. Westberg, H.; Bryngelsson, I.-L.; Marsh, G.; Buchanich, J.; Zimmerman, S.; Kennedy, K.; Esmen, N.; Svartengren, M. Mortality among hardmetal production workers: The Swedish cohort. *J. Occup. Environ. Med.* **2017**, *59*, e263–e274. [CrossRef]
63. Westberg, H.; Bryngelsson, I.-L.; Marsh, G.; Kennedy, K.; Buchanich, J.; Zimmerman, S.; Esmen, N.; Svartengren, M. Mortality among hardmetal production workers: Swedish measurement data and exposure assessment. *J. Occup. Environ. Med.* **2017**, *59*, e327–e341. [CrossRef]
64. Dunnick, J.K.; Elwell, M.R.; Radovsky, A.E.; Benson, J.M.; Hahn, F.F.; Nikula, K.J.; Barr, E.B.; Hobbs, C.H. Comparative carcinogenic effects of nickel subsulfide, nickel oxide, or nickel sulfate hexahydrate chronic exposures in the lung. *Cancer Res.* **1995**, *55*, 5251–5256.
65. Hueper, W.C. Experimental studies in metal cancerigenesis. IX. Pulmonary lesions in guinea pigs and rats exposed to prolonged inhalation of powdered metallic nickel. *Arch. Pathol.* **1958**, *65*, 600–607.
66. Pott, F.; Ziem, U.; Reiffer, F.J.; Huth, F.; Ernst, H.; Mohr, U. Carcinogenicity studies of fibres, metal compounds, and some other dusts in rats. *Exp. Pathol.* **1987**, *32*, 129–152. [CrossRef]
67. Wehner, A.P.; Dagle, G.E.; Milliman, E.M. Chronic inhalation exposure of hamsters to nickel-enriched fly ash. *Environ. Res.* **1981**, *26*, 195–216. [CrossRef]

68. National Toxicology Program (NTP). NTP toxicology studies of cobalt metal (CAS No. 7440-48-4) in F344/N rats and B6C3F1/N mice and toxicology and carcinogenesis studies of cobalt metal in F344/NTac rats and B6C3F1/N mice (inhalation studies). *Natl. Toxicol. Progr. Am. Tech. Rep. Ser.* **2014**, *581*, 1–308.
69. National Toxicology Program (NTP). NTP toxicology and carcinogenesis studies of cobalt sulfate heptahydrate (CAS No. 10026-24-1) in F344/N rats and B6C3F1/N mice (inhalation studies). *Natl. Toxicol. Progr. Am. Tech. Rep. Ser.* **1998**, *471*, 1–268.
70. International Agency for Research on Cancer (IARC). Chromium, nickel and welding. *IARC Monogr. Eval. Carcinog. Risks Hum.* **1990**, *49*, 1–648.
71. Uddin, A.N.; Burns, F.J.; Rossman, T.G.; Chen, H.; Kluz, T.; Costa, M. Dietary chromium and nickel enhance UV-carcinogenesis in skin of hairless mice. *Toxicol. Appl. Pharmacol.* **2007**, *221*, 329–338. [CrossRef]
72. Lee, J.E.; Ciccarelli, R.B.; Jennette, K.W. Solubilization of the carcinogen nickel subsulfide and its interaction with deoxyribonucleic acid and protein. *Biochemistry* **1982**, *21*, 771–778. [CrossRef]
73. Biggart, N.W.; Costa, M. Assessment of the uptake and mutagenicity of nickel chloride in Salmonella tester strains. *Mutat. Res.* **1986**, *175*, 209–215. [CrossRef]
74. Chen, H.; Giri, N.C.; Zhang, R.; Yamane, K.; Zhang, Y.; Maroney, M.; Costa, M. Nickel ions inhibit histone demethylase JMJD1A and DNA repair enzyme ABH2 by replacing the ferrous iron in the catalytic centers. *J. Biol. Chem.* **2010**, *285*, 7374–7383. [CrossRef]
75. Zhang, J.; Zhou, Y.; Ma, L.; Huang, S.; Wang, R.; Gao, R.; Wu, Y.; Shi, H.; Zhang, J. The alteration of miR-222 and its target genes in nickel-induced tumor. *Biol. Trace Elem. Res.* **2013**, *152*, 267–274. [CrossRef] [PubMed]
76. Zhang, J.; Zhou, Y.; Wu, Y.J.; Li, M.J.; Wang, R.J.; Huang, S.Q.; Gao, R.R.; Ma, L.; Shi, H.J.; Zhang, J. Hyper-methylated miR-203 dysregulates ABL1 and contributes to the nickel-induced tumorigenesis. *Toxicol. Lett.* **2013**, *223*, 42–51. [CrossRef] [PubMed]
77. Ji, W.; Yang, L.; Yuan, J.; Yang, L.; Zhang, M.; Qi, D.; Duan, X.; Xuan, A.; Zhang, W.; Lu, J.; et al. MicroRNA-152 targets DNA methyltransferase 1 in NiS-transformed cells via a feedback mechanism. *Carcinogenesis* **2013**, *34*, 446–453. [CrossRef]
78. Chen, C.-Y.; Wang, Y.-F.; Huang, W.-R.; Huang, Y.-T. Nickel induces oxidative stress and genotoxicity in human lymphocytes. *Toxicol. Appl. Pharmacol.* **2003**, *189*, 153–159. [CrossRef]
79. Åkerlund, E.; Cappellini, F.; Di Bucchianico, S.; Islam, S.; Skoglund, S.; Derr, R.; Wallinder, I.O.; Hendriks, G.; Karlsson, H.L. Genotoxic and mutagenic properties of Ni and NiO nanoparticles investigated by comet assay, γ-H2AX staining, Hprt mutation assay and ToxTracker reporter cell lines. *Environ. Mol. Mutagen.* **2018**, *59*, 211–222. [CrossRef] [PubMed]
80. Jia, J.; Chen, J. Chronic nickel-induced DNA damage and cell death: The protection role of ascorbic acid. *Environ. Toxicol.* **2008**, *23*, 401–406. [CrossRef] [PubMed]
81. Benson, J.M.; March, T.H.; Hahn, F.F.; Seagrave, J.C.; Divine, K.K.; Belinsky, S.A. *Short-Term Inhalation Study with Nickel Compounds*; Final Report to NiPERA, Inc.; Lovelace Respiratory Research Institute: Albuquerque, NM, USA, 2002.
82. Kawanishi, S.; Oikawa, S.; Inoue, S.; Nishino, K. Distinct mechanisms of oxidative DNA damage induced by carcinogenic nickel subsulfide and nickel oxides. *Environ. Health Perspect.* **2002**, *110*, 789–791. [CrossRef]
83. Saplakoglu, U.; Iscan, M.; Iscan, M. DNA single-strand breakage in rat lung, liver and kidney after single and combined treatments of nickel and cadmium. *Mutagen. Res.* **1997**, *394*, 133–140.
84. Werfel, U.; Langen, V.; Eickhoff, I.; Schoonbrood, J.; Vahrenholz, C.; Brauksiepe, A.; Popp, W.; Norpoth, K. Elevated DNA single-strand breakage frequencies in lymphocytes of welders exposed to chromium and nickel. *Carcinogenesis* **1998**, *19*, 413–418. [CrossRef]
85. Danadevi, K.; Rozati, R.; Banu, B.S.; Grover, P. Genotoxic evaluation of welders occupationally exposed to chromium and nickel using the comet and micronucleus assays. *Mutagenesis* **2004**, *19*, 35–41. [CrossRef]
86. Faccioni, F.; Franceschetti, P.; Cerpelloni, M.; Fracasso, M.E. In vivo study on metal release from fixed orthodontic appliances and DNA damage in oral mucosa cells. *Am. J. Orthod. Dentofac. Orthop.* **2003**, *124*, 687–694. [CrossRef]
87. Hafez, H.S.; Selim, E.M.N.; Kamel Eid, F.H.; Tawfik, W.A.; Al-Ashkar, E.A.; Mostafa, Y.A. Cytotoxity, genotoxicity, and metal release in patients with fixed orthodontic appliances: A longitudinal in vivo study. *Am. J. Orthod. Dentofac. Orthop.* **2011**, *140*, 298–308. [CrossRef] [PubMed]

88. Fernández-Miñano, E.; Ortiz, C.; Vicente, A.; Calvo, J.L.; Ortiz, A.J. Metallic ion content and damage to the DNA in oral mucosa cells of children with fixed orthodontic appliances. *Biometals* **2011**, *24*, 935–941. [CrossRef] [PubMed]
89. Hartwig, A.; Mullenders, L.H.F.; Schlepegrell, R.; Kasten, U.; Beyersmann, D. Nickel(II) interferes with the incision step in nucleotide excision repair in mammalian cells. *Cancer Res.* **1994**, *54*, 4045–4051. [PubMed]
90. Hartwig, A.; Schlepegrell, R.; Dally, H.; Hartmann, M. Interaction of carcinogenic metal compounds with deoxyribonucleic acid repair processes. *Ann. Clin. Lab. Sci.* **1996**, *26*, 31–38. [PubMed]
91. Hartwig, A.; Beyersmann, D. Enhancement of UV-induced mutagenesis and sister-chromatid exchanges by nickel ions in V79 cells: Evidence for inhibition of DNA repair. *Mutat. Res.* **1989**, *217*, 65–73. [CrossRef]
92. Mayer, C.; Klein, R.G.; Wesch, H.; Schmezer, P. Nickel subsulfide is genotoxic in vitro but shows no mutagenic potential in respiratory tract tissues of BigBlue^{TM} rats and Muta^{TM} Mouse mice in vivo after inhalation. *Mutat. Res.* **1998**, *420*, 85–98. [CrossRef]
93. Arrouijal, F.Z.; Hildebrand, H.F.; Vophi, H.; Marzin, D. Genotoxic activity of nickel subsulphide α-Ni$_3$S$_2$. *Mutagenesis* **1990**, *5*, 583–589. [CrossRef]
94. Fletcher, G.G.; Rossetto, F.E.; Turnbull, J.D.; Nieboer, E. Toxicity, uptake, and mutagenicity of particulate and soluble nickel compounds. *Environ. Health Perspect.* **1994**, *102* (Suppl. 3), 69–79.
95. Kargacin, B.; Klein, C.B.; Costa, M. Mutagenic responses of nickel oxides and nickel sulfides in Chinese Hamster V79 cell lines at the xanthine-guanine phosphoribosyl transferase locus. *Mutat. Res.* **1993**, *300*, 63–72. [CrossRef]
96. Danish Environmental Protection Agency (Danish EPA). *Nickel and Nickel Compounds: Background Document in Support of Individual Risk Assessment Reports of Nickel Compounds Prepared in Relation to Council Regulation (EEC) 793/93*; Danish Environmental Protection Agency: Copenhagen, Denmark, 2008.
97. Nishimura, M.; Umeda, M. Induction of chromosomal aberrations in cultured mammalian cells by nickel compounds. *Mutat. Res.* **1979**, *68*, 337–349. [CrossRef]
98. Sen, P.; Costa, M. Incidence and localization of sister chromatid exchanges induced by nickel and chromium compounds. *Carcinogenesis* **1986**, *7*, 1527–1533. [CrossRef] [PubMed]
99. Oller, A.R.; Erexson, G. Lack of micronuclei formation in bone marrow of rats after repeated oral exposure to nickel sulfate hexahydrate. *Mutat. Res.* **2007**, *626*, 102–110. [CrossRef] [PubMed]
100. Dhir, H.; Agarwal, K.; Sharma, A.; Talukder, G. Modifying role of Phyllanthus emblica and ascorbic acid against nickel clastogenicity in mice. *Cancer Lett.* **1991**, *59*, 9–18. [CrossRef]
101. Morita, T.; Asano, N.; Awogi, T.; Sasaki, Y.F.; Sato, S.; Shimada, H.; Sutou, S.; Suzuki, T.; Wakata, A.; Sofuni, T.; et al. Evaluation of the rodent micronucleus assay in the screening of IARC carcinogens (Groups 1, 2A and 2B), The summary report of the 6th collaborative study by CSGMT/JEMS.MMS. *Mutat. Res.* **1997**, *389*, 3–122. [CrossRef]
102. Waksvik, H.; Boysen, M.; Brogger, A.; Klepp, O. Chromosome aberrations and sister chromatid exchanges in persons occupationally exposed to mutagens/carcinogens. In *Chromosome Damage Repair*; NATO Advanced Study Institutes Series (Series A: Life Sciences); Seeberg, E., Kleppe, K., Eds.; Springer: New York, NY, USA, 1981; Volume 40, pp. 563–566.
103. Waksvik, H.; Boysen, M.; Hogetveit, C. Increased incidence of chromosomal aberrations in peripheral lymphocytes of retired nickel workers. *Carcinogenesis* **1984**, *5*, 1525–1527. [CrossRef]
104. Morán-Martínez, J.; Monreal-de Luna, K.D.; Betancourt-Martínez, N.D.; Carranza-Rosales, P.; Contreras-Martínez, J.G.; López-Meza, M.C.; Rodríguez-Villareal, O. Genotoxicity in oral epithelial cells in children caused by nickel in metal crowns. *Genet. Mol. Res.* **2013**, *12*, 3178–3185. [CrossRef]
105. Natarajan, M.; Padmanabhan, S.; Chiharajan, A.; Narasimhan, M. Evaluation of the genotoxic effects of fixed appliances on oral mucosal cells and the relationship to nickel and chromium concentrations: An in vivo study. *Am. J. Orthod. Dentofac. Orthop.* **2011**, *140*, 383–388. [CrossRef]
106. Heravi, F.; Abbaszadegan, M.R.; Merati, M.; Hasanzadeh, N.; Dadkhah, E.; Ahrari, F. DNA damage in oral mucosa cells of patients with fixed orthodontic appliances. *J. Dent.* **2013**, *10*, 494–500.
107. Westphalen, G.H.; Menezes, L.M.; Prá, D.; Garcia, G.G.; Schmitt, V.M.; Henriques, J.A.; Medina-Silva, R. In vivo determination of genotoxicity induced by metals from orthodontic appliances using micronucleus and comet assays. *Genet. Mol. Res.* **2008**, *7*, 1259–1266. [CrossRef]
108. Zhou, D.; Salnikow, K.; Costa, M. Cap43, a novel gene specifically induced by Ni^{2+} compounds. *Cancer Res.* **1998**, *58*, 2182–2189. [PubMed]

109. Salnikow, K.; An, W.G.; Melillo, G.; Blagosklonny, M.V.; Costa, M. Nickel-induced transformation shifts the balance between HIF-1 and p53 transcription factors. *Carcinogenesis* **1999**, *20*, 1819–1823. [CrossRef] [PubMed]
110. Salnikow, K.; Blagosklonny, M.V.; Ryan, H.; Johnson, R.; Costa, M. Carcinogenic nickel induces genes involved with hypoxic stress. *Cancer Res.* **2000**, *60*, 38–41. [PubMed]
111. Salnikow, K.; Davidson, T.; Costa, M. The role of hypoxia-inducible signaling pathway in nickel carcinogenesis. *Health Perspect.* **2002**, *110* (Suppl. 5), 831–834. [CrossRef]
112. Kang, Y.-T.; Hsu, W.-C.; Wu, C.-H.; Hsin, I.-L.; Wu, P.-R.; Yeh, K.-T.; Ko, J.-L. Metformin alleviates nickel-induced autophagy and apoptosis via inhibition of hexokinase-2, activating lipocalin-2, in human bronchial epithelial cells. *Oncotarget* **2017**, *8*, 105536–105552. [CrossRef] [PubMed]
113. Wong, V.C.; Morse, J.L.; Zhitkovich, A. p53 activation by Ni(II) is a HIF-1α independent response causing caspases 9/3-mediated apoptosis in human lung cells. *Toxicol. Appl. Pharmacol.* **2013**, *269*, 233–239. [CrossRef] [PubMed]
114. Green, S.E.; Luczak, W.W.; Morse, J.L.; DeLoughery, Z.; Zhitkovich, A. Uptake, p53 pathway activation, and cytotoxic responses for Co(II) and Ni(II) in human lung cells: Implications for carcinogenicity. *Toxicol. Sci.* **2013**, *136*, 467–477. [CrossRef] [PubMed]
115. Di Bucchianico, S.; Gliga, A.R.; Åkerlund, E.; Skoglund, S.; Wallinder, I.O.; Fadeel, B.; Karlsson, H.L. Calcium-dependent cyto-and genotoxicity of nickel metal and nickel oxide nanoparticles in human lung cells. *Part. Fibre Toxicol.* **2018**, *15*, 32. [CrossRef] [PubMed]
116. Bonin, S.; Larese, F.F.; Trevisan, G.; Avian, A.; Rui, F.; Stanta, G.; Bovenzi, M. Gene expression changes in peripheral blood mononuclear cells in occupational exposure to nickel. *Exp. Dermatol.* **2011**, *20*, 147–148. [CrossRef]
117. Chen, H.; Ke, Q.; Kluz, T.; Yan, Y.; Costa, M. Nickel ions increase histone H3 lysine 9 dimethylation and induce transgene silencing. *Mol. Cell. Biol.* **2006**, *26*, 3728–3737. [CrossRef]
118. Chen, H.; Kluz, T.; Zhang, R.; Costa, M. Hypoxia and nickel inhibit histone demethylase JMJD1A and repress Spry2 expression in human bronchial epithelial BEAS-2B cells. *Carcinogenesis* **2010**, *31*, 2136–2144. [CrossRef] [PubMed]
119. Ma, L.; Bai, Y.; Pu, H.; Gou, F.; Dai, M.; Wang, H.; He, J.; Zheng, T.; Cheng, N. Histone methylation in nickel-smelting industrial workers. *PLoS ONE* **2015**, *10*, e0140339. [CrossRef] [PubMed]
120. Arita, A.; Niu, J.; Qu, Q.; Zhao, N.; Ruan, Y.; Nadas, A.; Chervona, Y.; Wu, F.; Sun, H.; Hayes, R.B.; et al. Global levels of histone modifications in peripheral blood mononuclear cells of subjects with exposure to nickel. *Environ. Health Perspect.* **2012**, *120*, 198–203. [CrossRef] [PubMed]
121. Cantone, L.; Nordio, F.; Hou, L.; Apostoli, P.; Bonzini, M.; Tarantini, L.; Angelici, L.; Bollati, V.; Zanobetti, A.; Schwartz, J.; et al. Inhalable metal-rich air particles and histone H3K4 dimethylation and H3K9 acetylation in a cross-sectional study of steel workers. *Environ. Health Perspect.* **2011**, *119*, 964–969. [CrossRef] [PubMed]
122. Ke, Q.; Li, Q.; Ellen, T.P.; Sun, H.; Costa, M. Nickel compounds induce phosphorylation of histone H3 at serine 10 by activating JNK-MAPK pathway. *Carcinogenesis* **2008**, *29*, 1276–1281. [CrossRef]
123. Ke, Q.; Ellen, T.P.; Costa, M. Nickel compounds induce histone ubiquitination by inhibiting histone deubiquitinating enzyme activity. *Toxicol. Appl. Pharmacol.* **2008**, *228*, 190–199. [CrossRef] [PubMed]
124. Kang, J.; Zhang, Y.; Chen, J.; Chen, H.; Lin, C.; Wang, Q.; Ou, Y. Nickel-induced histone hypoacetylation: The role of reactive oxygen species. *Toxicol. Sci.* **2003**, *74*, 279–286. [CrossRef]
125. Kang, J.; Zhang, D.; Chen, J.; Lin, C.; Liu, Q. Involvement of histone hypoacetylation in Ni^{2+}-induced bcl-2 down-regulation and human hepatoma cell apoptosis. *J. Biol. Inorg. Chem.* **2004**, *9*, 713–723. [CrossRef]
126. Huang, X.; Frenkel, K.; Klein, C.B.; Costa, M. Nickel induces increased oxidants in intact cultured mammalian cells as detected by dichlorofluorescein fluorescence. *Toxicol. Appl. Pharmacol.* **1993**, *120*, 29–36.
127. Huang, X.; Klein, C.B.; Costa, M. Crystalline Ni_3S_2 specifically enhances the formation of oxidants in the nuclei of CHO cells as detected by dichlorofluorescein. *Carcinogenesis* **1994**, *15*, 545–548. [CrossRef]
128. Jadhav, S.H.; Sarkar, S.N.; Aggarwal, M.; Tripathi, H.C. Induction of oxidative stress in erythrocytes of male rats subchronically exposed to a mixture of eight metals found as groundwater contaminants in different parts of India. *Arch. Environ. Contam. Am. Toxicol.* **2007**, *52*, 145–151. [CrossRef] [PubMed]
129. El-Habit, O.H.; Moneim, A.E. Testing the genotoxicity, cytotoxicity, and oxidative stress of cadmium and nickel and their additive effect in male mice. *Biol. Trace Elem. Res.* **2014**, *159*, 364–372. [CrossRef] [PubMed]

130. Merzenich, H.; Hartwig, A.; Ahrens, W.; Beyersmann, D.; Schlepegrell, R.; Scholze, M.; Timm, J.; Jöckel, K.-H. Biomonitoring on carcinogenic metals and oxidative DNA damage in a cross-sectional study. *Cancer Epidemiol. Prev. Biomark.* **2001**, *10*, 515–522.
131. Chaschschin, V.P.; Artunina, G.P.; Norseth, T. Congenital defects, abortion and other health effects in nickel refinery workers. *Sci. Total Environ.* **1994**, *148*, 287–291. [CrossRef]
132. Vaktskjold, A.; Talykova, L.; Chashchin, V.; Odland, J.; Nieboer, E. Small-for-gestational age newborns of female refinery workers exposed to nickel. *Int. J. Occup. Med. Environ. Health* **2007**, *20*, 327–338. [CrossRef] [PubMed]
133. Vaktskjold, A.; Talykova, L.; Chashchin, V.; Odland, J.; Nieboer, E. Spontaneous abortions among nickel-exposed female refinery workers. *Int. J. Environ. Health Res.* **2008**, *18*, 99–115. [CrossRef] [PubMed]
134. Vaktskjold, A.; Talykova, L.; Chashchin, V.; Nieboer, E.; Thomassen, Y.; Odland, J. Genital malformations in newborns of female nickel-refinery workers. *Scand. J. Work Environ. Health* **2006**, *32*, 41–50. [CrossRef]
135. Vaktskjold, A.; Talykova, L.; Chashchin, V.; Odland, J.; Nieboer, E. Maternal nickel exposure and congenital musculoskeletal defects. *Am. J. Ind. Med.* **2008**, *51*, 825–833. [CrossRef]
136. Ebisu, K.; Bell, M.L. Airborne $PM_{2.5}$ chemical components and low birth weight in the Northeastern and Mid-Atlantic regions of the United States. *Environ. Health Perspect.* **2012**, *120*, 1746–1752. [CrossRef]
137. Ambrose, A.M.; Larson, P.S.; Borzelleca, J.F.; Hennigar, G.R., Jr. Long-term toxicologic assessment of nickel in rats and dogs. *J. Food Sci. Technol.* **1976**, *13*, 181–187.
138. Smith, M.K.; George, E.L.; Stober, J.A.; Feng, H.A.; Kimmel, G.A. Perinatal toxicity associated with nickel chloride. *Environ. Res.* **1993**, *61*, 200–211. [CrossRef] [PubMed]
139. Research Triangle Institute (RTI). *Fertility and Reproductive Performance of the P(0) Generation: Two-Generation Reproduction and Fertility Study of Nickel Chloride Administered to CD Rats in the Drinking Water*; Final Report, RTI Master Protocol No. 182, Study Code No. RT85-NICL.REPRO, Project No. 472U-3228-07, EPA No. 68-01-7075; Research Triangle Institute: Research Triangle Park, NC, USA, 1988.
140. Research Triangle Institute (RTI). *Fertility and Reproductive Performance of the F(1) Generation: Two-Generation Reproduction and Fertility Study of Nickel Chloride Administered to CD Rats in the Drinking Water*; Final Report; Research Triangle Institute: Research Triangle Park, NC, USA, 1988.
141. Springborn Laboratories, Inc. (SLI). *An Oral (Gavage) 1-Generation Reproduction Study of Nickel Sulfate Hexahydrate in Rats*; Study No. 3472.3; Springborn Laboratories, Inc.: Spencerville, OH, USA, 2000.
142. Springborn Laboratories, Inc. *An Oral (Gavage) Two-Generation Reproduction Toxicity Study in Sprague-Dawley Rats with Nickel Sulfate Hexahydrate*; Study No. 3472.4; Springborn Laboratories, Inc.: Spencerville, OH, USA, 2000.
143. Ishimatsu, S.; Kawamoto, T.; Matsuno, K.; Kodama, Y. Distribution of various nickel compounds in rat organs after oral administration. *Biol. Trace Elem. Res.* **1995**, *49*, 43–52. [CrossRef] [PubMed]
144. Classification, Labelling and Packaging (CLP). *Regulation (EC) No 1272/2008 of the European Parliament and of the Council of 16 December 2008 on Classification, Labelling and Packaging of Substances and Mixtures*; Amending and Repealing Directives 67/548/EEC and 1999/45/EC, and Amending Regulation (EC) No 1907/2006; European Union: Brussels, Belgium, 2008.
145. Haber, L.T.; Bates, H.K.; Allen, B.C.; Vincent, M.J.; Oller, A.R. Derivation of an oral toxicity reference value for nickel. *Regul. Toxicol. Pharmacol.* **2017**, *87* (Suppl. 1), S1–S18. [CrossRef]
146. Saini, S.; Nair, N.; Saini, M.R. Embryotoxic and teratogenic effects of nickel in Swiss albino mice during organogenetic period. *BioMed. Res. Int.* **2013**, *2013*, 701439. [CrossRef] [PubMed]
147. Akesson, B.; Skerfving, S. Exposure in welding of high nickel alloy. *Int. Arch. Occup. Environ. Health* **1985**, *56*, 111–117. [CrossRef] [PubMed]
148. Phillips, J.I.; Green, F.Y.; Davies, J.C.; Murray, J. Pulmonary and systemic toxicity following exposure to nickel nanoparticles. *Am. J. Ind. Med.* **2010**, *53*, 763–767. [CrossRef]
149. California Environmental Protection Agency (Cal EPA). *Nickel Reference Exposure Levels*; Office of Environmental Health Hazard Assessment: Sacramento, CA, USA, 2012.
150. Muir, D.C.F.; Julian, J.; Roberts, R.; Roos, J.; Chan, J.; Machle, W.; Morgan, W.K.C. Prevalence of small opacities in chest radiographs of nickel sinter plant workers. *Br. J. Ind. Med.* **1993**, *50*, 428–431. [CrossRef]
151. Berge, S.R.; Skyberg, K. Radiographic evidence of pulmonary fibrosis and possible etiologic factors at a nickel refinery in Norway. *J. Environ. Monit.* **2003**, *5*, 681–688. [CrossRef]

152. Miller, A.; Warshaw, R.; Nezamis, J. Diffusing capacity and forced vital capacity in 5003 asbestos-exposed workers: relationships to interstitial fibrosis (ILO profusion score) and pleural thickening. *Am. J. Ind. Med.* **2013**, *56*, 1383–1393. [CrossRef]
153. Sorahan, T. Mortality of workers at a plant manufacturing nickel alloys, 1958–2000. *Occup. Med.* **2004**, *54*, 28–34. [CrossRef]
154. Sorahan, T.; Williams, S.P. Mortality of workers at a nickel carbonyl refinery, 1958–2000. *Occup. Environ. Med.* **2005**, *62*, 80–85. [CrossRef] [PubMed]
155. Moulin, J.J.; Clavel, T.; Roy, D.; Danaché, B.; Marquis, N.; Févotte, J.; Fontana, J.M. Risk of lung cancer in workers producing stainless steel and metallic alloys. *Int. Arch. Occup. Environ. Health* **2000**, *73*, 171–180. [CrossRef] [PubMed]
156. Cragle, D.L.; Hollis, D.R.; Newport, T.H.; Shy, C.M. A Retrospective Cohort Mortality Study Aamon Workers Occupationally Exposed to Metallic Nickel Powder at the Oak Ridge Gaseous Diffusion Plant. In *Nickel in the Human Environment, Proceedings of a JoInt. Symposium, IARC Scientific Publications No. 53, Lyon, France, 6–8 October March 1983*; Sunderman, F.W., Jr., Ed.; International Agency for Research on Cancer: Lyon, France, 1984; pp. 57–63.
157. Block, G.T.; Yeung, M. Asthma induced by nickel. *JAMA* **1982**, *247*, 1600–1602. [CrossRef] [PubMed]
158. Estlander, T.; Kanerva, L.; Tupasela, O.; Keskinen, H.; Jolanki, R. Immediate and delayed allergy to nickel with contact urticaria, rhinitis, asthma and contact dermatitis. *Clin. Exp. Allergy* **1993**, *23*, 306–310. [CrossRef] [PubMed]
159. Bright, P.; Burge, P.S.; O'Hickey, S.P.; Gannon, P.F.G.; Robertson, A.S.; Boran, A. Occupational asthma due to chrome and nickel electroplating. *Thorax* **1997**, *52*, 28–32. [CrossRef] [PubMed]
160. Fernandez-Nieto, M.; Quirce, S.; Carnes, J.; Sastre, J. Occupational asthma due to chromium and nickel salts. *Int. Arch. Occup. Environ. Health* **2006**, *79*, 483–486. [CrossRef] [PubMed]
161. United States Environmental Protection Agency (US EPA). *Guidelines for Deriving Numerical National Water Quality Criteria for the Protection of Aquatic Organisms and Their Uses*; Office of Research and Development, Environmental Research Laboratories: Duluth, MN, USA; Narragansett, RI, USA; Corvallis, OR, USA, 1985; pp. 1–59.
162. Chau, Y.K.; Kulikovsky-Cordeiro, O.T.R. Occurrence of nickel in the Canadian environment. *Environ. Rev.* **1995**, *3*, 95–120. [CrossRef]
163. Ankley, G.T.; Di Toro, D.M.; Hansen, D.J.; Berry, W.J. Technical basis and proposal for deriving sediment quality criteria for metals. *Environ. Toxicol. Chem.* **1996**, *15*, 2056–2066. [CrossRef]
164. Merrington, G.; Peters, A.; Schlekat, C. Accounting for metal bioavailability in assessing water quality: A step change? *Environ. Toxicol. Chem.* **2016**, *35*, 257–265. [CrossRef]
165. Peters, A.; Merrington, G.; Schlekat, C.; de Schamphelaere, K.; Stauber, J.; Batley, G.; Harford, A.; van Dam, R.; Pease, C.; Mooney, T.; et al. Validation of the nickel biotic ligand model for locally relevant species in Australian freshwaters. *Environ. Toxicol. Chem.* **2018**, *37*, 2566–2574. [CrossRef]
166. Chapman, P.M. Environmental risks of inorganic metals and metalloids: A continuing, evolving scientific odyssey. *Hum. Ecol. Risk Assess.* **2008**, *14*, 5–40. [CrossRef]
167. Deleebeeck, N.; De Schamphelaere, K.; Janssen, C. A bioavailability model predicting the toxicity of nickel to rainbow trout (Oncorhynchus mykiss) and fathead minnow (Pimephales promelas) in synthetic and natural waters. *Ecotoxicol. Toxicol. Environ. Saf.* **2007**, *67*, 1–13. [CrossRef] [PubMed]
168. Deleebeeck, N.; De Schamphelaere, K.; Janssen, C. A novel method for predicting chronic nickel bioavailability and toxicity to Daphnia magna in artificial and natural waters. *Environ. Toxicol. Chem.* **2008**, *27*, 2097–2107. [CrossRef]
169. Deleebeeck, N.; De Schamphelaere, K.; Janssen, C. Effects of Mg^{2+} and H^+ on the toxicity of Ni^{2+} to the uni-cellular green alga Pseudokirchneriella subcapitata: Model development and validation with surface waters. *Sci. Total Environ.* **2009**, *407*, 1901–1914. [CrossRef] [PubMed]
170. Lock, K.; Van Eeckhout, H.; De Schamphelaere, K.; Criel, P.; Janssen, C. Development of a Biotic Ligand Model (BLM) predicting nickel toxicity to barley (Hordeum vulgare). *Chemosphere* **2007**, *66*, 1346–1352. [CrossRef] [PubMed]
171. Schlekat, C.; DeSchamphelaere, K.; Van Genderen, E.; Antunes, P.; Stubblefield, W.; Rogevich, E. Cross-species extrapolation of chronic nickel Biotic Ligand Models. *Sci. Total Environ.* **2010**, *408*, 6148–6157. [CrossRef] [PubMed]

172. Nys, C.; Janssen, C.; van Sprang, P.; de Schamphelaere, K. The effect of pH on chronic aquatic Ni toxicity is dependent on the pH itself: Extending the chronic Ni bioavailability models. *Environ. Toxicol. Chem.* **2016**, *35*, 1097–1106. [CrossRef] [PubMed]
173. Deleebeeck, N.M.E.; Muyssen, B.; De Lander, F.; De Schamphelaere, K.A.C.; Janssen, C.R. Comparison of nickel toxicity to cladocerans in soft versus hard surface waters. *Aquat. Toxicol.* **2007**, *84*, 223–235. [CrossRef]
174. Santore, R.C.; Di Toro, D.M.; Paquin, P.R.; Allen, H.E.; Meyer, J.S. Biotic ligand model of the acute toxicity of metals. 2. Application to acute copper toxicity in freshwater fish and Daphnia. *Environ. Toxicol. Chem.* **2001**, *20*, 2397–2402. [CrossRef]
175. Paquin, P.R.; Gorsuch, J.W.; Apte, S.; Batley, G.E.; Bowles, K.C.; Campbell, P.G.C.; Delos, C.G.; Di Toro, D.M.; Dwyer, R.L.; Galvez, F.; et al. The biotic ligand model: A historical overview. *Comp. BioChem. Physiol. C Toxicol. Pharmacol.* **2002**, *133*, 3–35. [CrossRef]
176. Nys, C.; Janssen, C.; De Schamphelaere, K. *Evaluation of Acute Ni Bioavailability Models for Model and Non-Model Species*; Faculty of Bioscience Engineering, Laboratory of Environmental Toxicology and Aquatic Ecology, Ghent University: Ghent, Belgium, 2015.
177. Blewett, T.A.; Dow, E.; Wood, C.M.; McGeer, J.C.; Smith, D.S. The role of dissolved organic carbon concentration and composition in ameliorating nickel toxicity to early life-stages of the blue mussel Mytilus edulis. *Ecotoxicol. Toxicol. Environ. Saf.* **2018**, *160*, 162–170. [CrossRef] [PubMed]
178. Sherman, S. Influence of Ligand Complexation on Nickel Toxicity, Speciation and Bioavailability in Marine Waters. Master's Thesis, Wilfrid Laurier University, Waterloo, ON, Canada, 2019.
179. Blewett, T.A.; Smith, D.S.; Wood, C.M.; Glover, C.N. Mechanisms of nickel toxicity in the highly sensitive embryos of the sea urchin Evechinus chloroticus, and the modifying effects of dissolved organic carbon. *Environ. Sci. Technol.* **2016**, *2*, 1595–1603. [CrossRef] [PubMed]
180. US EPA (United States Environmentmental Protection Agency). *The Aquatic Life Ambient Freshwater Criteria–Copper*; 2007 Revision (CAS Registry Number 7440-50-8); US Environmental Protection Agency, Office of Water, Office of Science and Technology: Washington, DC, USA, 2007.
181. Warne, M.; Batley, G.E.; van Dam, R.A.; Chapman, J.C.; Fox, D.R.; Hickey, C.W.; Stauber, J.L. *Revised Method for Deriving Australian and New Zealand Water Quality Guideline Values for Toxicants–Update of 2015 Version, Prepared for the Revision of the Australian and New Zealand Guidelines for Fresh and Marine Water Quality*; Australian and New Zealand Governments and Australian State and Territory Governments: Canberra, Australia, 2018.
182. Canadian Council of Ministers of the Environment (CCME). A Protocol for the Derivation of Water Quality Guidelines for the Protection of Aquatic Life 2007. In *Canadian Environmental Quality Guidelines, 1999*; Canadian Council of Ministers of the Environment: Winnipeg, MB, Canada, 2007.
183. European Commission (EC). *Nickel and Its Compounds (Final Revision Oct 12 2010) EQS Sheet*; Prepared by Denmark, Danish Environmental Protection Agency on behalf of the European Union; European Union: Brussels, Belgium, 2010.
184. European Commission (EC). *Common Implementation Strategy for the Water Framework Directive (2000/60/EC) Guidance Document No. 27 Technical Guidance for Deriving Environmental Quality Standards*; Technical Report for Approval by Water Directors-WD 2018-1-1; Office for Official Publications in the European Communities: Luxembourg, 2018.
185. Paller, M.H.; Knox, A.S. Bioavailability of metals in contaminated sediments. *EDP Sci.* **2013**, *1*, 2001. [CrossRef]
186. Besser, J.M.; Brumbaugh, W.G.; Ingersoll, C.G.; Ivey, C.D.; Kunz, J.L.; Kemble, N.E.; Schlekat, C.E.; Garman, E.R. Chronic Toxicity of Nickel-Spiked Freshwater Sediment: Variation in Toxicity Among Eight Invertebrate Taxa and Eight Sediments. *Environ. Toxicol. Chem.* **2013**, *32*, 2495–2506. [CrossRef] [PubMed]
187. Vangheluwe, M.; Verdonck, F.; Besser, J.M.; Brumbaugh, W.G.; Ingersoll, C.G.; Schlekat, C.E.; Garman, E.R. Improving sediment quality guidelines for nickel: Development and application of predictive bioavailability models to assess chronic toxicity of nickel in freshwater sediments. *Environ. Toxicol. Chem.* **2013**, *32*, 2507–2519. [CrossRef] [PubMed]
188. Di Toro, D.M.; Mahony, J.D.; Hansen, D.J.; Scott, K.J.; Carlson, A.R.; Ankley, G.T. Acid volatile sulfide predicts the acute toxicity of cadmium and nickel in sediments. *Environ. Sci. Technol.* **1992**, *26*, 96–101. [CrossRef]

189. Hansen, D.J.; Berry, W.J.; Mahony, J.D.; Boothman, W.S.; Di Toro, D.M.; Robson, D.L.; Ankley, G.T.; Ma, D.; Yan, Q.; Pesch, C.E. Predicting the toxicity of metal-contaminated field sediments using interstitial concentrations of metals and acid-volatile sulfide normalizations. *Environ. Toxicol. Chem.* **1996**, *15*, 2080–2094. [CrossRef]
190. Di Toro, D.M.; McGrath, J.A.; Hansen, D.J.; Berry, W.J.; Paquin, P.R.; Mathew, R.; Wu, K.B.; Santore, R.C. Predicting sediment metal toxicity using a sediment biotic ligand model: Methodology and initial application. *Environ. Toxicol. Chem.* **2005**, *24*, 2410–2427. [CrossRef]
191. Vangheluwe, M.; Nguyen, L. *Advanced Research on Nickel Toxicity in Sediments: Results Additional Species and Modelling*; Final Report to NiPERA; Nickel Producers Environmental Research Association, Inc.: Durham, NC, USA, 2015.
192. Costello, D.M.; Burton, G.A.; Hammerschmidt, C.R.; Rogevich, E.C.; Schlekat, C.E. Nickel phase partitioning and toxicity in field-deployed sediments. *Environ. Sci. Technol.* **2011**, *45*, 5798–5805. [CrossRef]
193. Lock, K.; Janssen, C. Ecotoxicity of nickel to Eisenia fetida, Enchytraeus albidus, and Folsomia candida. *Chemosphere* **2002**, *46*, 197–200. [CrossRef]
194. Smolders, E.; Oorts, K.; Van Sprang, P.; Schoeters, I.; Janssen, C.; McGrath, S.; McLaughlin, M. Toxicity of trace metals in soil as affected by soil type and aging after contamination: Using calibrated bioavailability models to set ecological soil standards. *Environ. Toxicol. Chem.* **2009**, *28*, 1633–1642. [CrossRef] [PubMed]
195. Rooney, C.; Zhao, F.J.; McGrath, S. Phytotoxicity of nickel in a range of European soils: Influence of soil properties, nickel solubility, and speciation. *Environ. Pollut.* **2007**, *145*, 596–605. [CrossRef] [PubMed]
196. European Commission (EC). *European Union Risk Assessment Report (EU RAR) of Nickel and Nickel Compounds*; Prepared by the Danish Environmental Protection Agency (DEPA); European Union: Brussels, Belgium, 2008.
197. Binet, M.T.; Adams, M.S.; Gissi, F.; Golding, L.A.; Schlekat, C.E.; Garman, E.R.; Merrington, G.; Stauber, J.L. Toxicity of nickel to tropical freshwater and sediment biota: A critical literature review and gap analysis. *Environ. Toxicol. Chem.* **2017**, *37*, 293–317. [CrossRef] [PubMed]
198. Gissi, F.; Stauber, J.L.; Binet, M.T.; Golding, L.A.; Adams, M.S.; Schlekat, C.E.; Garman, E.R.; Jolley, D.F. A review of nickel toxicity to marine and estuarine tropical biota with particular reference to the South East Asian and Melanesian region. *Environ. Pollut.* **2016**, *218*, 1308–1323. [CrossRef] [PubMed]
199. Klimisch, H.-J.; Andreae, M.; Tillmann, U. A systematic approach for evaluating the quality of experimental toxicological and ecotoxicological data. *Regul. Toxicol. Pharmacol.* **1997**, *25*, 1–5. [CrossRef] [PubMed]
200. Moermond, C.; Kase, R.; Korkaric, M.; Ågerstrand, M. CRED-Criteria for reporting and evaluating ecotoxicity data. *Environ. Toxicol. Chem.* **2015**, *35*, 1297–1309. [CrossRef] [PubMed]
201. Batley, G.E.; van Dam, R.; Warne, M.S.J.; Chapman, J.C.; Fox, D.R.; Hickey, C.W.; Stauber, J.L. *Technical Rationale for Changes to the Method for Deriving Australian and New Zealand Water Quality Guideline Values for Toxicants–Update of 2014 Version*; Prepared for the revision of the Australian and New Zealand Guidelines for Fresh and Marine Water Quality; Australian and New Zealand Governments and Australian State and Territory Governments: Canberra, Australia, 2018.
202. Schlekat, C.E.; Garman, E.R.; Vangheluwe, M.L.U.; Burton, G.A., Jr. Development of a bioavailability-based risk assessment approach for Ni in sediments. *Integr. Environ. Assess. Manag.* **2016**, *12*, 735–746. [CrossRef]
203. Aldenberg, T.; Jaworska, J.S. Uncertainty of the hazardous concentration and fraction affected for normal species sensitivity distributions. *Ecotoxicol. Toxicol. Environ. Saf.* **2000**, *46*, 1–18. [CrossRef]
204. Niyogi, S.; Brix, K.V.; Grosell, M. Effects of chronic waterborne nickel exposure on growth, ion homeostasis, acid-base balance, and nickel uptake in the freshwater pulmonate snail, Lymnaea stagnalis. *Aquat. Toxicol.* **2014**, *150*, 36–44. [CrossRef]
205. Peters, A.; Merrington, G.; Leverett, D.; Wilson, I.; Schlekat, C.; Garman, E. Comparison of the chronic toxicity of nickel to temperate and tropical freshwater species. *Environ. Toxicol. Chem.* **2019**. [CrossRef]
206. Hommen, U.; Knopf, B.; Rüdel, H.; Schäfers, C.; De Schamphelaere, K.; Schlekat, C.; Garman, E.R. A microcosm study to support aquatic risk assessment of nickel: Community-level effects and comparison with bioavailability-normalized species sensitivity distributions. *Environ. Toxicol. Chem.* **2016**, *35*, 1172–1182. [CrossRef] [PubMed]
207. Peters, A.; Simpson, P.; Merrington, G.; Schlekat, C.; Rogevich-Garman, E. Assessment of the effects of nickel on benthic macroinvertebrates in the field. *Environ. Sci. Pollut. Res. Int.* **2014**, *21*, 193–204. [CrossRef] [PubMed]

208. European Union (EU). Directive 2013/39/EU of the European Parliament and of the Council. *Off. J. Eur. Union* **2013**, *226*, 338–436.
209. Deforest, D.K.; Schlekat, C.E. Species sensitivity distribution evaluation for chronic nickel toxicity to marine organisms. *Integr. Environ. Assess. Manag.* **2013**, *9*, 580–589. [CrossRef] [PubMed]
210. Gissi, F.; Stauber, J.L.; Binet, M.T.; Trenfield, M.A.; Van Dam, J.W.; Jolley, D.F. Assessing the chronic toxicity of nickel to a tropical marine gastropod and two crustaceans. *Ecotoxicol. Toxicol. Environ. Saf.* **2018**, *159*, 284–292. [CrossRef] [PubMed]
211. Gissi, F.; Stauber, J.; Reichelt-Brushett, A.; Harrison, P.L.; Jolley, D.F. Inhibition in fertilisation of coral gametes following exposure to nickel and copper. *Ecotoxicol. Toxicol. Environ. Saf.* **2017**, *145*, 32–41. [CrossRef]
212. Gissi, F.; Reichelt-Brushett, A.J.; Chariton, A.A.; Stauber, J.L.; Greenfield, P.; Humphrey, C.; Salmon, M.; Stephenson, S.A.; Cresswell, T.; Jolley, D.F. The effect of dissolved nickel and copper on the adult coral *Acropora muricate* and its microbiome. *Environ. Pollut.* **2019**, *250*, 792–806. [CrossRef]
213. European Chemicals Agency (ECHA). *Guidance on Information Requirements and Chemical Safety Assessment, Chapter R.10: Characterisation of Dose [Concentration]-Response for Environment*; European Chemicals Agency: Helsinki, Finland, 2008.
214. Vandegehuchte, M.B.; Roman, Y.E.; Nguyen, L.T.; Janssen, C.R.; De Schamphelaere, K.A. Toxicological availability of nickel to the benthic oligochaete Lumbriculus variegatus. *Environ. Int.* **2007**, *33*, 736–742. [CrossRef]
215. Brumbaugh, W.G.; Besser, J.M.; Ingersoll, C.G.; May, T.W.; Ivey, C.D.; Schlekat, C.E.; Garman, E.R. Preparation and characterization of nickel-spiked freshwater sediments for toxicity tests: Toward more environmentally realistic testing conditions. *Environ. Toxicol. Chem.* **2013**, *32*, 2482–2494.
216. Chandler, G.T.; Schlekat, C.E.; Garman, E.R.; He, L.; Washburn, K.M.; Stewart, E.R.; Ferry, J.L. Sediment nickel bioavailability and toxicity to estuarine crustaceans of contrasting bioturbative behaviors—An evaluation of the SEM-AVS paradigm. *Environ. Sci. Technol.* **2014**, *48*, 12893–12901. [CrossRef]
217. Costello, D.M.; Burton, G.A.; Hammerschmidt, C.R.; Taulbee, W.K. Evaluating the performance of diffusive gradients in thin films (DGTs) for predicting Ni sediment toxicity. *Environ. Sci. Technol.* **2012**, *46*, 10239–10246. [PubMed]
218. Nguyen, L.T.; Burton, G.A., Jr.; Schlekat, C.E.; Janssen, C.R. Field measurement of nickel sediment toxicity: Role of acid volatile sulfide. *Environ. Toxicol. Chem.* **2011**, *30*, 162–172. [CrossRef] [PubMed]
219. Costello, D.M.; Burton, G.A. Response of stream ecosystem function and structure to sediment metal: Context dependency and variation among endpoints. *Elem. Sci. Anth.* **2014**, *2*, 30. [CrossRef]
220. Costello, D.M.; Hammerschmidt, C.R.; Burton, G.A. Nickel partitioning and toxicity in sediment during aging: Variation in toxicity related to stability of metal partitioning. *Environ. Sci. Technol.* **2016**, *50*, 11337–11345. [CrossRef] [PubMed]
221. Mendonca, R.M.; Daley, J.M.; Hudson, M.L.; Schlekat, C.E.; Burton, G.A.; Costello, D.M. Metal oxides in surface sediment control nickel bioavailability to benthic macroinvertebrates. *Environ. Sci. Technol.* **2017**, *51*, 13407–13416. [CrossRef] [PubMed]
222. Hale, B.; Gopalapillai, Y.; Pellegrino, A.; Jennett, T.; Kikkert, J.; Lau, W.; Schlekat, C.; McLaughlin, M.J. Validation of site-specific soil Ni toxicity thresholds with independent ecotoxicity and biogeochemistry data for elevated soil ni. *Environ. Pollut.* **2017**, *231 Pt 1*, 165–172. [CrossRef]
223. Brix, K.; Schlekat, C.; Garman, E. The mechanisms of nickel toxicity in aquatic environments: An adverse outcome pathway (AOP) analysis. *Environ. Toxicol. Chem.* **2017**, *36*, 1128–1137. [CrossRef]
224. He, E.; Qiu, H.; Dimitrova, K.; Van Gestel, C.A. A generic biotic ligand model quantifying the development in time of Ni toxicity to Enchytraeus crypticus. *Chemosphere* **2015**, *124*, 170–176. [CrossRef]
225. Luo, S.Q.; Plowman, M.C.; Hopfer, S.M.; Sunderman, F.W. $Mg^{(2+)}$-deprivation enhances and $Mg^{(2+)}$-supplementation diminishes the embryotoxic and teratogenic effects of Ni^{2+}, Co^{2+}, Zn^{2+} and Cd^{2+} for frog embryos in the FETAX assay. *Ann. Clin. Lab. Sci.* **1993**, *23*, 121–129.
226. Gopalapillai, Y.; Hale, B.; Vigneault, B. Effect of major cations (Ca^{2+}, Mg^{2+}, Na^+, K^+) and anions (SO_4^{2-}, Cl^-, NO_3^-) on Ni accumulation and toxicity in aquatic plant (*Lemna minor* L.): Implications for Ni risk assessment. *Environ. Toxicol. Chem.* **2013**, *32*, 810–821. [CrossRef]
227. Shahzad, B.; Tanveer, M.; Rehman, A.; Cheema, S.A.; Fahad, S.; Rehman, S.; Sharma, A. Nickel; whether toxic or essential for plants and environment—A review. *Plant Physiol. Biochem.* **2018**, *132*, 641–651. [CrossRef] [PubMed]

228. Schlekat, C.E.; McGeer, J.C.; Blust, R.; Borgmann, U.; Brix, K.; Bury, N.; Couillard, Y.; Dwyer, R.L.; Luoma, S.N.; Robertson, S.; et al. Bioaccumulation: Hazard identification of metals and inorganic metal substances. In *Assessing the Hazard of Metals and Inorganic Metal Substances in Aquatic and Terrestrial Systems*; Adams, W.J., Chapman, P.M., Eds.; SETAC Press: Pensacola, FL, USA, 2006; pp. 55–89.
229. DeForest, D.; Schlekat, C.; Brix, K.; Fairbrother, A. Secondary poisoning of nickel. *Integr. Environ. Assess. Manag.* **2011**, *8*, 107–119. [CrossRef] [PubMed]
230. Van der Ent, A.; Baker, A.J.M.; Reeves, R.D.; Pollard, A.J.; Schat, H. Hyperaccumulators of metal and metalloid trace elements: Facts and fiction. *Plant Soil* **2013**, *362*, 319–334. [CrossRef]
231. Meyer, J.S.; Farley, K.J.; Garman, E.R. Metal Mixtures Modeling Evaluation project: 1. Background. *Environ. Toxicol. Chem.* **2015**, *34*, 726–740. [CrossRef] [PubMed]
232. Farley, K.J.; Meyer, J.S. Metal Mixture Modeling Evaluation project: 3. Lessons learned and steps forward. *Environ. Toxicol. Chem.* **2015**, *34*, 821–832. [CrossRef] [PubMed]
233. Nys, C.; Janssen, C.R.; De Schamphelaere, K.A.C. Development and validation of a metal mixture bioavailability model (MMBM) to predict chronic toxicity of Ni–Zn–Pb mixtures to Ceriodaphnia dubia. *Environ. Pollut.* **2017**, *220 Pt B*, 1271–1281. [CrossRef]
234. Nys, C.; Van Regenmortel, T.; Janssen, C.R.; Oorts, K.; Smolders, E.; De Schamphelaere, K.A.C. A framework for ecological risk assessment of metal mixtures in aquatic systems. *Environ. Toxicol. Chem.* **2018**, *37*, 623–642. [CrossRef] [PubMed]
235. Lynch, N.R.; Hoang, T.C.; O'Brien, T.E. Acute toxicity of binary-metal mixtures of copper, zinc, and nickel to Pimephales promelas: Evidence of more-than-additive effect. *Environ. Toxicol. Chem.* **2015**, *35*, 446–457. [CrossRef]
236. Traudt, E.M.; Ranville, J.F.; Meyer, J.S. Effect of age on acute toxicity of cadmium, copper, nickel, and zinc in individual-metal exposures to Daphnia magna neonates. *Environ. Toxicol. Chem.* **2016**, *36*, 113–119. [CrossRef]
237. Traudt, E.M.; Ranville, J.F.; Meyer, J.S. Acute toxicity of ternary Cd–Cu–Ni and Cd–Ni–Zn mixtures to Daphnia magna: Dominant metal pairs change along a concentration gradient. *Environ. Sci. Technol.* **2017**, *51*, 4471–4481. [CrossRef]
238. Crémazy, A.; Brix, K.V.; Wood, C.M. Chronic toxicity of binary mixtures of six metals (Ag, Cu, Cd, Pb, Ni and Zn) to the great pond snail Lymnaea stagnalis. *Environ. Sci. Technol.* **2018**, *52*, 5979–5988. [CrossRef] [PubMed]
239. Crémazy, A.; Brix, K.V.; Wood, C.M. Using the Biotic Ligand Model framework to investigate binary metal interactions on the uptake of Ag, Cd, Cu, Ni, Pb and Zn in the freshwater snail Lymnaea stagnalis. *Sci. Total Environ.* **2019**, *647*, 1611–1625. [CrossRef] [PubMed]
240. Katsnelson, B.A.; Panov, V.G.; Minigaliyeva, I.A.; Varaksin, A.N.; Privalova, L.I.; Slyshkina, T.V.; Grebenkina, S.V. Further development of the theory and mathematical description of combined toxicity: An approach to classifying types of action of three-factorial combinations (a case study of manganese-chromium- nickel subchronic intoxication). *Toxicology* **2015**, *334*, 33–44. [CrossRef] [PubMed]
241. Minigaliyeva, I.A.; Katsnelson, B.A.; Privalova, L.I.; Gurvich, V.B.; Panov, V.G.; Varaksin, A.N.; Makeyev, O.H.; Sutunkova, M.P.; Loginova, N.V.; Kireyeva, E.P.; et al. Toxicodynamic and Toxicokinetic descriptors of combined Chromium (VI) and Nickel toxicity. *Int. J. Toxicol.* **2014**, *33*, 498–505. [CrossRef] [PubMed]
242. Kenston, S.S.F.; Su, H.; Li, Z.; Kong, L.; Wang, Y.; Song, X.; Gu, Y.; Barber, T.; Aldinger, J.; Hua, Q.; et al. The systemic toxicity of heavy metal mixtures in rats. *Toxicol. Res.* **2018**, *7*, 396–407. [CrossRef] [PubMed]
243. International Agency for Research on Cancer (IARC). Welding, Molybdenum Trioxide, and Indium Tin Oxide. *IARC Monogr. Eval. Carcinog. Risks Hum.* **2018**, *118*, 36–265.
244. Zimmer, A.T. The influence of metallurgy on the formation of welding aerosols. *J. Environ. Monit.* **2002**, *4*, 628–632. [CrossRef]
245. Zeidler-Erdely, P.C.; Meighan, T.G.; Erdely, A.; Battelli, L.A.; Kashon, M.L.; Keane, M.; Antonini, J.M. Lung tumor promotion by chromium-containing welding particulate matter in a mouse model. *Part. Fibre Toxicol.* **2013**, *10*, 45. [CrossRef]
246. Falcone, L.M.; Erdely, A.; Meighan, T.G.; Battelli, L.A.; Salmen, R.; McKinney, W.; Stone, S.; Cumpston, A.; Cumpston, J.; Andrews, R.N.; et al. Inhalation of gas metal arc-stainless steel welding fume promotes lung tumorigenesis in A/J mice. *Arch. Toxicol.* **2017**, *91*, 2953–2962. [CrossRef]

247. Zeidler-Erdely, P.C.; Falcone, L.M.; Antonini, J.M. Influence of welding fume metal composition on lung toxicity and tumor formation in experimental animal models. *J. Occup. Environ. Hyg.* **2019**, *16*, 372–377. [CrossRef]
248. Future Markets, Inc. *The Global Market for Metal and Metal Oxide Nanoparticles, 2010–2027*, 4th ed.; Future Markets, Inc.: Edinburgh, UK, 2017.
249. Biskos, G.; Schmidt-Ott, A. Airborne engineered nanoparticles: Potential risks and monitoring challenges for assessing their impacts on children. *Paediatr. Respir. Rev.* **2012**, *13*, 79–83. [CrossRef] [PubMed]
250. Munoz, A.; Costa, M. Elucidating the mechanisms of nickel compound uptake: A review of particulate and nano-nickel endocytosis and toxicity. *Toxicol. Appl. Pharmacol.* **2012**, *260*, 1–16. [CrossRef] [PubMed]
251. Magaye, R.; Zhao, J. Recent progress in studies of metallic nickel and nickel-based nanoparticles' genotoxicity and carcinogenicity. *Environ. Toxicol. Pharmacol.* **2012**, *34*, 644–650. [CrossRef] [PubMed]
252. Sutunkova, M.P.; Privalova, L.I.; Minigalieva, I.A.; Gurvich, V.B.; Panov, V.G.; Katsnelson, B.A. The most important inferences from the Ekaterinburg nanotoxicology team's animal experiments assessing adverse health effects of metallic and metal oxide nanoparticles. *Toxicol. Rep.* **2018**, *5*, 363–376. [CrossRef] [PubMed]
253. Sutunkova, M.P.; Solovyeva, S.N.; Minigalieva, I.A.; Gurvich, V.B.; Valamina, I.E.; Makeyev, O.H.; Shur, V.Y.; Shishkina, E.V.; Zubarev, I.V.; Saatkhudinova, R.R.; et al. Toxic effects of low-level long-term inhalation exposures of rats to nickel oxide nanoparticles. *Int. J. Mol. Sci.* **2019**, *20*, 1778. [CrossRef] [PubMed]
254. Horie, M.; Stowe, M.; Tabei, M.; Kuroda, E. Metal ion release of manufactured metal oxide nanoparticles is involved in the allergic response to inhaled ovalbumin in mice. *Occup. Dis. Environ.* **2016**, *4*, 17–26. [CrossRef]
255. Ispas, C.; Andreescu, D.; Patel, A.; Goia, D.V.; Andreescu, S.; Wallace, K.N. Toxicity and developmental defects of different sizes and shape nickel nanoparticles in zebrafish. *Environ. Sci. Technol.* **2009**, *43*, 6349–6356. [CrossRef]
256. Ogami, A.; Morimoto, Y.; Myojo, T.; Oyabu, T.; Murakami, M.; Todoroki, M.; Nishi, K.; Kadoya, C.; Yamamoto, M.; Tanaka, I. Pathological features of different sizes of nickel oxide following intratracheal instillation in rats. *Inhal. Toxicol.* **2009**, *21*, 812–818. [CrossRef]
257. Latvala, S.; Hedberg, J.; Di Bucchianico, S.; Möller, L.; Odnevall Wallinder, I.; Elihn, K.; Karlsson, H.L. Nickel release, ROS generation and toxicity of Ni and NiO micro- and nanoparticles. *PLoS ONE* **2016**, *11*, e0159684. [CrossRef]
258. Kong, L.; Tang, M.; Zhang, T.; Wang, D.; Hu, K.; Lu, W.; Wei, C.; Liang, G.; Pu, Y. Nickel nanoparticles exposure and reproductive toxicity in healthy adult rats. *Int. J. Mol. Sci.* **2014**, *15*, 21253–21269. [CrossRef]
259. Gallo, A.; Boni, R.; Buttino, I.; Tosti, E. Spermiotoxicity of nickel nanoparticles in the marine invertebrate Ciona intestinalis (Ascidians). *Nanotoxicology* **2016**, *10*, 1096–1104. [CrossRef] [PubMed]
260. Griffitt, R.J.; Luo, J.; Gao, J.; Bonzongo, J.-C.; Barber, D.S. Effects of particle composition and species on toxicity of metallic nanomaterials in aquatic organisms. *Environ. Toxicol. Chem.* **2008**, *27*, 1972–1978. [CrossRef] [PubMed]
261. Zhou, C.; Vitiello, V.; Casals, E.; Puntes, V.F.; Iamunno, F.; Pellegrini, D.; Changwen, W.; Benvenuto, G.; Buttino, I. Toxicity of nickel in the marine calanoid copepod Acartia tonsa: Nickel chloride versus nanoparticles. *Aquat. Toxicol.* **2016**, *170*, 1–12. [CrossRef] [PubMed]
262. Oukarroum, A.; Barhoumi, L.; Samadani, M.; Dewez, D. Toxic effects of nickel oxide bulk and nanoparticles on the aquatic plant Lemna gibba L. *BioMed. Res. Int.* **2015**, *2015*, 501326. [CrossRef]
263. Minigalieva, I.A.; Katsnelson, B.A.; Privalova, L.I.; Sutunkova, M.P.; Gurvich, V.B.; Shur, V.Y.; Shishkina, E.V.; Valamina, I.E.; Makeyev, O.H.; Panov, V.G.; et al. Attenuation of combined nickel(II) oxide and manganese(II, III) oxide nanoparticles' adverse effects with a complex of bioprotectors. *Int. J. Mol. Sci.* **2015**, *16*, 22555–22583. [CrossRef] [PubMed]
264. Minigalieva, I.; Bushueva, T.; Fröhlich, E.; Meindl, C.; Öhlinger, K.; Panov, V.; Varaksin, A.; Shur, V.; Shishkina, E.; Gurvich, V.; et al. Are in vivo and in vitro assessments of comparative and combined toxicity of the same metallic nanoparticles compatible, or contradictory, or both? A juxtaposition of data obtained in respective experiments with NiO and Mn$_3$O$_4$ nanoparticles. *Food Chem. Toxicol.* **2017**, *109 Pt 1*, 393–404. [CrossRef]
265. Katsnelson, B.A.; Minigaliyeva, I.A.; Panov, V.G.; Privalova, L.I.; Varaksin, A.N.; Gurvich, V.B.; Sutunkova, M.P.; Shur, V.Y.; Shishkina, E.V.; Valamina, I.E.; et al. Some patterns of metallic nanoparticles' combined subchronic toxicity as exemplified by a combination of nickel and manganese oxide nanoparticles. *Food Chem. Toxicol.* **2015**, *86*, 351–364. [CrossRef] [PubMed]

266. Ali, A.A. Evaluation of some biological, biochemical, and hematological aspects in male albino rats after acute exposure to the nano-structured oxides of nickel and cobalt. *Environ. Sci. Pollut. Res. Int.* **2019**, *26*, 17407–17417. [CrossRef]
267. Fischer, L.A.; Menné, T.; Johansen, J.D. Experimental nickel elicitation thresholds—A review focusing on occluded nickel exposure. *Contact Dermat.* **2005**, *52*, 57–64. [CrossRef]
268. Buekers, J.; De Brouwere, K.; Lefebvre, W.; Willems, H.; Vendenbroele, M.; Van Sprang, P.; Eliat-Eliat, M.; Hicks, K.; Schlekat, C.E.; Oller, A.R. Assessment of human exposure to environmental sources of nickel in Europe: Inhalation exposure. *Sci. Total Environ.* **2015**, *521–522*, 359–371. [CrossRef]

© 2019 by the authors. Licensee MDPI, Basel, Switzerland. This article is an open access article distributed under the terms and conditions of the Creative Commons Attribution (CC BY) license (http://creativecommons.org/licenses/by/4.0/).

Review

Role of Nickel in Microbial Pathogenesis

Robert J. Maier [1,2,*] **and Stéphane L. Benoit** [1,2]

1. Department of Microbiology, University of Georgia, Athens, GA 30602, USA
2. Center for Metalloenzyme Studies, University of Georgia, Athens, GA 30602, USA
* Correspondence: rmaier@uga.edu; Tel.: +1-706-542-2323

Received: 21 May 2019; Accepted: 20 June 2019; Published: 26 June 2019

Abstract: Nickel is an essential cofactor for some pathogen virulence factors. Due to its low availability in hosts, pathogens must efficiently transport the metal and then balance its ready intracellular availability for enzyme maturation with metal toxicity concerns. The most notable virulence-associated components are the Ni-enzymes hydrogenase and urease. Both enzymes, along with their associated nickel transporters, storage reservoirs, and maturation enzymes have been best-studied in the gastric pathogen *Helicobacter pylori*, a bacterium which depends heavily on nickel. Molecular hydrogen utilization is associated with efficient host colonization by the Helicobacters, which include both gastric and liver pathogens. Translocation of a *H. pylori* carcinogenic toxin into host epithelial cells is powered by H_2 use. The multiple [NiFe] hydrogenases of *Salmonella enterica* Typhimurium are important in host colonization, while ureases play important roles in both prokaryotic (*Proteus mirabilis* and *Staphylococcus* spp.) and eukaryotic (*Cryptoccocus* genus) pathogens associated with urinary tract infections. Other Ni-requiring enzymes, such as Ni-acireductone dioxygenase (ARD), Ni-superoxide dismutase (SOD), and Ni-glyoxalase I (GloI) play important metabolic or detoxifying roles in other pathogens. Nickel-requiring enzymes are likely important for virulence of at least 40 prokaryotic and nine eukaryotic pathogenic species, as described herein. The potential for pathogenic roles of many new Ni-binding components exists, based on recent experimental data and on the key roles that Ni enzymes play in a diverse array of pathogens.

Keywords: nickel; hydrogenase; urease; Ni-enzymes; pathogens

---

1. Introduction

Nickel (Ni) is well established as an essential cofactor for some pathogen virulence factors. While the majority of studies on pathogens' Ni enzymes relate to human pathogens, a sizable portion of animal pathogens (*Helicobacter hepaticus*, *Helicobacter mustelae*, *Ureaplasma diversum*, *Brucella* species, *Campylobacter* species) also use Ni-containing virulence factors (Table 1). In contrast, the literature on Ni-dependent plant pathogens is scarce. This discrepancy reflects a fundamental difference between plants and mammals: plants use nickel as a cofactor for urease, which they oftentimes make in abundance, therefore reducing its availability to pathogens. On the other hand, mammals do not synthesize any (known) Ni-requiring protein(s), hence (host) nickel is likely to be more available for Ni-utilizing pathogens. Still, the intestinal microflora of mammals is comprised of many Ni-utilizing members, such as urease-producing lactobacilli and *Bifidobacterium* species, or gut methanogens relying on nickel-dependent coenzyme M reductase [1]; those are likely to compete for nickel with pathogens. Interestingly, there is a strong link between nickel pools in plants (e.g., urease-bound) and mammals nickel pools, as plants are one of the main dietary sources of nickel for mammals.

The most notable virulence-associated components are the Ni-enzymes hydrogenase and urease, both of which have been shown to be important for pathogen virulence in various organisms [2–4]. Both hydrogenase and urease, along with their associated nickel transporters, storage reservoirs, maturation enzymes, and nickel-dependent regulators have been best-studied in the gastric pathogen *H. pylori*, so much of our review will discuss the nickel-metabolism factors in the gastric pathogen. Nevertheless, significant progress has been made towards understanding the role of [NiFe] hydrogenases in enteric pathogens, especially in S. Typhimurium; this microbe will thus be covered in our review as well. Also, urease enzymes play important roles in eukaryotic pathogens belonging to the *Cryptoccoccus* genus, as well as in prokaryotic pathogens such as *P. mirabilis* and *Staphylococcus* spp., which are causative agents of urinary tract infections (UTIs). Since the synthesis, structure, and catalytic activity of ureases and hydrogenases have been recently presented and discussed in comprehensive reviews [5–9], these aspects will not be covered in the present review. Likewise, *H. pylori* hydrogenase and urease maturation, as well as NikR-mediated gene regulation, have been extensively reviewed by our group and others [3,10–12], hence they will not be discussed herein.

Other Ni-requiring enzymes, such as Ni-acireductone dioxygenase (ARD) [6,13], Ni-superoxide dismutase (SOD) [14,15], and Ni-glyoxalase I (GloI) [6,16] play important metabolic or detoxifying roles in a few pathogens; little is known about their contribution to pathogenicity; nevertheless, their role will be briefly discussed. The hypothetical or demonstrated role of all Ni-enzymes in pathogens is summarized in Table 1.

Table 1. List of eukaryotic and prokaryotic pathogens with Ni-enzymes and their putative or demonstrated role in pathogenesis.

Pathogen	Ni-Enzyme *	Role in Pathogenesis (Reference)
EUKARYOTES		
Human fungi		
Cryptococcus neoformans	Ure	Virulence factor in experimental cryptococcosis [17]
		Required for microvascular sequestration and mouse brain invasion [18]
		Modulates phagolysosomal pH; important for mouse brain infection [19]
		Released via extracellular vesicles [20]
Cryptococcus gattii	Ure	Virulence factor in mice [21]
Coccidioides posadasii	Ure	Coccidioidomycosis in mice [22,23]
Histoplasma capsulatum	Ure	Released via extracellular vesicles [24]
Paracoccidioides brasiliensis	Ure	Up-expressed in mouse infection model [25]
Oomycetes		
Pythium insidiosum	Ure	Putative virulence factor for pythiosis [26]
Protists		
Leishmania major	Glo-I	Important for parasite metabolism: methylglyoxal detoxification [27]
Leishmania donovani	Glo-I	Essential for growth; suggested as drug target [28]
Trypanosoma cruzi	Ard	Important for parasite metabolism: methionine salvage pathway
	Glo-I	Important for parasite metabolism: methylglyoxal detoxification [29]
PROKARYOTES		
Actinobacteria		
Actinomyces naeslundii	Ure	Needed in acidic environment; promotes plaque formation [30]
Corynebacterium urealyticum	Ure	Plays an important role in urinary tract infection [31]
Mycobacterium tuberculosis	Hyc	Essential for optimal growth [32]
		Up-regulated during infection of human macrophage-like cells [33]
		Up-expressed in resting and active murine bone marrow macrophages [34]
	Ure	Important for survival under nitrogen-limited environment [35]
Streptomyces scabies	Sod	Important against oxidative stress encountered in host (plant)

Table 1. Cont.

Pathogen	Ni-Enzyme *	Role in Pathogenesis (Reference)
Firmicutes		
Clostridia	Glo-I	Important for metabolism: methylglyoxal detoxification [36]
Staphylococcus aureus	Ure	Increased expression of structural and accessory genes in biofilms [37]
		Required for acid response and persistent murine kidney infection [38]
		Decreased activity in mixed source (S. epidermidis) biofilms [39]
Staphylococcus epidermidis	Ure	Decreased activity in mixed source (S. aureus) biofilms [39]
Staphylococcus saprophyticus	Ure	Important for bladder infection and bladder stones in rats [40]
Streptococcus salivarius	Ure	Important as source of nitrogen and to combat acid stress [41]
Mollicutes		
Ureaplasma urealyticum	Ure	Role in human vaginal infection; used for diagnostic [42]
		Ammonia contributes to PMF-driven ATP synthesis [43]
		Ammonia generates struvite stone formation in rat bladders [44]
Ureaplasma parvum	Ure	Role in human vaginal infection; used for diagnostic [42]
Ureaplasma diversum	Ure	Role in vaginal infection of cattle and small ruminants [45]
Proteobacteria		
Alphaproteobacteria		
Brucella abortus	Ure	Needed for intestinal colonization in a murine model [46]
		Immunization with B. a. urease protects against B. abortus infection in mice [47]
Brucella melitensis	Ure	Immunization with B. a. urease protects against B. melitensis in mice [47]
Brucella suis	Ure	Required for intestinal colonization in a murine model [48]
		Immunization with B. a. urease protects against B. suis infection in mice [47]
Betaproteobacteria		
Neisseria meningitides	Glo-I	Important for methylglyoxal detoxification and potassium efflux (hypothesized)
Neisseria gonorrhoeae	Glo-I	Important for methylglyoxal detoxification and potassium efflux (hypothesized)
Gammaproteobacteria		
All γ-proteobacteria	Ard	Important for metabolism: methionine salvage pathway
All γ-proteobacteria	Glo-I	Important for methylglyoxal detoxification, potassium efflux
Acinetobacter baumannii	Ure	Virulence factor in worm and amoeba hosts [49]
Acinetobacter lwoffii	Ure	Needed to survive in the stomach [50]

Table 1. Cont.

Pathogen	Ni-Enzyme *	Role in Pathogenesis (Reference)
Actinobacillus pleuropneumoniae	Ure	Important for swine respiratory tract infection [51,52]
Escherichia coli	Hyd-1	Important for (PMF-driven) metabolism and motility
	Hyd-2	Important for (PMF-driven) metabolism and motility
	Hyc	Needed (as part of FHL) to dissipate formic acid-induced acidity [53]
E. coli (Shiga-toxin producing)	Ure	Needed for colonization in the murine model [54]
Edwardsiella tarda	Hyd	Hyd. accessory protein Sip2 essential for acid resistance and host infection [55]
Haemophilus influenzae	Ure	Important for acid resistance, expressed during human pulmonary infection [56]
Klebsiella pneumoniae	Ure	Required for colonization in murine intestinal model [57]
Morganella morganii	Ure	Needed for survival at low pH [58,59]
Proteus mirabilis	Hyd	Important for swarming motility [60]
	Ure	Role in persistence, urolithiasis, and acute pyelonephritis in a mouse model [61]
		Role in extracellular crystal stone cluster formation in the bladder [62]
		Induced in polymicrobial biofilms [63]
Providencia stuartii	Ure	Involved in crystal stones formation; induced in polymicrobial biofilms [59]
Pseudomonas aeruginosa	Glo-I	Important for methylglyoxal detoxification and potassium efflux (hypothesized)
Salmonella Typhimurium	Hyd-1	Important for acid resistance and macrophage colonization [64]
	Hyd-2	Most important hydrogenase for gut invasion [65,66]
	Hyd-5	Expressed under aerobic conditions and in macrophages [64,67,68]
	Hyc	Important for anaerobic acid resistance [69]
Shigella flexneri	Hyd	Important for acid resistance [70]
Vibrio parahaemolyticus	Ure	Important for pathogenicity [71]
Yersinia enterocolitica	Ure	Important for survival at low pH [58,72]
Yersinia pestis	Glo-I	Important for methylglyoxal detoxification and potassium efflux (hypothesized)
Deltaproteobacteria		
Bilophila wadsworthia	Hyd	H_2 used as energy source, optimal growth in presence of H_2 and taurine [73]
Epsilonproteobacteria		
Campylobacter jejuni	Hyd	Important for chicken cecum colonization [74]
		Essential for chicken colonization in absence of formate dehydrogenase [74]
		Required for in vitro interaction with human intestinal cells [75]

Table 1. *Cont.*

Pathogen	Ni-Enzyme *	Role in Pathogenesis (Reference)
Campylobacter concisus	Hyd	Essential for growth under microaerobic conditions [76]
Helicobacter hepaticus	Hyd	Role in amino-acid transport and causing liver lesions in mice [77]
	Ure	Promotes hepatic inflammation in mice [78]
Helicobacter mustelae	Ure	Essential for ferret stomach colonization [79]
Helicobacter pylori	Hyd	Needed for mouse stomach colonization [80]
		Role in CO_2 fixation [81]
		Role in CagA translocation [82]
	Ure	Cytotoxic effect on Caco-2 cells [83]
		Needed for nude mouse stomach colonization [84]
		Essential for gnotobiotic piglet stomach colonization [85]
		Activates human phagocytes and macrophages [86,87]
		Binds to class II MHC on gastric epithelial cells and induces their apoptosis [88]
		Essential for Mongolian gerbil colonization [89,90]
		Urease-produced CO_2 protects against host peroxynitrite [91]
		Urease-produced ammonia disrupts tight cell junction integrity [92]
		Dysregulates epithelial tight junctions through myosin activation [93]
		Activates blood platelets through a lipoxygenase-mediated pathway [94]
		Alters mucin gene expression in human gastric cells [95]
		Essential for chronic mouse infection [96]
		Role in angiogenesis, endothelial cells and chicken embryo models [97]
		Induces blood platelets inflammatory pathways [98]
		Non catalytic, oxidative stress-combatting role [99]

* Abreviations: **Ard**: acireductone dioxygenase; **Glo-I**: Glyoxalase I; **Hyc**: H_2-evolving hydrogenase; **Hyd**: H_2-uptake hydrogenase; **Sod**: superoxide dismutase; **Ure**: urease.

2. Nickel Availability to Pathogens and Host-Mediated Influences

Nickel, more so than for other required metals, presents both a difficult acquisition and a homeostasis problem for pathogens, ultimately due to its low availability within the host. Indeed, nickel is found at less than 5 ppm (µg/g of ash) in most human organs [100]. Compared to other metals, such as zinc, nickel is far less prevalent in organs: for instance it is found at a level of less than 1% of the amount of zinc in the brain, heart, lung, or muscle, and the amount of nickel is less than 0.1% of that of zinc measured in both the human liver and kidney [101].

To limit pathogen growth, animal hosts have developed metal sequestering strategies to abrogate the invading pathogen [102]. Indeed, this is thought to be a key antibacterial mechanism used to inhibit initial infection as well as to combat tissue-established ones. Metals are bound by the mammalian host mucosa (a process termed "nutritional immunity"), and this is known to involve mucosal-associated metal binding proteins [102,103]. Although we have significant knowledge of iron, zinc, and manganese sequestering by pathogenic bacteria, and the competition between host and pathogen for these metals is beginning to be understood [102,104,105], much less is known about these aspects with respect to cobalt, nickel, and copper [105]. Recruited neutrophils at inflammation sites express metal-binding proteins, such as calprotectin, lipocalin and lactoferrin. The role of the host defense protein calprotectin in zinc binding and subsequent pathogen inhibition is well-established [105]; however, a recent study from Nakashige and colleagues showed that coordination of Ni(II) at the hexahistidine site of calprotectin is preferred over that for Zn(II) [106]. In agreement with this finding, calprotectin was shown to sequester nickel away from two pathogens, *S. aureus* and *K. pneumoniae*, subsequently inhibiting their respective urease activity in bacterial culture [106].

Although the antibacterial effects of the multifunctional globular protein lactoferrin are attributed in part to its iron-binding capacity [107,108], the histidine and tyrosine ligands can bind other metals, including nickel; therefore, a nickel-sequestering effect of lactoferrin towards pathogens should not be ruled out. The same thinking might also apply to the peptide hormone hepcidin, a regulator of host iron homeostasis, which binds both Fe and Ni (II) [109]. While the role of hepcidin in starving pathogens of nickel has not been investigated, its antimicrobial activity via iron sequestering has been well established [110]. The list of siderophilic pathogens affected by hepcidin-driven iron chelation includes *E. coli*, *Staphylococcus* spp., including *S. epidermidis* and *S aureus*, group B *Streptococcus* bacteria, *Y. enterocolitica*, and *Candida albicans*. In particular, one demonstrated role of hepcidin is to affect metals levels within macrophages [111], so nickel availability may also be expected to be impacted for immune cell-engulfed pathogens. It is well known that macrophages use metal (iron, copper, zinc) sequestration to starve pathogens of essential metals [112]. Of relevance here is that one of the [Ni–Fe] H_2-uptake hydrogenases of *S.* Typhimurium (Hya or Hyd-1) is needed for survival within macrophages [64], an environment in which this hydrogenase, as well as another one, Hyd-5, are greatly up-expressed. Likewise, the *hyc* operon in *Mycobacterium tuberculosis* (encoding for a putative [Ni–Fe] H_2-evolving complex) is upregulated in human macrophage-like cells, as well as in resting or activated murine bone marrow macrophages [33,34]. Similarly, urease has been shown to be important for survival in macrophages of several pathogens, for instance that of *A. pleuropneumoniae* [51,52] or *C. neoformans* [19]. Thus, nickel starvation would be expected to be an immune cell strategy to attenuate pathogen growth.

Many more aspects of nickel restriction to pathogens by host metabolites need to be studied. For example, although not specifically studied with regard to Ni(II), the divalent metal ion transporter NRAMP1 can export metals out of the macrophage phagolysosome, thus restricting the metal availability to the engulfed or intracellular pathogen [113]. It seems likely that some of the antipathogen affects attributed to host iron restriction (or other metals starvation) actually employ nickel starvation as a goal as well. However, host-mediated metal restriction aimed at exacerbating pathogens also complicates metabolism for the host, as host processes are often metal-dependent as well. For example, calprotectin-mediated zinc and manganese starvation attenuates *S. aureus* abscess infection [114], but at the same time, the metal is needed for host enzymes and for normal immune processes [102].

Regarding nickel restriction, no major effect on the host metabolism is expected, as the mammalian hosts do not contain (known) Ni-dependent enzymes. This has caused several research groups to suggest nickel sequestrations as a possible therapeutic approach to combat Ni-requiring pathogens [10,115,116]. For instance, targeting nickel trafficking pathways to prevent proper maturation of both the H_2-uptake [Ni–Fe] hydrogenase and the urease in the gastric pathogen *H. pylori* has been proposed by several groups, including ours [10,116]. Similarly, a recent study identified the nickel requirement for *C. neoformans*'s urease as the fungus's "Achilles' heels" [117]. However, one has to keep in mind that hosts also contain Ni-requiring prokaryotic and eukaryotic microorganisms as part of their (healthy) microbiota [1], thus multiple aspects of host physiology would be affected. Indeed, disruption of nickel homeostasis in these microflora would be expected to lead to dysbiosis and subsequent health consequences for the host. In this particular case, Ni-enzymes do not play a direct role in pathogenesis, but rather they can be considered a "health-related factor" [118].

3. Ureases

Ureases, which catalyze the hydrolytic decomposition of urea into bicarbonate and ammonia, play a dual role: (i) the ammonia produced by the enzyme is an important source of nitrogen for microorganisms and (ii) both ammonia and bicarbonate can be used to neutralize the pH, thus allowing urease-containing organisms to survive and even thrive in acidic environments [7]. Indeed, mammalian pathogens encounter acid in transit through the stomach, on skin, within abscesses, and inside host cells; the nickel-containing urease activity is thus an enzyme that is critical to many bacteria for surviving acidic environments, for instance, *H. pylori* in the stomach. The acidic environment the pathogen encounters may be while in transit (such as when temporarily residing in the gastric milieu) whereby the pathogen is "surviving and seeking" a more hospitable host area, or may be encountered while inside the (acidified) phagolysosome. Before its concentration in the kidneys, urea is present in the bloodstream, thus in blood-rich organs. It is abundant in the blood, with levels estimated to range between 2.5 and 7.1 mM; urea is also found in sweat, saliva, and gastric juices [38,119]. Saliva levels in healthy individuals vary from 3 to 10 mM, albeit they can reach up to 15 mM in patients with renal disease [120]. Urea is also present at high levels in the lungs (2–4 mM) and can be used by urease-positive lung pathogens (e.g., *H. influenzae*, *M. tuberculosis*, *K. pneumoniae*, or *C. neoformans*) as a nitrogen source and/or as a way to neutralize acidic pH; in fact, urease can be used as breath test diagnosis of lung pathogens [121].

As stated above, in addition to its acid-neutralizing properties, urease hydrolyzes urea into ammonium, an important nitrogen source for many pathogens, especially urinary tract pathogens such as *P. stuartii*, *Morganella* spp., *Pseudomonas* spp., *U. urealyticum*, *Klebsiella* spp., *P. mirabilis*, and others (see Table 1) [16]. Urea is extremely abundant in human urine; although urea levels fluctuate widely, the average concentration is around 400 mM [119]. The formation of urinary stones is a direct result of alkalinization of the urinary tract by these pathogens' urease activity [63]. In this case, Ni-related pathogenesis can be viewed as having several notable outcomes. While exacerbating the host excretory system, the crystalline stones also provide a surface for the pathogen(s) to build biofilms and augment its (their) growth [63]. Furthermore, synergistic induction of urease activity in polymicrobial populations (belonging to the species listed above) leads to an increased incidence of urolithiasis and bacteremia [59]. *S. salivarius* uses salivary urea both as a source of nitrogen and to combat acid stress [41]. In *S. aureus*, a pathogen causing significant morbidity due to both acute and chronic infections, the transcription of urease-associated (e.g., structural and accessory/maturation) genes is up-expressed during bacterial biofilm growth [37]. This up-expression is considered to be one component of its acid response network [37,122].

While most ureases are Ni-enzymes, there are a few exceptions. Interestingly, some gastric *Helicobacter* species, such as *H. mustelae* (ferret), *H. felis* (big cats) and *H. acinonychis* (cheetah), possess two distinct urease gene loci, *ureABIEFGH* and *ureA2B2* [123,124]. The former encodes for a nickel-containing urease (similar to that found in *H. pylori*) while the latter encodes for a nickel-free, iron-containing isoform [125]. Transcription of the *ureABIEFGH* operon is induced by the addition of nickel, whereas transcription of *ureA2B2* is upregulated by iron and downregulated by nickel; the nickel-responsive transcriptional regulator NikR is involved in this dual Ni-dependent control [124]. The role of the iron-containing urease is not clear; however, the fact that it is only present in species inhabiting the stomach of carnivores may reflect an evolutionary adaptation, according to the authors of these studies [123–125]. Carnivores encounter an iron-rich, nickel-scarce diet [126] so infection by these *Helicobacter* species could be limited if they were to possess only Ni-ureases. The flexibility to produce either a Ni- or Fe-urease allows these *Helicobacter* species to colonize the gastric mucosa regardless of their host's diets.

We review here the role in pathogenesis of some of the best characterized microbial urease systems, emphasizing perspectives on the most recently published findings.

3.1. H. pylori

H. pylori must first survive the harsh environment of the human stomach and then survive a prolonged immune response that includes bombardment with oxidative radicals and oxidizing acids. The most severe *H. pylori*-mediated disease is due to long term infection. Urease, which comprises up to 10% of the total *H. pylori* proteome [127], is essential for the in vivo survival of *H. pylori*, as the buffering molecules from urea hydrolysis are essential to maintain the pathogen's cytoplasmic pH close to neutral. The constant production of ammonia in *H. pylori*-infected patients with cirrhosis can lead to blood hyperammonemia, which has been linked to a condition named minimal hepatic encephalopathy (MHE). As expected, anti-*H. pylori* therapy led to a reduction in blood ammonia levels, with subsequent improvement in MHE [128]. Besides the well-established acid-combatting role, many additional roles have been attributed to *H. pylori* urease over the last 25 years (see Table 1). The Ni-containing active form of urease is clearly required for initial colonization; however, several studies suggest that non-neutralizing roles also exist for urease, and those may not even require the Ni-form. For instance, a urease negative strain was unable to colonize a pH neutral pig stomach [85]. A recent study from Debowski et al. found that urease is needed for persistence in the mouse gastric mucosa, where pH approaches neutrality [96]. This is at first puzzling, since the main role of urease (to survive the low pH) requires considerable expense in terms of number of maturation/accessory enzymes and energy (in the form of GTP hydrolysis); this expenditure and Ni-drain should not occur (i.e., should not be needed) when the bacterium occupies neutral pH environments. However, urease-expressing bacteria are favored for survival/colonization in vivo over long time periods, while the pathogen resides in the mucosa, so urease seems to be needed for chronic and persistent infection [96]. In addition, urease is linked to gastric carcinoma incidence via its ability to promote angiogenesis [97]; the enzyme has been shown to induce proinflammatory cytokines, stimulate chemotaxis of neutrophils and monocytes, and to induce apoptosis in gastric endothelial cells [88,98].

As previously stated, urease (UreAB) is the most abundant enzyme synthesized in *H. pylori*; however, it appears most of it is never active as a ureolytic enzyme; this is probably due to limiting extracellular (host) and intracellular nickel levels. Indeed, a study from Stingl and De Reuse estimated that the fraction of nickel-activated urease ranged from 2% to 25% depending on growth conditions [129]. Therefore, this observation raises a question: why does *H. pylori* synthesize so much urease, if the bulk of it is not activated and therefore is not useful to combat acidity? Perhaps much of the above can be explained at the molecular level by a study indicating inactive urease plays a large role in *H. pylori* survival, due simply to its amino acid residue composition. Indeed, a new role for urease as an antioxidant or reactive oxygen-combating protein has been recently unveiled [99] (Figure 1). Catalytically inactive urease was able to protect the pathogen from oxidative damage, via a Met residue oxidant quenching mechanism. This mechanism does not require nickel, but requires surface Met residues, that cycle between oxidized and reduced forms, yet it would seem that nickel is important to facilitate the overall process, as the nickel containing version would be more stable to proteolysis within the cell, and urease synthesis is up-expressed by nickel [130,131]. The UreAB heterodimer contains 25 Met residues; 11 of these are subject to oxidation and subsequent methionine sulfoxide reductase (MSR) repair through a Met/Met-sulfoxide cycle [99]. In summary, both the catalytic and the noncatalytic role(s) of urease are important for *H. pylori* initial infection and long-term persistence in the host, as depicted in Figure 1.

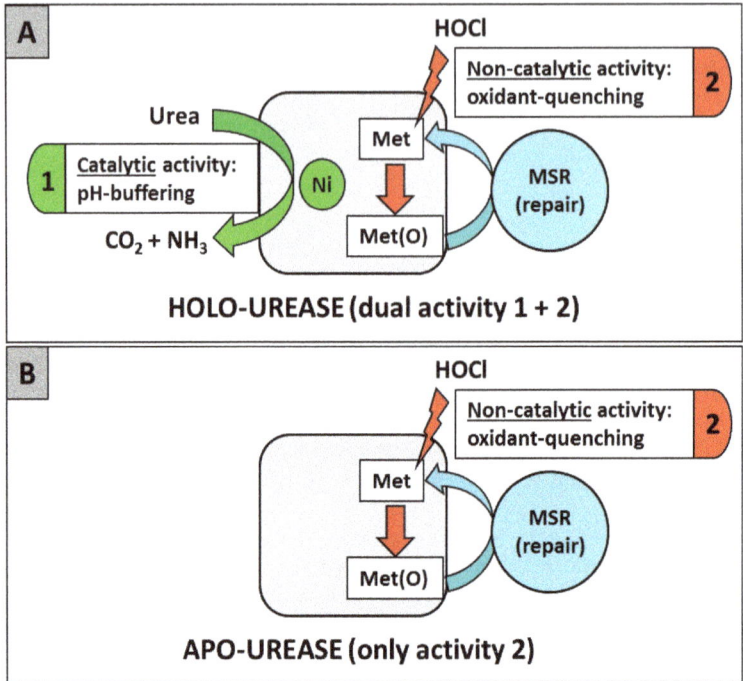

Figure 1. Dual role for *H. pylori* urease. (**A**) Holo (Ni-bound) urease can convert urea into ammonia and carbon dioxide (catalytic activity, #1) and can protect against oxidative stress (nonspecific quenching through Msr-repairable Met, activity #2). (**B**) Apo-urease (Ni-free) is non ureolytic yet it retains the nonspecific oxidative stress combatting activity (#2).

3.2. S. aureus

S. aureus is both a commensal bacterium and a human pathogen. It colonizes approximately 30% of the population asymptomatically; however, it can also cause infections ranging from mild skin and soft tissue infections to invasive infections, including sepsis and pneumonia [132]. A recent study by Zhou and coworkers examined the *S. aureus* urease response and its roles in host persistence [38]. The authors note that *S. aureus* combats many different host environments, and urease may have very different roles in different pH conditions, in part due to the largely underappreciated differing mode of action of strong versus weak acids on the cells. For example, weak acids such as acetate enter the cytoplasm more easily than strong acids that fully dissociate in water. Therefore, the weak acids cause macromolecular damage via intracellular proton release. Using a mutant strain analysis approach, it was concluded that urease activity is important to cell viability under weak acid stress conditions [38]. Kidney colonization was compared in a mouse bacteremia model; kidneys infected with the *S. aureus* Δure mutant strain had significantly lower bacterial burden in the longer term (12 and 19-day) infections than did the wild type strain. The host immune response, as assessed by leukocyte populations, did not differ between the two infected groups, so the absence of urease did not seem to influence the immune response (i.e., to augment clearance). These authors postulate that urease is important for host skin survival, where *S. aureus* resides; indeed, human skin is a major reservoir of this pathogen (for instance in sweat glands and hair follicles) and the authors reported that sweat contains 22 mM urea and the skin pH ranges from 4 to 6.

Biofilms are a significant contributor to host colonization and subsequent virulence by many pathogens, including *S. aureus*. Interestingly, a study from Resch et al. found increased expression of some of the urease structural and accessory genes (in correlation with increased urease activity) in biofilm-embedded *S. aureus* cells compared to planktonic cells of the same strain [37]. Connections between staphylococcal biofilms, urease production, and antibiotic resistance were further analyzed in a recent study [39]. When mixed biofilms composed of two species of *Staphyloccoccus* (*S. aureus* and *S. epidermidis*) were compared to monospecies biofilms, the urease subunit genes as well as the accessory protein genes were downexpressed in the mixed source. Since each species did not influence survival of the other, and the initial ratios used in biofilm formation were maintained, it was therefore determined that the cospecies influence on urease gene expression was specific to biofilm cultures [39]. According to the authors, *S. epidermidis* inhibits metabolic activity of *S. aureus*, leading to less acid production. As a consequence, less urease activity is required to compensate for low pH. Importantly, the two species used, *S. epidermis* and *S. aureus*, are oftentimes coisolated from biofilms on indwelling medical devices [133].

3.3. P. mirabilis

P. mirabilis is a major cause of urinary stones and it also forms resilient crystalline biofilms on catheters [63]. The initial formation of large clusters of the bacteria in the bladder lumen may be the etiology of stone formation [62], and urease is considered to be one of the two most important virulence factors in the initiation of cluster development [62,134]. In catheter-associated UTIs, urease-produced ammonium and carbon dioxide bind with Mg^{2+} and Ca^{2+}, respectively, found in the urine. These minerals precipitate, forming crystalline deposits on catheters or/and aggregates that evolve into macroscopic stones within the urinary system [135]. Adherent bacteria grow and the crystalline biofilm enlarges, so that bacteria and crystals become tightly associated [134]. Later, bacteria can become dissociated and begin the crystallization process elsewhere. *P. mirabilis* mutant strains that lack urease are unable to form crystalline biofilms [134]. Note that the urea level in human urine is reportedly about 400 mM [119], the average nickel level is approximately 1.7 µg/L (ranging from 0.1 to 20 µg/L) [136] and that the urease activity of *P. mirabilis* is especially robust.

Not much is known about *P. mirabilis*' ability to import and sequester nickel. Based on genome sequence analysis, *P. mirabilis* has two predicted Ni-transporters (*nikAB* and *yntABCD*); both transcriptional units are induced in experimentally-infected mice compared with laboratory-grown cells [63]. While it does not contain Ni-storage proteins, such as the ones found in *H. pylori* (Hpn, Hpn-like and HspA); nevertheless, *P. mirabilis* contains two His-rich accessory proteins: UreE, a putative urease accessory protein, contains a His-rich C-terminus domain (eight His residues out of nine residues); HypB, a putative hydrogenase accessory protein, is unusually His-rich, as it contains 17% His; almost all His residues are located in the N-terminus part of *Pm*HypB, bringing the percentage of His residues to an astonishing 39% (Figure 2). It is possible HypB plays a dual role in both hydrogenase and urease maturation in *P. mirabilis*, as previously demonstrated for both *H. pylori* [137] and *H. hepaticus* [138]. In addition, the maturation accessory factor could serve as nickel storage or as a sensor of Ni homeostasis status for the urinary tract pathogen but this remains to be shown.

```
MCSTCGCGEGNVSIEGVAPHSHDHHHHSHDHDHHDHGHHHHGHHHHHGHHHGHDHHHEHNATPANTVHKYIDKSE
QKHKHNYETHGQPIIIHHHHYHNSGDVHLHFYHDAQQNEAQVFHEHHHGHDDHHAHSHEHTHSHEHEHSHDHEHSHE
HEEQFSPVIDNDNMHYGQGEAGTHAPGISQKRMLKIEMDVLDKNNRIAVHNREHFAQQNVLALNLVSSPGSGKTTLLTQT
LKQLTQRVPCAVIEGDQQTTNDADRIRETGVAAIQVNTGKGCHLDAQMVHDATHQLGLKDNSILFIENVGNLVCPASFDLGE
KHKVAILSVTEGEDKPLKYPHMFAAADLMIINKIDLVPHLNIDVQACIESARRVNPNIEIIALSATTGEGMEEWLAWLESRLCA
```

Figure 2. The *P. mirabilis* HypB hydrogenase accessory protein is exceptionally His-rich. The sequence shown was translated from the *hypB* nucleotide sequence, as found in genome sequence of *P. mirabilis* strain HI4320 [139]. His residues are shown in red. The His-rich sequence suggests *Pm*HypB might play additional roles in Ni-enzyme maturation and/or Ni-storage besides its expected role as hydrogenase accessory protein.

The extracellular cluster of urease-containing *P. mirabilis* in the bladder lumen leads to a robust immune response [62]. Adjacent to the *P. mirabilis* clusters at the bacteria–bladder interface in the mouse model of UTI, neutrophil marker characteristics consistent with antimicrobial peptides in neutrophil extracellular traps/webs and phagocytosed bacteria were all observed. Therefore, the bacterium is likely to be subjected to significant oxidative stress *in vivo*, and it seems thus conceivable that alternative roles of urease (such as the one recently described for *H. pylori* urease [99]), could occur at these sites.

3.4. Ureaplasma spp.

Ureaplasma spp. include *U. urealyticum*, *U. parvum*, and *U. diversum*. The first two species are responsible for vaginal infections in humans while the later species cause urogenital tract infection in cattle and small ruminants [42,45]. As the name implies, *Ureaplasma* species are urease positive, to the extent that the enzyme is actually used as genus-specific diagnostic [42]. At least in *U. urealyticum*, and probably in other members of this Mollicute class, urease fulfils several roles: (i) the ammonia released in the cytosol contributes to PMF-driven ATP synthesis [43]; (ii) the ammonia also increases the urinary pH, leading to Mg precipitation and subsequent struvite stone formation, as shown in rat bladders [44].

3.5. Eukaryotic Pathogens

In contrast to metals such as copper, zinc, and iron, little is known of the roles of nickel in fungal pathogenesis. Still, ureases play important roles in fungal pathogens, for example in *C. neoformans* and in *C. immitis* (Table 1) [140]. In *C. neoformans* (responsible for human meningoencephalitis), Ni-urease is an important factor for brain invasion, as shown in several independent studies [18,19]. The enzyme can be found in extracellular vesicles, apparently used by the fungus to colonize host tissues [20]. Urease maturation components, sometimes referred to as accessory proteins or maturation chaperones, largely resemble their bacterial counterparts [141]. Although the cryptococcal genome lacks *ureE* and *ureG* homologs, one accessory protein, named Ure7, combines the nickel incorporation functions normally assigned to both UreE and UreG [141]. In *C. posadasii*, the causative agent of San Joaqin Valley fever, the extracellular ammonia generated (by urease) at sites of pulmonary infection contributes to severity of the respiratory disease [23], and urease mutants are less virulent in a mouse intranasal challenge [22]. A nickel permease homolog is present in the *Aspergillus fumigatus* genome, but the role it plays is not known [140]. It is worth noting, however, that many fungal pathogens apparently have no need for nickel, relying instead on a non-Ni, biotin-requiring urease to metabolize urea.

4. Hydrogenases

Hydrogenases are found in bacteria, archaea, and in some eukarya. They catalyze the conversion of molecular hydrogen (H_2) into protons and electrons and the reverse reaction, the generation of H_2 [142]. Three classes of hydrogenases have been defined, based on the metallic content of their active site: [NiFe], [FeFe], and [Fe] hydrogenases. Several [NiFe] hydrogenases, (especially of the

H$_2$-uptake type) have been shown to be key to colonization and virulence in various organisms such as *H. pylori* [80] or *S.* Typhimurium [143]. We review here the major findings on [NiFe]-mediated H$_2$ use by pathogenic bacteria.

4.1. H. pylori

H$_2$-uptake hydrogenase activity was first measured in whole cells of microaerobically grown *H. pylori*, using an amperometric assay and various artificial and natural electron acceptors, including oxygen [144]. The activity was subsequently shown to be specifically associated with membrane fractions and within these membranes, the hydrogenase enzyme was shown to be poised at a redox potential to oxidize H$_2$ rather than to evolve the gas [144]. Based on genome sequence analysis, *H. pylori* contains only one hydrogenase, of the H$_2$-uptake type (*hydABCDE* operon). Transcription of the *hyd* operon is controlled by various regulatory proteins in response to distinct stimuli (iron, nickel, pH, H$_2$); for instance, *hyd* genes are transcriptionally repressed by the apo (iron-free) form of Fur, the ferric uptake regulator [145]. Furthermore, the transcription of each of the *hydABC* structural genes is repressed in wild-type cells grown in nickel-supplemented medium; however, this repression is not observed in a Δ*nikR* mutant [146,147]. Finally, H$_2$ supplementation increases both *hydA* transcription and H$_2$-uptake hydrogenase activity in *H. pylori* [80]; however, neither the H$_2$-sensing mechanism, the H$_2$-responding regulatory mechanism, nor the global H$_2$-responsive proteome, has been characterized in *H. pylori*.

A unique particularity of nickel trafficking in *H. pylori* is the interplay between the urease and hydrogenase maturation pathways. Indeed, Olson et al. found that two of the hydrogenase accessory enzymes, HypA and HypB, are required not only for hydrogenase maturation, but also for urease maturation [137]. Additional studies from various groups (including ours) provided further evidence of the interconnectivity between both maturation pathways. Indeed, HypA was shown to physically interact with the urease accessory protein UreE [148,149], and a HypA-(UreE)$_2$ heterotrimeric complex able to bind nickel has been characterized [150]. Furthermore, nickel transfer between both proteins (from HypA to UreE) was demonstrated [151]. Finally, HypB was also found to be physically associated with another urease maturation protein, UreG [152].

Since the *H. pylori* hydrogenase K_m for H$_2$ is approximately 1.8 μM and the concentration of dissolved H$_2$ in animal and human stomachs is in the high micromolar-low millimolar range, the enzyme is predicted to be chronically saturated with H$_2$ [80,153,154]. The [Ni–Fe] hydrogenase is important for virulence: the *hydB* mutant colonized only 24% of mouse stomachs, while 100% of stomachs inoculated with the parent strain were colonized [80]. Recent studies have showed that the energy (proton motive force, PMF) derived from H$_2$ respiration can drive various important cellular mechanisms in *H. pylori*. Firstly, a link between H$_2$ utilization and CO$_2$ fixation (in the form of HCO$_3^-$) was established in *H. pylori* [81]. It is interesting to note that this H$_2$-stimulated CO$_2$ fixation (also referred to as "H$_2$-stimulated mixotrophy") is a growth mode that has never been described for a human pathogen [81]. Secondly, the hydrogenase-mediated H$_2$ respiration can fuel CagA (cytotoxin-associated gene A) translocation into host cells. CagA-positive strains have increased adenocarcinoma incidence [155]. A carcinogenic derivative strain that had greater ability to translocate CagA was found to have higher hydrogenase activity than its noncarcinogenic parent strain [82]. In agreement with this result, a *H. pylori* Δ*hyd* hydrogenase deletion mutant was unable to translocate CagA into human gastric epithelial AGS cells and the strain did not induce gastric cancer in gerbils [82]. By contrast, 50% of gerbils infected with the wild-type strain (hydrogenase positive, CagA translocating) developed gastric cancers [82]. Finally, albeit a limited strain set was studied, a significantly higher hydrogenase activity was measured in *H. pylori* strains isolated from cancer patients, compared to those measured in strains isolated from gastritis patients [82]. Taken together, these results suggest a correlation between the *H. pylori* Ni-hydrogenase and (CagA-mediated) cancer.

4.2. H. hepaticus

H. hepaticus has been shown to induce liver disease in mice, as well as colitis, colorectal cancer, inflammatory bowel disease (IBD), and prostate cancer [156,157]. Based on genome sequence, H. hepaticus possesses only one [NiFe] H_2-uptake membrane-bound hydrogenase [158]. Similar to what was reported in H. pylori, hyp hydrogenase accessory genes are present, and mutations in either hypA or hypB abolish both the hydrogenase and the urease activities [138]. Whole cells of H. hepaticus are able to couple H_2 oxidation to O_2 uptake [159]. H_2 concentrations measured in the livers of live adult mice are above 50 µM, which means that H. hepaticus hydrogenase, with an apparent K_m of approximately 2.5 µM, is saturated with H_2 [159]. Mehta and colleagues showed that the energy derived from H_2-oxidation can be used for amino acid uptake, eventually enhancing cell growth; this dual phenotype was observed with the WT strain but not in a ΔhyaB mutant strain [77]. While there was no significant difference in bacterial count numbers between WT and ΔhyaB mutant strains in the liver or cecum of mice, various liver lesions were observed with the WT but not with the mutant [77]. To summarize, the H. hepaticus [NiFe] hydrogenase provides energy (in the form of PMF) to the cell, aiding amino acid transport, bolstering growth and eventually contributing to liver pathogenesis, at least in the established murine model.

4.3. S. Typhimurium

Similar to E. coli, the enteric pathogen S. Typhimurium contains four different [Ni–Fe] hydrogenases: Hya (Hyd-1), Hyb (Hyd-2), Hyc (Hyd-3), and Hyd (Hyd-5) [67]. However, in contrast to E. coli that contains two H_2-uptake and two H_2-evolving hydrogenases, S. Typhimurium possesses three respiratory (H_2-uptake) enzymes (Hya, Hyb, and Hyd) and only one H_2-synthesizing enzyme (Hyc); the latter forms the formate-hydrogen-lyase (FHL) system together with the formate dehydrogenase-H (FDH-H), coupling H_2 production to formate oxidation, similar to what has been described in E.coli [160]. Each of the three respiratory hydrogenases is coupled to a respiratory pathway that can use O_2 as the terminal electron acceptor [143,161]. However, S. Typhimurium can use many terminal acceptors and it can be expected that they could all be coupled to H_2 oxidation. The role of each respiratory enzyme as well as their specific expression in various environments (murine macrophages, human polymorphonuclear leukocyte (PMN)-like cells, and mice) was studied using a mutagenesis approach, combined with RIVET (Resolvase In Vivo Expression Technology) [64]. The hya mutant was expressed at low levels in all (mouse) locations tested (e.g., the ileum, the liver and the spleen) and its survival in macrophages was decreased (compared to the WT), a phenotype attributed to the higher acid sensitivity observed for this mutant [64]. The hyd (Hyd-5) gene was found to be highly expressed in the liver and spleen, and weakly expressed in the ileum, at early stages of infection. In the late stages of infection, hyd was expressed at high levels in all organs tested [64]. Expression of the hyb (Hyd-2) gene could not be studied, due to a lack of stability of the hyb RIVET construct.

The role of each enzyme in physiology and virulence was assessed by constructing a series of markerless mutants and testing them using the typhoid fever-mouse model [143]. Double-mutant strains expressing only Hya (Δhyb Δhyd) or only Hyd (Δhya Δhyb) had lower virulence compared to the WT. In contrast, the Δhya Δhyd double mutant strain retaining Hyb activity was almost as virulent as the WT strain, suggesting Hyb is the most important hydrogenase for S. Typhimurium virulence [66,162]. Interestingly, the triple mutant (Δhya Δhyb Δhyd) was found to be avirulent (100% survival in the typhoid fever mouse model) [143]. This was confirmed by an independent study [163].

Based on the analysis of the S. enterica Typhi genome sequence, it appears the causative agent of typhoid fever in humans has the same set of hydrogenases as S. Typhimurium. Given the results of mouse studies with S. Typhimurium, it is expected that one or several of the [Ni–Fe] respiratory hydrogenases of S. Typhi could play an important role in the pathogenicity of typhoid fever in humans.

4.4. C. jejuni

C. jejuni, a leading cause of human diarrheal disease, is a microaerophilic bacterium that possesses a unique, energy-conserving, membrane-bound [Ni–Fe] uptake-type hydrogenase [74,164]. The enzyme is important for both *C. jejuni*'s growth and virulence. Indeed, in addition to carbon sources formate and fumarate, the respiratory reductant H_2 has been found to enhance growth of *C. jejuni* [74,165]. Disruption of the *hydB* gene led to abolition of hydrogenase activity, as expected, and the Δ*hydB* mutant showed severe colonization deficiency of the chicken cecum (compared to the WT) but only in the context of a Δ*fdhA* (formate dehydrogenase) mutant background [74]. Both Δ*fdhA* and Δ*hydB* single mutants showed only modest reduced colonization compared to WT. Finally, the *C. jejuni* Δ*hydB* is impaired in cell division (scanning electron microscopy revealed a filamentous phenotype) and is unable to interact with either human intestinal cell lines (INT-407) or with primary chicken intestinal epithelial cells [75]. Thus, similar to what has been observed in *H. pylori* (which belongs to the same phylogenetic group, the ε-proteobacteria), the [NiFe] H_2-uptake hydrogenase plays an important role in *C. jejuni* metabolism and pathogenesis.

4.5. C. concisus

C. concisus has been found throughout the entire human oral-gastrointestinal tract. The bacterium is associated with various ailments and diseases, such as gingivitis, periodontitis, inflammatory bowel disease, including Crohn's disease [166]. *C. concisus* contains genes encoding for two distinct Ni-containing hydrogenase complexes: a H_2-uptake type hydrogenase ("Hyd") similar to those found in other pathogenic ε-proteobacteriae (such as *H. pylori* or *C. jejuni*) and a H_2-evolving type hydrogenase similar to Hyd-3 (Hyc) and Hyd-4 (Hyf) complexes found in *E. coli* [167]. The former appears essential, as it is possible to disrupt components of the Hyf complex (*hyfB*), whereas attempts to generate *hyd* mutants were unsuccessful [167]. Furthermore, *C. concisus* has the highest H_2-uptake hydrogenase activity reported so far among pathogenic bacteria [167]. In agreement with these observations, H_2 was found to be needed for optimal growth under anaerobic conditions, and required for growth under microaerobic conditions, highlighting the importance of the H_2-uptake hydrogenase in the pathogen's metabolism [76,168].

4.6. S. flexneri

Shigella spp. including *S. flexneri*, *S. boydii*, *S. sonnei*, and *S. dysenteriae*, cause shigellosis (also called bacillary dysentery). *Shigella* spp. are responsible for approximately 165 million illness episodes worldwide, leading to an estimated 164,000 diarrhoeal deaths annually [169]. Based on genome sequences, *Shigella* spp. have four predicted unidirectional hydrogenases: two H_2-uptake enzymes, Hya and Hyb, and two H_2-evolving enzymes, Hyc and Hyf, although the role of the latter remains elusive. McNorton and Maier used a targeted mutagenesis approach to address each enzyme's respective role in *S. flexneri* [70]. Both H_2-uptake hydrogenases in *S. flexneri*, and more specifically Hya, can combat severe acid stress through generation of abundant periplasmic proton pools that are hypothesized to act as a barrier against proton influx from the outside [70]. Based on mutant strain analysis, much of the H_2 oxidation was attributed to the Hya hydrogenase: its activity was three-fold activated within minutes of acid exposure. This acid activation phenomena has clear pathogen survival consequences, as the Hya enzyme is the hydrogenase shown (in *S.* Typhimurium) to combat or to resist phagolysosome killing, and a primary method of such killing by immune cells is acidification [170].

5. Other Ni-Dependent Enzymes

Besides urease and hydrogenase, three other Ni-dependent enzymes can be found in a few pathogens: these are the Ni-activated forms of acireductone dioxygenase (ARD) [6,13], glyoxalase I (GloI) [16], and superoxide dismutase (SOD) [15]. Although it could be argued that neither of these three enzymes directly contributes to pathogenesis, the first two (ARD and GloI) play important roles in metabolism of their respective host, while the third (SOD) is a key contributor to oxidative stress resistance in bacteria. Thus, all three Ni-enzymes are expected to play (to a certain degree) a role in metabolism, growth, and virulence of their bacterial hosts. In support of this, heterozygous *glo-I* mutants of *L. donovani* (causative agent of visceral leishmaniasis) were found to exhibit reduced methylglyoxal detoxification, and *glo-I* null mutants were not viable, illustrating the importance of Ni-GloI for this parasite [28].

5.1. Acireductone Dioxygenase (ARD)

The ARD enzyme has two different activities depending on whether it uses Fe^{2+} or Ni^{2+} as cofactor [6,13]. The enzyme, which is part of the methionine salvage pathway, uses the same substrates (1,2-dihydroxy-3-keto-5-methylthiopent-1-ene (acireductone) and O_2) regardless of the bound metal (Fe^{2+} or Ni^{2+}); however formate and the ketoacid precursor of methionine, 2-keto-4-methylthiobutyrate are produced in presence of Fe^{2+}, whereas methylthiopropionate, carbon monoxide and formate are produced in presence of Ni^{2+} [13]. Based on genome sequence analysis, the Ni-containing form of ARD is expected to be found in all pathogenic γ-proteobacteriaceae, as well as in *A. baumannii*, *P. aeruginosa*, and *S. pneumoniae* (Table 1). There is no known Ni-ARD in eukaryotes. The structure of the *K. pneumoniae* Ni-ARD was revealed by Pochapsky and coworkers, using NMR and X-ray absorption spectroscopy [171].

5.2. Ni-Glyoxalase I

The glyoxalase I (GloI) enzyme, also called lactoylglutathione lyase, is part of a three-component system aimed at detoxifying methylglyoxal, a chemical that forms adducts with DNA; besides GloI, the system involves the thioesterase glyoxalase II (GloII) and reduced glutathione (GSH), the final product of the detoxification pathway being D-lactate [6,16]. There are two distinct classes of GloI: a Zn^{2+}-dependent class and a Co^{2+}/Ni^{2+}-dependent class, both of which can be found in a variety of eukaryotic and prokaryotic organisms [16]. The former (Zn-GloI) includes *Homo sapiens*, *Saccharomyces cerevisiae*, and *Pseudomonas putida*, while the latter (Ni-GloI) was originally described in *E. coli* [172]. Since then, nickel has been shown to be the preferred cofactor of GloI in various prokaryotic pathogens such as *P. aeruginosa*, *N. meningitidis*, and *Y. pestis* [173]. In addition, *Clostridium acetobutylicum* GloI co-crystallized with nickel [36]. Based on sequence analysis, the authors of the study hypothesize other clostridial GloI to be also Ni-activated, including those of pathogenic *C. botulinum*, *C. perfringens*, and *C. tetani* [36]. Since S-lactoylglutathione, a product of GloI, has been shown to play an important role in potassium efflux in *E. coli* [174], a similar role can be expected for pathogenic *E. coli* species, as well as for the (Ni) GloI-containing bacteria cited above. In fact, based on genome sequence analysis, the Ni-containing isoform of GloI is widespread among bacterial species. For instance, all Enterobacteriaceae (including *E. coli*, *Enterobacter* spp., *Klebsiella* spp., *Morganella* spp., *Proteus* spp., *Providencia* spp., *Serratia* spp., *Salmonella* spp) are expected to have the Ni-GloI type. Finally, Ni-GloI can also be found in protozoan parasites [175], including *L. major* [27] and *T. cruzi* [29]. Likewise, the GloI homolog from *L. donovani* is also expected to be Ni-dependent, based on genome sequence analysis. As stated above, the *gloI* gene is essential in *L. donovani*, leading the authors to identify GloI as a potential drug target [28].

5.3. Ni-Superoxide Dismutase (Ni-SOD)

Superoxide dismutases (SOD), which catalyze the dismutation of superoxide radicals ($O_2^{\bullet -}$) into molecular oxygen (O_2) or hydrogen peroxide (H_2O_2), can be found in all domains of life. Three distinct groups of SODs have been defined, based on amino acid sequence homology and preferred metallic cofactors: Cu-Zn-SOD, Fe-SOD and Mn-SOD, and Ni-SOD [15]. The Ni-SOD are seldom encountered; they were first described in a few species of the genus *Streptomyces* [176], including phytopathogenic species such as *S. scabies*, *S. acidiscabies*, and other related species. More recently, the Ni-SOD gene (*sodN*) has been found in cyanobacteria, marine γ-proteobacteria species, and in a marine eukaryote [177].

6. Nickel Transport and Nickel Metallophores

Pathogens must provide soluble Ni(II) [15] to mobilize the metal into the key nickel enzymes amongst an environment where this metal is in low availability (~0.5 nM) in the host [178]. A number of Ni-binding strategies are used by the pathogens, and the transporters vary in subunit composition, in Ni-binding affinity, and in chelating mechanism and chemistry. Several recent reviews have extensively covered nickel import by bacteria, including in human pathogens [11,179,180]. Therefore, we will only present the latest findings on nickel transport, and limit this to pathogens.

Like other transition metals, nickel needs to be first scavenged and imported from the extracellular environment. These so-called "nickelophores" (by analogy to iron siderophores) are small molecules which can chelate nickel ions before delivering it to specific transporters. Several recent studies have deciphered the structure and the specificity of these metallophores. In the case of nickel, L-His and its derivatives could play such roles [11,179]: for instance, in *E. coli*, a Ni–(L-His)$_2$ complex with NikA has been revealed by X-ray crystallography [181]. Likewise, *S. aureus* produces a nicotianamine-like metallophore called staphylopine (StP) to acquire metals under metal-limited conditions [182]. At first, StP was thought to be mostly zinc specific, but recent studies have demonstrated that it can also bind nickel [183,184]. In *P. aeruginosa*, an organism best known for its high affinity siderophores pyochelin and pyoverdin (the latter being a virulence factor), a recent study from Lhospice et al. has shown that a staphylopine-like metallophore named pseudopaline is able to import nickel in metal scarce environment [185].

In Gram-negative bacteria, the TonB/ExbB/ExbD machinery is needed to energize TonB-dependent transporters, allowing them to transport metals, including nickel, across the outer membrane (OM). For instance, the TonB-dependent FrpB4 protein has been shown to transport nickel through the OM in *H. pylori* [186].

Two main types of high affinity transporters are used by bacteria to transport Ni(II) across the cytoplasmic membrane. These are the ATP-binding cassette (ABC)-type transporters and the "secondary" nickel/cobalt NiCoT transporters [180]. In addition, a subclass of ABC transporters has been identified: the energy-coupling factor (ECF) transporter, which also requires ATP but uses a membrane-embedded solute binding protein; instead, ABC transporters rely on a soluble periplasmic binding protein (reviewed in [187]). So far, nine nickel ABC importers have been experimentally shown to import nickel in vivo, and only three nickel-binding proteins from human pathogens have been characterized: these are the *C. jejuni* NikZ, the *B. suis* NikA and the *Y. pestis* YntA [179].

Most prokaryotic pathogens mentioned in this review use both types of Ni-transporters. For instance, *H. pylori* possess both the NiuBDE transporter (ABC-type) and the NixA (NiCoT-type) [188,189]. Both NiuBDE and NixA function as nickel transporters independently of each other, and they are the sole nickel transporters [189]. Although both NiuBDE and NixA participate in nickel acquisition for urease activation, NiuBDE is the only transporter that can operate at both acidic and neutral pH. Furthermore, NiuBDE is also able to transport cobalt or bismuth (this is important, as bismuth is currently used in *H. pylori* eradication therapy), whereas NixA only transports nickel [189]. Finally, *H. pylori nixA* mutants retained some colonization ability in two different murine models, in contrast to *niuBDE* mutants. The latter strains are unable to colonize mouse stomachs, indicating that NiuBDE is required in vivo, but NixA is not [189,190].

In *S. aureus*, there are three distinct nickel transporters; however, they also fall into the two classes discussed above. Indeed, *S. aureus* possesses two canonical ABC-importers, the NikABCDE and the (recently discovered) CntABCDF systems, as well as the NixA system that belongs to the NiCoT family [191–193]. The Nik system functions in metal-replete medium and is required for urease activity as well as for urinary tract colonization [192]. In contrast to Nik, the multi-cation transporter Cnt is expressed under zinc-depleted conditions. However Cnt also plays an essential role in *S. aureus* virulence, as it contributes to colonization of the bladder and kidneys in an ascending urinary tract infection model, as well as in systemic infections in mice [193]. Extracytoplasmic nickel-binding components for the Nik and the Cnt ABC-type systems are *Sa*NikA and *Sa*CntA, respectively. A recent study, combining crystallography and mass spectrometry approaches, defined each protein's substrate specificity: SaNikA is able to bind either a Ni–(L-His)$_2$ complex or a Ni–(L-His) (2-methyl-thiazolidine dicarboxylate) complex (depending on their availability), while SaCntA binds Ni(II) via a different histidine-dependent chelator; however, it cannot bind Ni–(L-His)$_2$ [194].

In *E. coli* and other enterobacteria, only the ABC-type transporter Nik system is present. As stated above, a Ni–(L-His)$_2$:NikA complex was identified a few year ago [181]. It is worth noting however that transport of Ni–(L-His)$_2$ in *E. coli* is not a TonB-dependent process, since *E. coli* ΔtonB mutants are still able to transport it inside the cell [195]. Interestingly, the *nikABCDE* gene cluster of uropathogenic *E. coli* (UPEC) is up-expressed in urine samples isolated from UTI patients, as compared to the same UPEC strain cultured in urine from healthy volunteers or grown in lysogeny broth, suggesting that nickel transport is a key fitness factor for the bacteria during human UTI [196]. In agreement with this hypothesis, Δ*nik* mutants were shown to be compromised in fitness in the mouse model of UTI [196]. Even though enterobacteria do not possess the second Ni-transport system (NiCoT-type), it seems they have developed alternate strategies to import metals, including nickel. Indeed, enterobacteria that encode the *Yersinia* high pathogenicity island (HPI), including strains of *E. coli*, *Klebsiella*, and *Y. pestis*, secrete the metallophore Yersiniabactin (Ybt). Originally shown to chelate iron ions during infection, Ybt can also bind extracellular nickel in UPEC [197]. Ni-Ybt complexes are internalized, then metal-free Ybt is recycled outside the cell while the captured nickel is liberated for use by Ni-requiring enzymes (i.e., hydrogenases and/or urease depending upon the bacterial species). The authors hypothesize the Ybt system can chelate nickel ions that appear to be otherwise inaccessible to the NikABCDE permease [197]. In *Mycobacterium avium* subsp. *paratuberculosis*, expression of the *dppA* gene encoding for a Nickel/dipeptide transporter (ABC type) increased during early infection in an epithelium-macrophage co-culture system [198].

Finally, while nickel importers play a major role in providing Ni-requiring enzymes with the metallic cofactor, nickel exporters are equally important, as they ensure that intracellular nickel levels do not reach toxic concentration. Several nickel export systems have been characterized. For instance, the CznABC (cobalt zinc nickel) export pump of *H. pylori* was shown to play a critical role in both nickel homeostasis and in vivo stomach colonization: *czn* mutants had higher urease activities, yet they were unable to colonize in a Mongolian gerbil stomach animal model [199]. Likewise, a *P. mirabilis* putative nickel export transporter (PMI1518) was found to be essential for CAUTI, in single-species kidney colonization as well as in bladder and kidney colonization coinfection with *P. stuartii* [200].

7. Nickel Storage, Toxicity, and Metabolism

Among bacterial pathogens, mechanisms used to sequester and store nickel, as well as to remove it due to its toxic properties on macromolecules, are best known for the gastric pathogen *H. pylori*. This is not surprising as this bacterium contains two nickel enzymes (see above) that are key to the pathogen's in vivo survival. As such, its demand for nickel is great, but along with the high demand necessarily goes risk for potential toxicity. Much of the bacterium's regulatory mechanisms are based on sensing nickel levels, thus on nickel-binding proteins. Nickel overload inside the cell is countered by efflux mechanisms and by repression of nickel transport factors, while intracellular levels as well as storage of the metal for later use in Ni-enzyme manufacture is mediated by proteins known within the field as nickel storage proteins. The latter system involves two histidine-rich proteins (termed Hpn and Hpn-like) as well as at least one chaperone (HspA) that plays multiple roles. Most likely, the multiple Ni-binding proteins needed for eventual Ni-enzyme maturation have caused this bacterium to develop complex and unique mechanisms for dealing with nickel. The physiological challenge for the bacterium must then include requirements that minimally must encompass a critical need to discriminately deliver nickel to at least two Ni-enzymes and probably to other proteins, dealing with a very high nickel demand, recognizing nickel amongst a variety of (sometimes competing) metal cations, and balancing metal need with toxicity due to the fluctuating reservoirs of the metal both intracellularly and extracellularly.

7.1. Hpn and Hpn-Like Proteins

H. pylori possess two proteins with remarkably high histidine content. These small His-rich proteins, named Hpn and Hpn-like (referred to herein as Hpnl), contain 47% and 25% His, respectively. Interestingly, both are confined to the gastric colonizing types of Helicobacters [201]. However, Hpn is present in every gastric *Helicobacter* species, and Hpnl is restricted to *H. pylori* and its closely related species *H. acinonychis*, originally isolated from cheetah [202].

These small proteins have apparently redundant functions to one another regarding roles in nickel storage for urease manufacture. Recombinant Hpn exists primarily as a 20-mer with each monomer binding five Ni (II) with a Kd of 7.1 µM [203]. Hpnl monomers bind two Ni (II) with a Kd of 3.8 µM, but it also forms multimeric structures of more than 20 subunits [204]. While the His residues are directly involved in nickel binding, as expected, the additional multiple Gln residues (in Hpnl) are thought to improve stability of the metal complexes [205]. Strains carrying mutations in *hpn* and *hpnl* are more sensitive to nickel toxicity, and they influence active urease maturation in a nickel-dependent manner [206]. Nickel release from Hpn and Hpnl is observed under acidic conditions, suggesting that these proteins may supply nickel when urease is also needed (e.g., to combat acidity, see [207]). Expression of the storage proteins confers nickel resistance to *E. coli*, confirming their ability to sequester excess nickel [201]. Based on studies with pure proteins or on *H. pylori* mutant strain in-lab phenotypes, the initial suggestions that these proteins might play nickel storage roles in vivo was supported by use of mice maintained under strict nickel-limited conditions [208] or NMRI-specific pathogen-free mice [202]. *H. pylori* mutant strains lacking either Hpn, or Hpnl, or both storage proteins, were poorer colonizers than their wild type counterparts, when the hosts (C57/Bl6 mice) were subjected to nickel-deficient diets [208].

Purified Hpn was shown to interact with the UreA subunit of urease, while Hpnl interacted most strongly uniquely with the HypA and HypB hydrogenase maturation proteins [202]. Also, Hpn and Hpnl together impact intracellular nickel trafficking, and influence urease activity. However, the authors of the study concluded that Hpn is the primary nickel sequestering reservoir, and that the two storage proteins compete for nickel under low nickel conditions [202]. The result in low nickel is "restricted activation" or basal levels of urease. In high nickel, where both storage proteins are saturated, their nickel delivery roles would ensure Ni-activated levels of urease. In the suggested model, Hpnl would thus not play a nickel detoxification role, due to its more limited capacity to bind nickel and its lower abundance [202].

In contrast to the rather limited roles assigned to the two storage proteins by Vinella et al. [202], the most recent work on the two nickel storage proteins [207] indicates they both play a much broader role than previously reported. Considering their importance in virulence, it seemed reasonable they may play roles in nickel-sensing, regulation, or delivery of nickel; this would likely require many protein–protein interactions. An affinity pulldown approach was used, whereby cross-linking to each storage protein was followed by a Ni-based purification/enrichment; the results indicate the storage proteins interact with a wide array of proteins [207]. The storage proteins interacted with known nickel delivery systems involved in urease and hydrogenase maturation, and hydrogenase activity was severely diminished in a Δ*hpn* Δ*hpnl* double mutant strain when nickel was limited. Interestingly, both storage proteins play roles in ammonia production independent of urease activity, i.e., via amide hydrolysis. Indeed, Hpn and Hpnl were shown to synergistically suppress aliphatic amidase (AmiE) activity [207]. This role makes sense from a physiological view, as the complementary ammonia-producing enzyme (urease) is known to be made in large amounts when cells are provided with nickel [130,131], and the aliphatic amidase AmiE plays a role in acid resistance especially in the absence of urease [209]. In addition, interactions between Hpn and the aminopeptidase PepA were observed [207]. PepA can accept a variety of divalent cations for its activation, and the study implicates the storage proteins role in peptide salvage processes. The interactions between Hpn and Hpnl and AmiE or PepA, respectively, were further confirmed by using purified proteins and a tryptophan (Trp) fluorescence-based method, taking advantage of the concomitant presence of Trp residues in both

AmiE and PepA and the lack of Trp in both storage proteins [210] (Figure 3). As expected, neither Hpn nor Hpnl had measurable fluorescence, while both AmiE and PepA, as pure proteins, had observable fluorescence profiles (Figure 3). The maximum fluorescence for both AmiE and PepA fell within the previously established range for Trp fluorescence maxima [211]. Upon addition of either purified Hpn or purified Hpnl, the fluorescence profile of AmiE and PepA shifted markedly (Figure 3). The observed shift suggests that the storage proteins have either altered the microenvironment of the Trp residues within the peptidase and amidase, or that they have caused conformational changes in the target proteins that altered their fluorescence profiles [212]. Bovine serum albumin (BSA), a protein that contains three Trp residues, was used as a negative control. Incubation of either Hpn or Hpnl had no effect on the fluorescence profile of BSA (data not shown), confirming the specificity of the (AmiE and PepA) interactions described above [210].

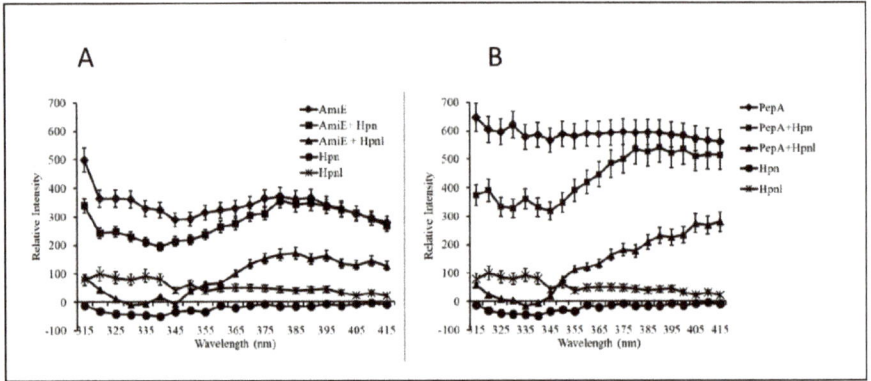

Figure 3. Tryptophan fluorescence as a way to detect Hpn and Hpnl protein interactions with (**A**) AmiE or (**B**) PepA. Purified proteins were incubated overnight at 4 °C with 1 mM DMS (crosslinker). Tryptophan fluorescence was recorded on a BioTek SynergyMx spectrophotometer (excitation at 295 nm and emission at 315–415 nm). Samples were normalized against their respective (buffer-only) controls and plotted as relative fluorescence against wavelength. See [210].

Ni-acquisition and peptide salvage, maturation of Ni-enzymes, and oxidative stress-combating enzymes are some of the enzymes that may be impacted by interaction with the storage proteins (199). Whether or not nickel delivery to or from these enzymes is involved is not known, and only a few interactions were characterized [207]. Still, the cross-linking results supported that each protein transiently but intimately interacts with perhaps 100 or more proteins. This may at first seems to be a gross overestimate, but considering that (i) nickel metabolism plays a central role in *H. pylori*; (ii) both storage proteins can bind other metals; (iii) both proteins apparently represent the major nickel reservoirs in the cell; thus, a variety of sensing and delivery roles for Hpn and Hpnl seems more possible. We must also remember that metal binding proteins may comprise a much larger fraction of the total bacterial proteomes than previously appreciated [213].

7.2. HspA

HspA is a nickel-binding protein that has homology to the highly-conserved and essential heat shock protein GroES. However, the *H. pylori* version has a unique His-rich C-terminus that binds nickel. Although a strain lacking HspA is not recoverable (i.e., lethal), a strain lacking only the C-terminal extension of HspA is viable, and it has been tested for virulence, i.e., mouse colonization capacity [214]. Although the strain had diminished hydrogenase activity and low tolerance to exogenous nickel, the strain was normal in colonization capacity. Of course, this strain still had other nickel-binding proteins. It would be interesting to perform the colonization assays in mice that have reduced nickel levels, to assess HspA's role in nickel storage when the metal is limiting. Also, comparing a strain lacking both the C-terminus His-rich extension of HspA as well as Hpn and Hpnl, to a strain lacking only the two storage proteins could be an approach to evaluate the (additive) role of HspA in the overall Ni-storage budget. HspA has long been considered to be a candidate for use as anti-*H. pylori* vaccine. Partial protection against *H. pylori* in mice was observed after intranasal administration of HspA [215]), while the interesting goal of expressing HpHspA in a probiotic bacterium (*Lactococcus lactis*) was unfortunately not successful [216].

8. Conclusions

As a required cofactor for some key enzymes, most notably hydrogenase and urease, nickel clearly plays roles in microbial pathogenesis. Still, new information on the role of the metal in pathogens that have other Ni-enzymes is needed, and it is hypothesized that more proteins that use nickel, or respond to fluctuating nickel levels, will be identified. In addition to uncovering the molecular mechanisms of nickel trafficking and homeostasis in the Ni-requiring pathogens, large gaps in our knowledge on nickel in vivo availability exist. These include understanding the dietary nickel sources, host metabolic factors that modulate accessibility of nickel to the pathogens and how nickel availability in the host is impacted by the (Ni-utilizing) host intestinal microbiota composition. These represent just some research areas that are in need of development. Availability of the metal within the host certainly varies, and it is important that we know how generally accessible the metal is within specific host organs, specific tissue types, and within specific host cell (e.g., epithelial cell, immune cell, blood cell) types. Finally, one promising avenue for future nickel-related research is the fact that numerous mammalian pathogens (at least 39 prokaryotes and nine eukaryotes) require the metal (for various enzymes, e.g., Ni-ARD, Ni-GloI, [Fe-Ni] hydrogenase, urease, and Ni-SOD), while their host do not; this presents an opportunity to specifically target pathogens via nickel sequestration. It seems such sequestration naturally occurs in higher plants, since plants use nickel (in the form of Ni-urease), while at the same time, the number of nickel-utilizing plant pathogens is very limited: indeed, only *S. scabies* and a few other related *Streptomyces* species are known plant pathogens that contain a Ni-enzyme (Ni-SOD). This interesting two-kingdom conundrum (plants use nickel/few Ni-requiring pathogens; mammals do not require nickel/many Ni-utilizing pathogens) not only gives us insights on the evolution of host/pathogens competition for nickel, but can perhaps provide us with a roadmap for future projects aimed at inhibiting or eradicating the nickel-requiring human or mammal pathogens.

Funding: This research received no external funding.

Conflicts of Interest: The authors declare no conflict of interest.

References

1. Zambelli, B.; Uversky, V.N.; Ciurli, S. Nickel impact on human health: An intrinsic disorder perspective. *Biochim. Biophys. Acta (BBA) Proteins Proteom.* **2016**, *1864*, 1714–1731. [CrossRef]
2. Rutherford, J.C. The emerging role of urease as a general microbial virulence factor. *PLoS Pathog.* **2014**, *10*, e1004062. [CrossRef]
3. Benoit, S.L.; Maier, R.J. Hydrogen and nickel metabolism in *Helicobacter* species. *Ann. N. Y. Acad. Sci.* **2008**, *1125*, 242–251. [CrossRef]
4. Maier, R.J. Use of molecular hydrogen as an energy substrate by human pathogenic bacteria. *Biochem. Soc. Trans.* **2005**, *33*, 83–85. [CrossRef]
5. Boer, J.L.; Mulrooney, S.B.; Hausinger, R.P. Nickel-dependent metalloenzymes. *Arch. Biochem. Biophys.* **2014**, *544*, 142–152. [CrossRef]
6. Maroney, M.J.; Ciurli, S. Nonredox nickel enzymes. *Chem. Rev.* **2013**, *114*, 4206–4228. [CrossRef]
7. Mazzei, L.; Musiani, F.; Ciurli, S. Urease. In *The Biological Chemistry of Nickel*; The Royal Society of Chemistry: Cambridge, UK, 2017; pp. 60–97.
8. Tai, H.; Higuchi, Y.; Hirota, S. Comprehensive reaction mechanisms at and near the Ni–Fe active sites of [NiFe] hydrogenases. *Dalton Trans.* **2018**, *47*, 4408–4423. [CrossRef]

9. Ogata, H.; Lubitz, W.; Higuchi, Y. Structure and function of [NiFe] hydrogenases. *J. Biochem.* **2016**, *160*, 251–258. [CrossRef]
10. De Reuse, H.; Vinella, D.; Cavazza, C. Common themes and unique proteins for the uptake and trafficking of nickel, a metal essential for the virulence of *Helicobacter pylori*. *Front. Cell. Infect. Microbiol.* **2013**, *3*, 94. [CrossRef]
11. Zeer-Wanklyn, C.J.; Zamble, D.B. Microbial nickel: Cellular uptake and delivery to enzyme centers. *Curr. Opin. Chem. Biol.* **2017**, *37*, 80–88. [CrossRef]
12. Gaddy, J.A.; Haley, K.P. Metalloregulation of *Helicobacter pylori* physiology and pathogenesis. *Front. Microbiol.* **2015**, *6*, 911.
13. Deshpande, A.R.; Pochapsky, T.C.; Ringe, D. The Metal Drives the Chemistry: Dual Functions of Acireductone Dioxygenase. *Chem. Rev.* **2017**, *117*, 10474–10501. [CrossRef]
14. Miller, A.-F. Superoxide dismutases: Ancient enzymes and new insights. *FEBS Lett.* **2012**, *586*, 585–595. [CrossRef]
15. Ryan, K.C.; Guce, A.I.; Johnson, O.E.; Brunold, T.C.; Cabelli, D.E.; Garman, S.C.; Maroney, M.J. Nickel superoxide dismutase: Structural and functional roles of His1 and its H-bonding network. *Biochemistry* **2015**, *54*, 1016–1027. [CrossRef]
16. Honek, J.F. Nickel Glyoxalase I. In *The Biological Chemistry of Nickel*; Zamble, D., Rowinska-Zyrek, M., Kozlowski, H., Eds.; Royal Society of Chemistry: Cambridge, UK, 2017.
17. Cox, G.M.; Mukherjee, J.; Cole, G.T.; Casadevall, A.; Perfect, J.R. Urease as a virulence factor in experimental cryptococcosis. *Infect. Immun.* **2000**, *68*, 443–448. [CrossRef]
18. Olszewski, M.A.; Noverr, M.C.; Chen, G.-H.; Toews, G.B.; Cox, G.M.; Perfect, J.R.; Huffnagle, G.B. Urease expression by *Cryptococcus neoformans* promotes microvascular sequestration, thereby enhancing central nervous system invasion. *Am. J. Pathol.* **2004**, *164*, 1761–1771. [CrossRef]
19. Fu, M.S.; Coelho, C.; De Leon-Rodriguez, C.M.; Rossi, D.C.P.; Camacho, E.; Jung, E.H.; Kulkarni, M.; Casadevall, A. *Cryptococcus neoformans* urease affects the outcome of intracellular pathogenesis by modulating phagolysosomal pH. *PLoS Pathog.* **2018**, *14*, e1007144. [CrossRef]
20. Rodrigues, M.L.; Nakayasu, E.S.; Oliveira, D.L.; Nimrichter, L.; Nosanchuk, J.D.; Almeida, I.C.; Casadevall, A. Extracellular vesicles produced by *Cryptococcus neoformans* contain protein components associated with virulence. *Eukaryot. Cell* **2008**, *7*, 58–67. [CrossRef]
21. Feder, V.; Kmetzsch, L.; Staats, C.C.; Vidal-Figueiredo, N.; Ligabue-Braun, R.; Carlini, C.R.; Vainstein, M.H. *Cryptococcus gattii* urease as a virulence factor and the relevance of enzymatic activity in cryptococcosis pathogenesis. *FEBS J.* **2015**, *282*, 1406–1418. [CrossRef]
22. Mirbod-Donovan, F.; Schaller, R.; Hung, C.Y.; Xue, J.; Reichard, U.; Cole, G.T. Urease produced by *Coccidioides posadasii* contributes to the virulence of this respiratory pathogen. *Infect. Immun.* **2006**, *74*, 504–515. [CrossRef]
23. Wise, H.Z.; Hung, C.-Y.; Whiston, E.; Taylor, J.W.; Cole, G.T. Extracellular ammonia at sites of pulmonary infection with *Coccidioides posadasii* contributes to severity of the respiratory disease. *Microb. Pathog.* **2013**, *59*, 19–28. [CrossRef]
24. Baltazar, L.M.; Zamith-Miranda, D.; Burnet, M.C.; Choi, H.; Nimrichter, L.; Nakayasu, E.S.; Nosanchuk, J.D. Concentration-dependent protein loading of extracellular vesicles released by *Histoplasma capsulatum* after antibody treatment and its modulatory action upon macrophages. *Sci. Rep.* **2018**, *8*, 8065. [CrossRef]
25. Costa, M.; Borges, C.L.; Bailao, A.M.; Meirelles, G.V.; Mendonça, Y.A.; Dantas, S.F.; de Faria, F.P.; Felipe, M.S.; Molinari-Madlum, E.N.E.; Mendes-Giannini, M.J. Transcriptome profiling of *Paracoccidioides brasiliensis* yeast-phase cells recovered from infected mice brings new insights into fungal response upon host interaction. *Microbiology* **2007**, *153*, 4194–4207. [CrossRef]
26. Rujirawat, T.; Patumcharoenpol, P.; Lohnoo, T.; Yingyong, W.; Kumsang, Y.; Payattikul, P.; Tangphatsornruang, S.; Suriyaphol, P.; Reamtong, O.; Garg, G.; et al. Probing the Phylogenomics and Putative Pathogenicity Genes of *Pythium insidiosum* by Oomycete Genome Analyses. *Sci. Rep.* **2018**, *8*, 4135. [CrossRef]
27. Ariza, A.; Vickers, T.J.; Greig, N.; Armour, K.A.; Dixon, M.J.; Eggleston, I.M.; Fairlamb, A.H.; Bond, C.S. Specificity of the trypanothione-dependent *Leishmania major* glyoxalase I: Structure and biochemical comparison with the human enzyme. *Mol. Microbiol.* **2006**, *59*, 1239–1248. [CrossRef]
28. Chauhan, S.C.; Madhubala, R. Glyoxalase I gene deletion mutants of *Leishmania donovani* exhibit reduced methylglyoxal detoxification. *PLoS ONE* **2009**, *4*, e6805. [CrossRef]

29. Greig, N.; Wyllie, S.; Vickers, T.J.; Fairlamb, A.H. Trypanothione-dependent glyoxalase I in *Trypanosoma cruzi*. *Biochem. J.* **2006**, *400*, 217–223. [CrossRef]
30. Morou-Bermudez, E.; Burne, R.A. Genetic and physiologic characterization of urease of Actinomyces naeslundii. *Infect. Immun.* **1999**, *67*, 504–512.
31. Salem, N.; Salem, L.; Saber, S.; Ismail, G.; Bluth, M.H. *Corynebacterium urealyticum*: A comprehensive review of an understated organism. *Infect. Drug Resist.* **2015**, *8*, 129–145.
32. Sassetti, C.M.; Rubin, E.J. Genetic requirements for mycobacterial survival during infection. *Proc. Natl. Acad. Sci. USA* **2003**, *100*, 12989–12994. [CrossRef]
33. Fontán, P.; Aris, V.; Ghanny, S.; Soteropoulos, P.; Smith, I. Global transcriptional profile of *Mycobacterium tuberculosis* during THP-1 human macrophage infection. *Infect. Immun.* **2008**, *76*, 717–725. [CrossRef]
34. Schnappinger, D.; Ehrt, S.; Voskuil, M.I.; Liu, Y.; Mangan, J.A.; Monahan, I.M.; Dolganov, G.; Efron, B.; Butcher, P.D.; Nathan, C. Transcriptional adaptation of *Mycobacterium tuberculosis* within macrophages: Insights into the phagosomal environment. *J. Exp. Med.* **2003**, *198*, 693–704. [CrossRef]
35. Lin, W.; Mathys, V.; Ang, E.L.; Koh, V.H.; Martinez Gomez, J.M.; Ang, M.L.; Zainul Rahim, S.Z.; Tan, M.P.; Pethe, K.; Alonso, S. Urease activity represents an alternative pathway for *Mycobacterium tuberculosis* nitrogen metabolism. *Infect. Immun.* **2012**, *80*, 2771–2779. [CrossRef]
36. Suttisansanee, U.; Lau, K.; Lagishetty, S.; Rao, K.N.; Swaminathan, S.; Sauder, J.M.; Burley, S.K.; Honek, J.F. Structural Variation in Bacterial Glyoxalase I Enzymes Investigation of the Metalloenzyme Glyoxalase I from *Clostridium acetobutylicum*. *J. Biol. Chem.* **2011**, *286*, 38367–38374. [CrossRef]
37. Resch, A.; Rosenstein, R.; Nerz, C.; Gotz, F. Differential gene expression profiling of *Staphylococcus aureus* cultivated under biofilm and planktonic conditions. *Appl. Environ. Microbiol.* **2005**, *71*, 2663–2676. [CrossRef]
38. Zhou, C.; Bhinderwala, F.; Lehman, M.K.; Thomas, V.C.; Chaudhari, S.S.; Yamada, K.J.; Foster, K.W.; Powers, R.; Kielian, T.; Fey, P.D. Urease is an essential component of the acid response network of *Staphylococcus aureus* and is required for a persistent murine kidney infection. *PLoS Pathog.* **2019**, *15*, e1007538. [CrossRef]
39. Vandecandelaere, I.; Van Nieuwerburgh, F.; Deforce, D.; Coenye, T. Metabolic activity, urease production, antibiotic resistance and virulence in dual species biofilms of *Staphylococcus epidermidis* and *Staphylococcus aureus*. *PLoS ONE* **2017**, *12*, e0172700. [CrossRef]
40. Gatermann, S.; John, J.; Marre, R. *Staphylococcus saprophyticus* urease: Characterization and contribution to uropathogenicity in unobstructed urinary tract infection of rats. *Infect. Immun.* **1989**, *57*, 110–116.
41. Chen, Y.Y.; Weaver, C.A.; Burne, R.A. Dual functions of *Streptococcus salivarius* urease. *J. Bacteriol.* **2000**, *182*, 4667–4669. [CrossRef]
42. Kokkayil, P.; Dhawan, B. Ureaplasma: Current perspectives. *Indian J. Med. Microbiol.* **2015**, *33*, 205–214.
43. Smith, D.; Russell, W.; Ingledew, W.; Thirkell, D. Hydrolysis of urea by *Ureaplasma urealyticum* generates a transmembrane potential with resultant ATP synthesis. *J. Bacteriol.* **1993**, *175*, 3253–3258. [CrossRef]
44. Grenabo, L.; Hedelin, H.; Pettersson, S. Urinary infection stones caused by *Ureaplasma urealyticum*: A review. *Scand. J. Infect. Dis. Suppl.* **1988**, *53*, 46–49.
45. Silva, J.; Marques, L.; Timenetsky, J.; de Farias, S.T. *Ureaplasma diversum* protein interaction networks: Evidence of horizontal gene transfer and evolution of reduced genomes among the Mollicutes. *Can. J. Microbiol.* **2019**. [CrossRef]
46. Sangari, F.J.; Seoane, A.; Rodriguez, M.C.; Aguero, J.; Garcia Lobo, J.M. Characterization of the urease operon of *Brucella abortus* and assessment of its role in virulence of the bacterium. *Infect. Immun.* **2007**, *75*, 774–780. [CrossRef]
47. Abkar, M.; Amani, J.; Sahebghadam Lotfi, A.; Nikbakht Brujeni, G.; Alamian, S.; Kamali, M. Subcutaneous immunization with a novel immunogenic candidate (urease) confers protection against *Brucella abortus* and *Brucella melitensis* infections. *APMIS* **2015**, *123*, 667–675. [CrossRef]
48. Bandara, A.B.; Contreras, A.; Contreras-Rodriguez, A.; Martins, A.M.; Dobrean, V.; Poff-Reichow, S.; Rajasekaran, P.; Sriranganathan, N.; Schurig, G.G.; Boyle, S.M. *Brucella suis* urease encoded by ure 1 but not ure 2 is necessary for intestinal infection of BALB/c mice. *BMC Microbiol.* **2007**, *7*, 57. [CrossRef]
49. Smith, M.G.; Gianoulis, T.A.; Pukatzki, S.; Mekalanos, J.J.; Ornston, L.N.; Gerstein, M.; Snyder, M. New insights into *Acinetobacter baumannii* pathogenesis revealed by high-density pyrosequencing and transposon mutagenesis. *Genes Dev.* **2007**, *21*, 601–614. [CrossRef]
50. Rathinavelu, S.; Zavros, Y.; Merchant, J.L. *Acinetobacter lwoffii* infection and gastritis. *Microbes Infect.* **2003**, *5*, 651–657. [CrossRef]

51. Bossé, J.T.; MacInnes, J.I. Urease activity may contribute to the ability of Actinobacillus pleuropneumoniae to establish infection. *Can. J. Vet. Res.* **2000**, *64*, 145.
52. Klitgaard, K.; Friis, C.; Jensen, T.K.; Angen, O.; Boye, M. Transcriptional portrait of *Actinobacillus pleuropneumoniae* during acute disease—Potential strategies for survival and persistence in the host. *PLoS ONE* **2012**, *7*, e35549. [CrossRef]
53. Pinske, C.; Sawers, R.G. Anaerobic Formate and Hydrogen Metabolism. *EcoSal Plus* **2016**, *7*. [CrossRef]
54. Steyert, S.R.; Kaper, J.B. Contribution of urease to colonization by Shiga toxin-producing *Escherichia coli*. *Infect. Immun.* **2012**, *80*, 2589–2600. [CrossRef]
55. Li, M.F.; Sun, L. *Edwardsiella tarda* Sip2: A Serum-Induced Protein That Is Essential to Serum Survival, Acid Resistance, Intracellular Replication, and Host Infection. *Front. Microbiol.* **2018**, *9*, 1084. [CrossRef]
56. Murphy, T.F.; Brauer, A.L. Expression of urease by *Haemophilus influenzae* during human respiratory tract infection and role in survival in an acid environment. *BMC Microbiol.* **2011**, *11*, 183. [CrossRef]
57. Maroncle, N.; Rich, C.; Forestier, C. The role of *Klebsiella pneumoniae* urease in intestinal colonization and resistance to gastrointestinal stress. *Res. Microbiol.* **2006**, *157*, 184–193. [CrossRef]
58. Young, G.M.; Amid, D.; Miller, V.L. A bifunctional urease enhances survival of pathogenic *Yersinia enterocolitica* and *Morganella morganii* at low pH. *J. Bacteriol.* **1996**, *178*, 6487–6495. [CrossRef]
59. Armbruster, C.E.; Smith, S.N.; Yep, A.; Mobley, H.L. Increased incidence of urolithiasis and bacteremia during *Proteus mirabilis* and *Providencia stuartii* coinfection due to synergistic induction of urease activity. *J. Infect. Dis.* **2014**, *209*, 1524–1532. [CrossRef]
60. Alteri, C.J.; Himpsl, S.D.; Engstrom, M.D.; Mobley, H.L. Anaerobic respiration using a complete oxidative TCA cycle drives multicellular swarming in *Proteus mirabilis*. *MBio* **2012**, *3*, e00365-12. [CrossRef]
61. Johnson, D.; Russell, R.; Lockatell, C.; Zulty, J.; Warren, J.; Mobley, H. Contribution of *Proteus mirabilis* urease to persistence, urolithiasis, and acute pyelonephritis in a mouse model of ascending urinary tract infection. *Infect. Immun.* **1993**, *61*, 2748–2754.
62. Schaffer, J.N.; Norsworthy, A.N.; Sun, T.T.; Pearson, M.M. *Proteus mirabilis* fimbriae- and urease-dependent clusters assemble in an extracellular niche to initiate bladder stone formation. *Proc. Natl. Acad. Sci. USA* **2016**, *113*, 4494–4499. [CrossRef]
63. Armbruster, C.E.; Mobley, H.L.T.; Pearson, M.M. Pathogenesis of *Proteus mirabilis* Infection. *EcoSal Plus* **2018**, *8*. [CrossRef] [PubMed]
64. Zbell, A.L.; Maier, S.E.; Maier, R.J. *Salmonella enterica* serovar Typhimurium NiFe uptake-type hydrogenases are differentially expressed in vivo. *Infect. Immun.* **2008**, *76*, 4445–4454. [CrossRef] [PubMed]
65. Maier, L.; Barthel, M.; Stecher, B.; Maier, R.J.; Gunn, J.S.; Hardt, W.-D. *Salmonella* Typhimurium strain ATCC14028 requires H_2-hydrogenases for growth in the gut, but not at systemic sites. *PLoS ONE* **2014**, *9*, e110187. [CrossRef] [PubMed]
66. Lam, L.H.; Monack, D.M. Intraspecies competition for niches in the distal gut dictate transmission during persistent *Salmonella* infection. *PLoS Pathog.* **2014**, *10*, e1004527. [CrossRef] [PubMed]
67. Zbell, A.L.; Benoit, S.L.; Maier, R.J. Differential expression of NiFe uptake-type hydrogenase genes in *Salmonella enterica* serovar Typhimurium. *Microbiology* **2007**, *153*, 3508–3516. [CrossRef] [PubMed]
68. Parkin, A.; Bowman, L.; Roessler, M.M.; Davies, R.A.; Palmer, T.; Armstrong, F.A.; Sargent, F. How *Salmonella* oxidises H_2 under aerobic conditions. *FEBS Lett.* **2012**, *586*, 536–544. [CrossRef] [PubMed]
69. Lamichhane-Khadka, R.; Benoit, S.L.; Miller-Parks, E.F.; Maier, R.J. Host hydrogen rather than that produced by the pathogen is important for *Salmonella enterica* serovar Typhimurium virulence. *Infect. Immun.* **2015**, *83*, 311–316. [CrossRef]
70. McNorton, M.M.; Maier, R.J. Roles of H_2 uptake hydrogenases in *Shigella flexneri* acid tolerance. *Microbiology* **2012**, *158*, 2204–2212. [CrossRef]
71. Cai, Y.; Ni, Y. Purification, characterization, and pathogenicity of urease produced by *Vibrio parahaemolyticus*. *J. Clin. Lab. Anal.* **1996**, *10*, 70–73. [CrossRef]
72. De Koning-Ward, T.F.; Robins-Browne, R.M. Contribution of urease to acid tolerance in *Yersinia enterocolitica*. *Infect. Immun.* **1995**, *63*, 3790–3795.
73. Da Silva, S.M.; Venceslau, S.S.; Fernandes, C.L.; Valente, F.M.; Pereira, I.A. Hydrogen as an energy source for the human pathogen *Bilophila wadsworthia*. *Antonie Leeuwenhoek* **2008**, *93*, 381–390. [CrossRef] [PubMed]
74. Weerakoon, D.R.; Borden, N.J.; Goodson, C.M.; Grimes, J.; Olson, J.W. The role of respiratory donor enzymes in *Campylobacter jejuni* host colonization and physiology. *Microb. Pathog.* **2009**, *47*, 8–15. [CrossRef] [PubMed]

75. Kassem, I.; Khatri, M.; Esseili, M.A.; Sanad, Y.M.; Saif, Y.M.; Olson, J.W.; Rajashekara, G. Respiratory proteins contribute differentially to *Campylobacter jejuni*'s survival and in vitro interaction with hosts' intestinal cells. *BMC Microbiol.* **2012**, *12*, 258. [CrossRef] [PubMed]
76. Benoit, S.L.; Maier, R.J. Site-directed mutagenesis of *Campylobacter concisus* respiratory genes provides insight into the pathogen's growth requirements. *Sci. Rep.* **2018**, *8*, 14203. [CrossRef] [PubMed]
77. Mehta, N.S.; Benoit, S.; Mysore, J.V.; Sousa, R.S.; Maier, R.J. *Helicobacter hepaticus* hydrogenase mutants are deficient in hydrogen-supported amino acid uptake and in causing liver lesions in A/J mice. *Infect. Immun.* **2005**, *73*, 5311–5318. [CrossRef] [PubMed]
78. Ge, Z.; Lee, A.; Whary, M.T.; Rogers, A.B.; Maurer, K.J.; Taylor, N.S.; Schauer, D.B.; Fox, J.G. *Helicobacter hepaticus* urease is not required for intestinal colonization but promotes hepatic inflammation in male A/JCr mice. *Microb. Pathog.* **2008**, *45*, 18–24. [CrossRef]
79. Andrutis, K.A.; Fox, J.G.; Schauer, D.B.; Marini, R.P.; Murphy, J.C.; Yan, L.; Solnick, J.V. Inability of an isogenic urease-negative mutant stain of *Helicobacter mustelae* to colonize the ferret stomach. *Infect. Immun.* **1995**, *63*, 3722–3725.
80. Olson, J.W.; Maier, R.J. Molecular hydrogen as an energy source for *Helicobacter pylori*. *Science* **2002**, *298*, 1788–1790. [CrossRef]
81. Kuhns, L.G.; Benoit, S.L.; Bayyareddy, K.; Johnson, D.; Orlando, R.; Evans, A.L.; Waldrop, G.L.; Maier, R.J. Carbon fixation driven by Molecular Hydrogen Results in Chemolithoautotrophically Enhanced Growth of *Helicobacter pylori*. *J. Bacteriol.* **2016**, *198*, 1423–1428. [CrossRef]
82. Wang, G.; Romero-Gallo, J.; Benoit, S.L.; Piazuelo, M.B.; Dominguez, R.L.; Morgan, D.R.; Peek, R.M., Jr.; Maier, R.J. Hydrogen metabolism in *Helicobacter pylori* plays a role in gastric carcinogenesis through facilitating CagA translocation. *MBio* **2016**, *7*, e01022-16. [CrossRef]
83. Segal, E.D.; Shon, J.; Tompkins, L.S. Characterization of *Helicobacter pylori* urease mutants. *Infect. Immun.* **1992**, *60*, 1883–1889. [PubMed]
84. Tsuda, M.; Karita, M.; Morshed, M.G.; Okita, K.; Nakazawa, T. A urease-negative mutant of *Helicobacter pylori* constructed by allelic exchange mutagenesis lacks the ability to colonize the nude mouse stomach. *Infect. Immun.* **1994**, *62*, 3586–3589. [PubMed]
85. Eaton, K.A.; Krakowka, S. Effect of gastric pH on urease-dependent colonization of gnotobiotic piglets by *Helicobacter pylori*. *Infect. Immun.* **1994**, *62*, 3604–3607. [PubMed]
86. Harris, P.R.; Ernst, P.B.; Kawabata, S.; Kiyono, H.; Graham, M.F.; Smith, P.D. Recombinant *Helicobacter pylori* urease activates primary mucosal macrophages. *J. Infect. Dis.* **1998**, *178*, 1516–1520. [CrossRef] [PubMed]
87. Harris, P.; Mobley, H.; Perez-Perez, G.; Blaser, M.; Smith, P. *Helicobacter pylori* urease is a potent stimulus of mononuclear phagocyte activation and inflammatory cytokine production. *Gastroenterology* **1996**, *111*, 419–425. [CrossRef] [PubMed]
88. Fan, X.; Gunasena, H.; Cheng, Z.; Espejo, R.; Crowe, S.E.; Ernst, P.B.; Reyes, V.E. *Helicobacter pylori* urease binds to class II MHC on gastric epithelial cells and induces their apoptosis. *J. Immunol.* **2000**, *165*, 1918–1924. [CrossRef] [PubMed]
89. Kavermann, H.; Burns, B.P.; Angermuller, K.; Odenbreit, S.; Fischer, W.; Melchers, K.; Haas, R. Identification and characterization of *Helicobacter pylori* genes essential for gastric colonization. *J. Exp. Med.* **2003**, *197*, 813–822. [CrossRef]
90. Wirth, H.P.; Beins, M.H.; Yang, M.; Tham, K.T.; Blaser, M.J. Experimental infection of Mongolian gerbils with wild-type and mutant *Helicobacter pylori* strains. *Infect. Immun.* **1998**, *66*, 4856–4866.
91. Kuwahara, H.; Miyamoto, Y.; Akaike, T.; Kubota, T.; Sawa, T.; Okamoto, S.; Maeda, H. *Helicobacter pylori* urease suppresses bactericidal activity of peroxynitrite via carbon dioxide production. *Infect. Immun.* **2000**, *68*, 4378–4383. [CrossRef]
92. Lytton, S.D.; Fischer, W.; Nagel, W.; Haas, R.; Beck, F.X. Production of ammonium by *Helicobacter pylori* mediates occludin processing and disruption of tight junctions in Caco-2 cells. *Microbiology* **2005**, *151*, 3267–3276. [CrossRef]
93. Wroblewski, L.E.; Shen, L.; Ogden, S.; Romero-Gallo, J.; Lapierre, L.A.; Israel, D.A.; Turner, J.R.; Peek, R.M., Jr. *Helicobacter pylori* dysregulation of gastric epithelial tight junctions by urease-mediated myosin II activation. *Gastroenterology* **2009**, *136*, 236–246. [CrossRef] [PubMed]

94. Wassermann, G.E.; Olivera-Severo, D.; Uberti, A.F.; Carlini, C.R. *Helicobacter pylori* urease activates blood platelets through a lipoxygenase-mediated pathway. *J. Cell. Mol. Med.* **2010**, *14*, 2025–2034. [CrossRef] [PubMed]
95. Perrais, M.; Rousseaux, C.; Ducourouble, M.-P.; Courcol, R.; Vincent, P.; Jonckheere, N.; Van Seuningen, I. *Helicobacter pylori* urease and flagellin alter mucin gene expression in human gastric cancer cells. *Gastric Cancer* **2014**, *17*, 235–246. [CrossRef] [PubMed]
96. Debowski, A.W.; Walton, S.M.; Chua, E.G.; Tay, A.C.; Liao, T.; Lamichhane, B.; Himbeck, R.; Stubbs, K.A.; Marshall, B.J.; Fulurija, A.; et al. *Helicobacter pylori* gene silencing in vivo demonstrates urease is essential for chronic infection. *PLoS Pathog.* **2017**, *13*, e1006464. [CrossRef] [PubMed]
97. Olivera-Severo, D.; Uberti, A.F.; Marques, M.S.; Pinto, M.T.; Gomez-Lazaro, M.; Figueiredo, C.; Leite, M.; Carlini, C.R. A new role for *Helicobacter pylori* urease: Contributions to angiogenesis. *Front. Microbiol.* **2017**, *8*, 1883. [CrossRef] [PubMed]
98. Scopel-Guerra, A.; Olivera-Severo, D.; Staniscuaski, F.; Uberti, A.F.; Callai-Silva, N.; Jaeger, N.; Porto, B.N.; Carlini, C.R. The impact of *Helicobacter pylori* urease upon platelets and consequent contributions to inflammation. *Front. Microbiol.* **2017**, *8*, 2447. [CrossRef] [PubMed]
99. Schmalstig, A.A.; Benoit, S.L.; Misra, S.K.; Sharp, J.S.; Maier, R.J. Noncatalytic antioxidant role for *Helicobacter pylori* urease. *J. Bacteriol.* **2018**, *200*, e00124-18. [CrossRef] [PubMed]
100. Iyengar, G.V.; Kollmer, W.E.; Bowen, H.J.M. *The Elemental Composition of Human Tissues and Body Fluids: A Compilation of Values for Adults*; Verlag Chemie: Weinheim, Germany, 1978.
101. Maret, W. Metalloproteomics, metalloproteomes, and the annotation of metalloproteins. *Metallomics* **2010**, *2*, 117–125. [CrossRef] [PubMed]
102. Diaz-Ochoa, V.E.; Jellbauer, S.; Klaus, S.; Raffatellu, M. Transition metal ions at the crossroads of mucosal immunity and microbial pathogenesis. *Front. Cell. Infect. Microbiol.* **2014**, *4*, 2. [CrossRef]
103. Zackular, J.P.; Chazin, W.J.; Skaar, E.P. Nutritional Immunity: S100 Proteins at the host-pathogen interface. *J. Biol. Chem.* **2015**, *290*, 18991–18998. [CrossRef]
104. Choby, J.E.; Mike, L.A.; Mashruwala, A.A.; Dutter, B.F.; Dunman, P.M.; Sulikowski, G.A.; Boyd, J.M.; Skaar, E.P. A small-molecule inhibitor of iron-sulfur cluster assembly uncovers a link between virulence regulation and metabolism in *Staphylococcus aureus*. *Cell Chem. Biol.* **2016**, *23*, 1351–1361. [CrossRef] [PubMed]
105. Palmer, L.D.; Skaar, E.P. Transition Metals and Virulence in Bacteria. *Annu. Rev. Genet.* **2016**, *50*, 67–91. [CrossRef] [PubMed]
106. Nakashige, T.G.; Zygiel, E.M.; Drennan, C.L.; Nolan, E.M. Nickel sequestration by the host-defense protein human calprotectin. *J. Am. Chem. Soc.* **2017**, *139*, 8828–8836. [CrossRef] [PubMed]
107. Kanwar, J.R.; Roy, K.; Patel, Y.; Zhou, S.F.; Singh, M.R.; Singh, D.; Nasir, M.; Sehgal, R.; Sehgal, A.; Singh, R.S.; et al. Multifunctional iron bound lactoferrin and nanomedicinal approaches to enhance its bioactive functions. *Molecules* **2015**, *20*, 9703–9731. [CrossRef] [PubMed]
108. Ganz, T. Iron and infection. *Int. J. Hematol.* **2018**, *107*, 7–15. [CrossRef]
109. Kulprachakarn, K.; Chen, Y.L.; Kong, X.; Arno, M.C.; Hider, R.C.; Srichairatanakool, S.; Bansal, S.S. Copper(II) binding properties of hepcidin. *J. Biol. Inorg. Chem.* **2016**, *21*, 329–338. [CrossRef]
110. Stefanova, D.; Raychev, A.; Deville, J.; Humphries, R.; Campeau, S.; Ruchala, P.; Nemeth, E.; Ganz, T.; Bulut, Y. Hepcidin protects against lethal *Escherichia coli* sepsis in mice inoculated with isolates from septic patients. *Infect. Immun.* **2018**, *86*, e00253-18. [CrossRef]
111. Stefanova, D.; Raychev, A.; Arezes, J.; Ruchala, P.; Gabayan, V.; Skurnik, M.; Dillon, B.J.; Horwitz, M.A.; Ganz, T.; Bulut, Y.; et al. Endogenous hepcidin and its agonist mediate resistance to selected infections by clearing non-transferrin-bound iron. *Blood* **2017**, *130*, 245–257. [CrossRef]
112. Stafford, S.L.; Bokil, N.J.; Achard, M.E.; Kapetanovic, R.; Schembri, M.A.; McEwan, A.G.; Sweet, M.J. Metal ions in macrophage antimicrobial pathways: Emerging roles for zinc and copper. *Biosci. Rep.* **2013**, *33*, e00049. [CrossRef]
113. Cellier, M.F.M. Developmental Control of NRAMP1 (SLC11A1) Expression in Professional Phagocytes. *Biology* **2017**, *6*, 28. [CrossRef]
114. Corbin, B.D.; Seeley, E.H.; Raab, A.; Feldmann, J.; Miller, M.R.; Torres, V.J.; Anderson, K.L.; Dattilo, B.M.; Dunman, P.M.; Gerads, R.; et al. Metal chelation and inhibition of bacterial growth in tissue abscesses. *Science* **2008**, *319*, 962–965. [CrossRef] [PubMed]

115. Rowinska-Zyrek, M.; Zakrzewska-Czerwinska, J.; Zawilak-Pawlik, A.; Kozlowski, H. Ni^{2+} chemistry in pathogens–a possible target for eradication. *Dalton Trans.* **2014**, *43*, 8976–8989. [CrossRef] [PubMed]
116. Maier, R.J. Availability and use of molecular hydrogen as an energy substrate for *Helicobacter* species. *Microbes Infect.* **2003**, *5*, 1159–1163. [CrossRef] [PubMed]
117. Morrow, C.A.; Fraser, J.A. Is the nickel-dependent urease complex of *Cryptococcus* the pathogen's Achilles' heel? *MBio* **2013**, *4*, e00408-13. [CrossRef] [PubMed]
118. Mora, D.; Arioli, S. Microbial urease in health and disease. *PLoS Pathog.* **2014**, *10*, e1004472. [CrossRef] [PubMed]
119. Huang, C.-T.; Chen, M.-L.; Huang, L.-L.; Mao, I.-F. Uric acid and urea in human sweat. *Chin. J. Physiol.* **2002**, *45*, 109–116. [PubMed]
120. Lasisi, T.J.; Raji, Y.R.; Salako, B.L. Salivary creatinine and urea analysis in patients with chronic kidney disease: A case control study. *BMC Nephrol.* **2016**, *17*, 10. [CrossRef] [PubMed]
121. Bishai, W.; Timmins, G. Potential for breath test diagnosis of urease positive pathogens in lung infections. *J. Breath Res.* **2019**, *13*, 032002. [CrossRef] [PubMed]
122. Anderson, K.L.; Roux, C.M.; Olson, M.W.; Luong, T.T.; Lee, C.Y.; Olson, R.; Dunman, P.M. Characterizing the effects of inorganic acid and alkaline shock on the *Staphylococcus aureus* transcriptome and messenger RNA turnover. *FEMS Immunol. Med. Microbiol.* **2010**, *60*, 208–250. [CrossRef] [PubMed]
123. Pot, R.G.; Stoof, J.; Nuijten, P.J.; De Haan, L.A.; Loeffen, P.; Kuipers, E.J.; Van Vliet, A.H.; Kusters, J.G. UreA2B2: A second urease system in the gastric pathogen *Helicobacter felis*. *FEMS Immunol. Med. Microbiol.* **2007**, *50*, 273–279. [CrossRef] [PubMed]
124. Stoof, J.; Breijer, S.; Pot, R.G.; van der Neut, D.; Kuipers, E.J.; Kusters, J.G.; van Vliet, A.H. Inverse nickel-responsive regulation of two urease enzymes in the gastric pathogen *Helicobacter mustelae*. *Environ. Microbiol.* **2008**, *10*, 2586–2597. [CrossRef] [PubMed]
125. Carter, E.L.; Tronrud, D.E.; Taber, S.R.; Karplus, P.A.; Hausinger, R.P. Iron-containing urease in a pathogenic bacterium. *Proc. Natl. Acad. Sci. USA* **2011**, *108*, 13095–13099. [CrossRef] [PubMed]
126. Solomons, N.W.; Viteri, F.; Shuler, T.R.; Nielsen, F.H. Bioavailability of nickel in man: Effects of foods and chemically-defined dietary constituents on the absorption of inorganic nickel. *J. Nutr.* **1982**, *112*, 39–50. [CrossRef] [PubMed]
127. Bauerfeind, P.; Garner, R.; Dunn, B.; Mobley, H. Synthesis and activity of *Helicobacter pylori* urease and catalase at low pH. *Gut* **1997**, *40*, 25–30. [CrossRef] [PubMed]
128. Agrawal, A.; Gupta, A.; Chandra, M.; Koowar, S. Role of *Helicobacter pylori* infection in the pathogenesis of minimal hepatic encephalopathy and effect of its eradication. *Indian J. Gastroenterol.* **2011**, *30*, 29–32. [CrossRef] [PubMed]
129. Stingl, K.; De Reuse, H. Staying alive overdosed: How does *Helicobacter pylori* control urease activity? *Int. J. Med. Microbiol.* **2005**, *295*, 307–315. [CrossRef] [PubMed]
130. van Vliet, A.H.; Poppelaars, S.W.; Davies, B.J.; Stoof, J.; Bereswill, S.; Kist, M.; Penn, C.W.; Kuipers, E.J.; Kusters, J.G. NikR mediates nickel-responsive transcriptional induction of urease expression in *Helicobacter pylori*. *Infect. Immun.* **2002**, *70*, 2846–2852. [CrossRef]
131. van Vliet, A.H.; Kuipers, E.J.; Waidner, B.; Davies, B.J.; de Vries, N.; Penn, C.W.; Vandenbroucke-Grauls, C.M.; Kist, M.; Bereswill, S.; Kusters, J.G. Nickel-responsive induction of urease expression in *Helicobacter pylori* is mediated at the transcriptional level. *Infect. Immun.* **2001**, *69*, 4891–4897. [CrossRef] [PubMed]
132. Tong, S.Y.; Davis, J.S.; Eichenberger, E.; Holland, T.L.; Fowler, V.G. *Staphylococcus aureus* infections: Epidemiology, pathophysiology, clinical manifestations, and management. *Clin. Microbiol. Rev.* **2015**, *28*, 603–661. [CrossRef] [PubMed]
133. Molina-Manso, D.; del Prado, G.; Ortiz-Perez, A.; Manrubia-Cobo, M.; Gomez-Barrena, E.; Cordero-Ampuero, J.; Esteban, J. In vitro susceptibility to antibiotics of staphylococci in biofilms isolated from orthopaedic infections. *Int. J. Antimicrob. Agents* **2013**, *41*, 521–523. [CrossRef]
134. Norsworthy, A.N.; Pearson, M.M. From Catheter to Kidney Stone: The uropathogenic lifestyle of *Proteus mirabilis*. *Trends Microbiol.* **2017**, *25*, 304–315. [CrossRef] [PubMed]
135. Bichler, K.H.; Eipper, E.; Naber, K.; Braun, V.; Zimmermann, R.; Lahme, S. Urinary infection stones. *Int. J. Antimicrob. Agents* **2002**, *19*, 488–498. [CrossRef]
136. Heitland, P.; Köster, H.D. Biomonitoring of 30 trace elements in urine of children and adults by ICP-MS. *Clin. Chim. Acta* **2006**, *365*, 310–318. [CrossRef] [PubMed]

137. Olson, J.W.; Mehta, N.S.; Maier, R.J. Requirement of nickel metabolism proteins HypA and HypB for full activity of both hydrogenase and urease in *Helicobacter pylori*. *Mol. Microbiol.* **2001**, *39*, 176–182. [CrossRef] [PubMed]
138. Benoit, S.L.; Zbell, A.L.; Maier, R.J. Nickel enzyme maturation in *Helicobacter hepaticus*: Roles of accessory proteins in hydrogenase and urease activities. *Microbiology* **2007**, *153*, 3748–3756. [CrossRef] [PubMed]
139. Pearson, M.M.; Sebaihia, M.; Churcher, C.; Quail, M.A.; Seshasayee, A.S.; Luscombe, N.M.; Abdellah, Z.; Arrosmith, C.; Atkin, B.; Chillingworth, T.; et al. Complete genome sequence of uropathogenic *Proteus mirabilis*, a master of both adherence and motility. *J. Bacteriol.* **2008**, *190*, 4027–4037. [CrossRef] [PubMed]
140. Gerwien, F.; Skrahina, V.; Kasper, L.; Hube, B.; Brunke, S. Metals in fungal virulence. *FEMS Microbiol. Rev.* **2018**, *42*. [CrossRef] [PubMed]
141. Singh, A.; Panting, R.J.; Varma, A.; Saijo, T.; Waldron, K.J.; Jong, A.; Ngamskulrungroj, P.; Chang, Y.C.; Rutherford, J.C.; Kwon-Chung, K.J. Factors required for activation of urease as a virulence determinant in *Cryptococcus neoformans*. *MBio* **2013**, *4*, e00220-13. [CrossRef]
142. Lubitz, W.; Ogata, H.; Rüdiger, O.; Reijerse, E. Hydrogenases. *Chem. Rev.* **2014**, *114*, 4081–4148. [CrossRef]
143. Maier, R.J.; Olczak, A.; Maier, S.; Soni, S.; Gunn, J. Respiratory hydrogen use by *Salmonella enterica* serovar Typhimurium is essential for virulence. *Infect. Immun.* **2004**, *72*, 6294–6299. [CrossRef]
144. Maier, R.J.; Fu, C.; Gilbert, J.; Moshiri, F.; Olson, J.; Plaut, A.G. Hydrogen uptake hydrogenase in *Helicobacter pylori*. *FEMS Microbiol. Lett.* **1996**, *141*, 71–76. [CrossRef] [PubMed]
145. Ernst, F.D.; Bereswill, S.; Waidner, B.; Stoof, J.; Mader, U.; Kusters, J.G.; Kuipers, E.J.; Kist, M.; van Vliet, A.H.; Homuth, G. Transcriptional profiling of *Helicobacter pylori* Fur- and iron-regulated gene expression. *Microbiology* **2005**, *151*, 533–546. [CrossRef] [PubMed]
146. Muller, C.; Bahlawane, C.; Aubert, S.; Delay, C.M.; Schauer, K.; Michaud-Soret, I.; De Reuse, H. Hierarchical regulation of the NikR-mediated nickel response in *Helicobacter pylori*. *Nucleic Acids Res.* **2011**, *39*, 7564–7575. [CrossRef] [PubMed]
147. Contreras, M.; Thiberge, J.M.; Mandrand-Berthelot, M.A.; Labigne, A. Characterization of the roles of NikR, a nickel-responsive pleiotropic autoregulator of *Helicobacter pylori*. *Mol. Microbiol.* **2003**, *49*, 947–963. [CrossRef] [PubMed]
148. Benoit, S.L.; Mehta, N.; Weinberg, M.V.; Maier, C.; Maier, R.J. Interaction between the *Helicobacter pylori* accessory proteins HypA and UreE is needed for urease maturation. *Microbiology* **2007**, *153*, 1474–1482. [CrossRef] [PubMed]
149. Benoit, S.L.; McMurry, J.L.; Hill, S.A.; Maier, R.J. *Helicobacter pylori* hydrogenase accessory protein HypA and urease accessory protein UreG compete with each other for UreE recognition. *Biochim. Biophys. Acta (BBA)-Gen. Subj.* **2012**, *1820*, 1519–1525. [CrossRef] [PubMed]
150. Hu, H.Q.; Huang, H.-T.; Maroney, M.J. The *Helicobacter pylori* HypA· UreE2 Complex Contains a Novel High-Affinity Ni(II)-Binding Site. *Biochemistry* **2018**, *57*, 2932–2942. [CrossRef] [PubMed]
151. Yang, X.; Li, H.; Cheng, T.; Xia, W.; Lai, Y.-T.; Sun, H. Nickel translocation between metallochaperones HypA and UreE in *Helicobacter pylori*. *Metallomics* **2014**, *6*, 1731–1736. [CrossRef]
152. Stingl, K.; Schauer, K.; Ecobichon, C.; Labigne, A.; Lenormand, P.; Rousselle, J.-C.; Namane, A.; de Reuse, H. In vivo interactome of *Helicobacter pylori* urease revealed by tandem affinity purification. *Mol. Cell. Proteom.* **2008**, *7*, 2429–2441. [CrossRef]
153. Levitt, M.D. Production and excretion of hydrogen gas in man. *N. Engl. J. Med.* **1969**, *281*, 122–127. [CrossRef]
154. Kanazuru, T.; Sato, E.F.; Nagata, K.; Matsui, H.; Watanabe, K.; Kasahara, E.; Jikumaru, M.; Inoue, J.; Inoue, M. Role of hydrogen generation by *Klebsiella pneumoniae* in the oral cavity. *J. Microbiol.* **2010**, *48*, 778–783. [CrossRef] [PubMed]
155. Blaser, M.J.; Perez-Perez, G.I.; Kleanthous, H.; Cover, T.L.; Peek, R.M.; Chyou, P.H.; Stemmermann, G.N.; Nomura, A. Infection with *Helicobacter pylori* strains possessing cagA is associated with an increased risk of developing adenocarcinoma of the stomach. *Cancer Res.* **1995**, *55*, 2111–2115. [PubMed]
156. Fox, J.G.; Ge, Z.; Whary, M.T.; Erdman, S.E.; Horwitz, B.H. *Helicobacter hepaticus* infection in mice: Models for understanding lower bowel inflammation and cancer. *Mucosal Immunol.* **2011**, *4*, 22–30. [CrossRef] [PubMed]
157. Poutahidis, T.; Cappelle, K.; Levkovich, T.; Lee, C.W.; Doulberis, M.; Ge, Z.; Fox, J.G.; Horwitz, B.H.; Erdman, S.E. Pathogenic intestinal bacteria enhance prostate cancer development via systemic activation of immune cells in mice. *PLoS ONE* **2013**, *8*, e73549. [CrossRef] [PubMed]

158. Suerbaum, S.; Josenhans, C.; Sterzenbach, T.; Drescher, B.; Brandt, P.; Bell, M.; Droge, M.; Fartmann, B.; Fischer, H.P.; Ge, Z.; et al. The complete genome sequence of the carcinogenic bacterium *Helicobacter hepaticus*. *Proc. Natl. Acad. Sci. USA* **2003**, *100*, 7901–7906. [CrossRef] [PubMed]
159. Maier, R.J.; Olson, J.; Olczak, A. Hydrogen-oxidizing capabilities of *Helicobacter hepaticus* and in vivo availability of the substrate. *J. Bacteriol.* **2003**, *185*, 2680–2682. [CrossRef] [PubMed]
160. McDowall, J.S.; Murphy, B.J.; Haumann, M.; Palmer, T.; Armstrong, F.A.; Sargent, F. Bacterial formate hydrogenlyase complex. *Proc. Natl. Acad. Sci. USA* **2014**, *111*, E3948–E3956. [CrossRef] [PubMed]
161. Zbell, A.L.; Maier, R.J. Role of the Hya hydrogenase in recycling of anaerobically produced H_2 in *Salmonella enterica* serovar Typhimurium. *Appl. Environ. Microbiol.* **2009**, *75*, 1456–1459. [CrossRef]
162. Maier, L.; Vyas, R.; Cordova, C.D.; Lindsay, H.; Schmidt, T.S.; Brugiroux, S.; Periaswamy, B.; Bauer, R.; Sturm, A.; Schreiber, F.; et al. Microbiota-derived hydrogen fuels *Salmonella typhimurium* invasion of the gut ecosystem. *Cell Host Microbe* **2013**, *14*, 641–651. [CrossRef]
163. Craig, M.; Sadik, A.Y.; Golubeva, Y.A.; Tidhar, A.; Slauch, J.M. Twin-arginine translocation system (tat) mutants of *Salmonella* are attenuated due to envelope defects, not respiratory defects. *Mol. Microbiol.* **2013**, *89*, 887–902. [CrossRef]
164. Parkhill, J.; Wren, B.W.; Mungall, K.; Ketley, J.M.; Churcher, C.; Basham, D.; Chillingworth, T.; Davies, R.M.; Feltwell, T.; Holroyd, S.; et al. The genome sequence of the food-borne pathogen *Campylobacter jejuni* reveals hypervariable sequences. *Nature* **2000**, *403*, 665–668. [CrossRef]
165. Carlone, G.M.; Lascelles, J. Aerobic and anaerobic respiratory systems in *Campylobacter fetus* subsp. *jejuni* grown in atmospheres containing hydrogen. *J. Bacteriol.* **1982**, *152*, 306–314.
166. Istivan, T.; Ward, P.; Coloe, P. *Current Research, Technology and Education Topics in Applied Microbiology and Microbial Biotechnology*; Méndez-Vilas, A., Ed.; FORMATEX: Badajoz, Spain, 2010; pp. 626–634.
167. Benoit, S.L.; Holland, A.A.; Johnson, M.K.; Maier, R.J. Iron-sulfur protein maturation in *Helicobacter pylori*: Identifying a Nfu-type cluster carrier protein and its iron-sulfur protein targets. *Mol. Microbiol.* **2018**, *108*, 379–396. [CrossRef]
168. Lee, H.; Ma, R.; Grimm, M.C.; Riordan, S.M.; Lan, R.; Zhong, L.; Raftery, M.; Zhang, L. Examination of the anaerobic growth of *Campylobacter concisus* strains. *Int. J. Microbiol.* **2014**, *2014*, 476047. [CrossRef]
169. Kotloff, K.L.; Riddle, M.S.; Platts-Mills, J.A.; Pavlinac, P.; Zaidi, A.K. Shigellosis. *Lancet* **2018**, *391*, 801–812. [CrossRef]
170. Kissing, S.; Saftig, P.; Haas, A. Vacuolar ATPase in phago (lyso) some biology. *Int. J. Med. Microbiol.* **2018**, *308*, 58–67. [CrossRef]
171. Pochapsky, T.C.; Pochapsky, S.S.; Ju, T.; Mo, H.; Al-Mjeni, F.; Maroney, M.J. Modeling and experiment yields the structure of acireductone dioxygenase from *Klebsiella pneumoniae*. *Nat. Struct. Mol. Biol.* **2002**, *9*, 966. [CrossRef]
172. Clugston, S.L.; Barnard, J.F.; Kinach, R.; Miedema, D.; Ruman, R.; Daub, E.; Honek, J.F. Overproduction and characterization of a dimeric non-zinc glyoxalase I from *Escherichia coli*: Evidence for optimal activation by nickel ions. *Biochemistry* **1998**, *37*, 8754–8763. [CrossRef]
173. Sukdeo, N.; Clugston, S.L.; Daub, E.; Honek, J.F. Distinct classes of glyoxalase I: Metal specificity of the *Yersinia pestis*, *Pseudomonas aeruginosa* and *Neisseria meningitidis* enzymes. *Biochem. J.* **2004**, *384*, 111–117. [CrossRef]
174. Ozyamak, E.; Black, S.S.; Walker, C.A.; Maclean, M.J.; Bartlett, W.; Miller, S.; Booth, I.R. The critical role of S-lactoylglutathione formation during methylglyoxal detoxification in *Escherichia coli*. *Mol. Microbiol.* **2010**, *78*, 1577–1590. [CrossRef]
175. Deponte, M. Glyoxalase diversity in parasitic protists. *Biochem. Soc. Trans.* **2014**, *42*, 473–478. [CrossRef]
176. Youn, H.D.; Kim, E.J.; Roe, J.H.; Hah, Y.C.; Kang, S.O. A novel nickel-containing superoxide dismutase from *Streptomyces* spp. *Biochem. J.* **1996**, *318 Pt 3*, 889–896. [CrossRef]
177. Dupont, C.; Neupane, K.; Shearer, J.; Palenik, B. Diversity, function and evolution of genes coding for putative Ni-containing superoxide dismutases. *Environ. Microbiol.* **2008**, *10*, 1831–1843. [CrossRef]
178. Zambelli, B.; Ciurli, S. Nickel and human health. In *Interrelations between Essential Metal Ions and Human Diseases*; Springer: Dordrecht, The Netherlands, 2013; pp. 321–357.
179. Lebrette, H.; Brochier-Armanet, C.; Zambelli, B.; de Reuse, H.; Borezee-Durant, E.; Ciurli, S.; Cavazza, C. Promiscuous nickel import in human pathogens: Structure, thermodynamics, and evolution of extracytoplasmic nickel-binding proteins. *Structure* **2014**, *22*, 1421–1432. [CrossRef]

180. Chivers, P.T. Nickel recognition by bacterial importer proteins. *Metallomics* **2015**, *7*, 590–595. [CrossRef]
181. Lebrette, H.; Iannello, M.; Fontecilla-Camps, J.C.; Cavazza, C. The binding mode of Ni-(L-His) 2 in NikA revealed by X-ray crystallography. *J. Inorg. Biochem.* **2013**, *121*, 16–18. [CrossRef]
182. Ghssein, G.; Brutesco, C.; Ouerdane, L.; Fojcik, C.; Izaute, A.; Wang, S.; Hajjar, C.; Lobinski, R.; Lemaire, D.; Richaud, P. Biosynthesis of a broad-spectrum nicotianamine-like metallophore in *Staphylococcus aureus*. *Science* **2016**, *352*, 1105–1109. [CrossRef]
183. Grim, K.P.; San Francisco, B.; Radin, J.N.; Brazel, E.B.; Kelliher, J.L.; Solórzano, P.K.P.; Kim, P.C.; McDevitt, C.A.; Kehl-Fie, T.E. The metallophore staphylopine enables *Staphylococcus aureus* to compete with the host for zinc and overcome nutritional immunity. *MBio* **2017**, *8*, e01281-17. [CrossRef]
184. Song, L.; Zhang, Y.; Chen, W.; Gu, T.; Zhang, S.-Y.; Ji, Q. Mechanistic insights into staphylopine-mediated metal acquisition. *Proc. Natl. Acad. Sci. USA* **2018**, *115*, 3942–3947. [CrossRef]
185. Lhospice, S.; Gomez, N.O.; Ouerdane, L.; Brutesco, C.; Ghssein, G.; Hajjar, C.; Liratni, A.; Wang, S.; Richaud, P.; Bleves, S. *Pseudomonas aeruginosa* zinc uptake in chelating environment is primarily mediated by the metallophore pseudopaline. *Sci. Rep.* **2017**, *7*, 17132. [CrossRef]
186. Schauer, K.; Gouget, B.; Carrière, M.; Labigne, A.; De Reuse, H. Novel nickel transport mechanism across the bacterial outer membrane energized by the TonB/ExbB/ExbD machinery. *Mol. Microbiol.* **2007**, *63*, 1054–1068. [CrossRef]
187. Zhang, P. Structure and mechanism of energy-coupling factor transporters. *Trends Microbiol.* **2013**, *21*, 652–659. [CrossRef]
188. Bauerfeind, P.; Garner, R.M.; Mobley, L. Allelic exchange mutagenesis of *nixA* in *Helicobacter pylori* results in reduced nickel transport and urease activity. *Infect. Immun.* **1996**, *64*, 2877–2880.
189. Fischer, F.; Robbe-Saule, M.; Turlin, E.; Mancuso, F.; Michel, V.; Richaud, P.; Veyrier, F.J.; De Reuse, H.; Vinella, D. Characterization in *Helicobacter pylori* of a nickel transporter essential for colonization that was acquired during evolution by gastric *Helicobacter* species. *PLoS Pathog.* **2016**, *12*, e1006018. [CrossRef]
190. Nolan, K.J.; McGee, D.J.; Mitchell, H.M.; Kolesnikow, T.; Harro, J.M.; O'Rourke, J.; Wilson, J.E.; Danon, S.J.; Moss, N.D.; Mobley, H.L. In vivo behavior of a *Helicobacter pylori* SS1 *nixA* mutant with reduced urease activity. *Infect. Immun.* **2002**, *70*, 685–691. [CrossRef]
191. Eitinger, T.; Suhr, J.; Moore, L.; Smith, J.A.C. Secondary transporters for nickel and cobalt ions: Theme and variations. *Biometals* **2005**, *18*, 399–405. [CrossRef]
192. Hiron, A.; Posteraro, B.; Carriere, M.; Remy, L.; Delporte, C.; La Sorda, M.; Sanguinetti, M.; Juillard, V.; Borezee-Durant, E. A nickel ABC-transporter of *Staphylococcus aureus* is involved in urinary tract infection. *Mol. Microbiol.* **2010**, *77*, 1246–1260. [CrossRef]
193. Remy, L.; Carrière, M.; Derré-Bobillot, A.; Martini, C.; Sanguinetti, M.; Borezée-Durant, E. The *Staphylococcus aureus* Opp1 ABC transporter imports nickel and cobalt in zinc-depleted conditions and contributes to virulence. *Mol. Microbiol.* **2013**, *87*, 730–743. [CrossRef]
194. Lebrette, H.; Borezée-Durant, E.; Martin, L.; Richaud, P.; Erba, E.B.; Cavazza, C. Novel insights into nickel import in *Staphylococcus aureus*: The positive role of free histidine and structural characterization of a new thiazolidine-type nickel chelator. *Metallomics* **2015**, *7*, 613–621. [CrossRef]
195. Chivers, P.T.; Benanti, E.L.; Heil-Chapdelaine, V.; Iwig, J.S.; Rowe, J.L. Identification of Ni-(L-His)$_2$ as a substrate for NikABCDE-dependent nickel uptake in *Escherichia coli*. *Metallomics* **2012**, *4*, 1043–1050. [CrossRef]
196. Subashchandrabose, S.; Hazen, T.H.; Brumbaugh, A.R.; Himpsl, S.D.; Smith, S.N.; Ernst, R.D.; Rasko, D.A.; Mobley, H.L. Host-specific induction of *Escherichia coli* fitness genes during human urinary tract infection. *Proc. Natl. Acad. Sci. USA* **2014**, *111*, 18327–18332. [CrossRef]
197. Robinson, A.E.; Lowe, J.E.; Koh, E.I.; Henderson, J.P. Uropathogenic enterobacteria use the yersiniabactin metallophore system to acquire nickel. *J. Biol. Chem.* **2018**, *293*, 14953–14961. [CrossRef]
198. Lamont, E.A.; Xu, W.W.; Sreevatsan, S. Host-*Mycobacterium avium* subsp. *paratuberculosis* interactome reveals a novel iron assimilation mechanism linked to nitric oxide stress during early infection. *BMC Genom.* **2013**, *14*, 694. [CrossRef]
199. Stahler, F.N.; Odenbreit, S.; Haas, R.; Wilrich, J.; Van Vliet, A.H.; Kusters, J.G.; Kist, M.; Bereswill, S. The novel *Helicobacter pylori* CznABC metal efflux pump is required for cadmium, zinc, and nickel resistance, urease modulation, and gastric colonization. *Infect. Immun.* **2006**, *74*, 3845–3852. [CrossRef]

200. Armbruster, C.E.; Forsyth-DeOrnellas, V.; Johnson, A.O.; Smith, S.N.; Zhao, L.; Wu, W.; Mobley, H.L.T. Genome-wide transposon mutagenesis of *Proteus mirabilis*: Essential genes, fitness factors for catheter-associated urinary tract infection, and the impact of polymicrobial infection on fitness requirements. *PLoS Pathog.* **2017**, *13*, e1006434. [CrossRef]
201. De Reuse, H. Nickel and Virulence in Bacterial Pathogens. In *The Biological Chemistry of Nickel*; Zamble, D., Rowińska-Żyrek, M., Kozlowski, H., Eds.; The Royal Society of Chemistry: Cambridge, UK, 2017; pp. 339–356.
202. Vinella, D.; Fischer, F.; Vorontsov, E.; Gallaud, J.; Malosse, C.; Michel, V.; Cavazza, C.; Robbe-Saule, M.; Richaud, P.; Chamot-Rooke, J.; et al. Evolution of *Helicobacter*: Acquisition by gastric species of two histidine-rich proteins essential for colonization. *PLoS Pathog.* **2015**, *11*, e1005312. [CrossRef]
203. Ge, R.; Watt, R.M.; Sun, X.; Tanner, J.A.; He, Q.Y.; Huang, J.D.; Sun, H. Expression and characterization of a histidine-rich protein, Hpn: Potential for Ni^{2+} storage in *Helicobacter pylori*. *Biochem. J.* **2006**, *393*, 285–293. [CrossRef]
204. Zeng, Y.-B.; Zhang, D.-M.; Li, H.; Sun, H. Binding of Ni^{2+} to a histidine-and glutamine-rich protein, Hpn-like. *JBIC J. Biol. Inorg. Chem.* **2008**, *13*, 1121. [CrossRef]
205. Chiera, N.M.; Rowinska-Zyrek, M.; Wieczorek, R.; Guerrini, R.; Witkowska, D.; Remelli, M.; Kozlowski, H. Unexpected impact of the number of glutamine residues on metal complex stability. *Metallomics* **2013**, *5*, 214–221. [CrossRef]
206. Seshadri, S.; Benoit, S.L.; Maier, R.J. Roles of His-rich hpn and hpn-like proteins in *Helicobacter pylori* nickel physiology. *J. Bacteriol.* **2007**, *189*, 4120–4126. [CrossRef]
207. Saylor, Z.; Maier, R. *Helicobacter pylori* nickel storage proteins: Recognition and modulation of diverse metabolic targets. *Microbiology* **2018**, *164*, 1059–1068. [CrossRef]
208. Benoit, S.L.; Miller, E.F.; Maier, R.J. *Helicobacter pylori* stores nickel to aid its host colonization. *Infect. Immun.* **2013**, *81*, 580–584. [CrossRef]
209. Skouloubris, S.; Labigne, A.; De Reuse, H. The AmiE aliphatic amidase and AmiF formamidase of *Helicobacter pylori*: Natural evolution of two enzyme paralogues. *Mol. Microbiol.* **2001**, *40*, 596–609. [CrossRef]
210. Saylor, Z.J. *Helicobacter pylori* nickel storage proteins: Recognition and modulation of diverse metabolic targets. Ph.D. Thesis, The University of Georgia, Athens, GA, USA, 2018.
211. Lopez, A.J.; Martinez, L. Parametric models to compute tryptophan fluorescence wavelengths from classical protein simulations. *J. Comput. Chem.* **2018**, *39*, 1249–1258. [CrossRef]
212. Deshayes, S.; Divita, G. Fluorescence technologies for monitoring interactions between biological molecules in vitro. *Prog. Mol. Biol. Transl. Sci.* **2013**, *113*, 109–143.
213. Cvetkovic, A.; Menon, A.L.; Thorgersen, M.P.; Scott, J.W.; Poole, F.L., II; Jenney, F.E., Jr.; Lancaster, W.A.; Praissman, J.L.; Shanmukh, S.; Vaccaro, B.J. Microbial metalloproteomes are largely uncharacterized. *Nature* **2010**, *466*, 779. [CrossRef]
214. Schauer, K.; Muller, C.; Carrière, M.; Labigne, A.; Cavazza, C.; De Reuse, H. The *Helicobacter pylori* GroES cochaperonin HspA functions as a specialized nickel chaperone and sequestration protein through its unique C-terminal extension. *J. Bacteriol.* **2010**, *192*, 1231–1237. [CrossRef]
215. Zhang, X.; Zhang, J.; Yang, F.; Wu, W.; Sun, H.; Xie, Q.; Si, W.; Zou, Q.; Yang, Z. Immunization with heat shock protein A and gamma-glutamyl transpeptidase induces reduction on the *Helicobacter pylori* colonization in mice. *PLoS ONE* **2015**, *10*, e0130391.
216. Zhang, X.J.; Feng, S.Y.; Li, Z.T.; Feng, Y.M. Expression of *Helicobacter pylori hspA* Gene in *Lactococcus lactis* NICE System and Experimental Study on Its Immunoreactivity. *Gastroenterol. Res. Pract.* **2015**, *2015*, 750932. [CrossRef]

© 2019 by the authors. Licensee MDPI, Basel, Switzerland. This article is an open access article distributed under the terms and conditions of the Creative Commons Attribution (CC BY) license (http://creativecommons.org/licenses/by/4.0/).

Review

Human Acireductone Dioxygenase (HsARD), Cancer and Human Health: Black Hat, White Hat or Gray?

Xinyue Liu [1] and Thomas C. Pochapsky [1,2,*]

1. Department of Chemistry, Brandeis University, Waltham, MA 02454, USA
2. Department of Biochemistry, Brandeis University, Waltham, MA 02454, USA
* Correspondence: pochapsk@brandeis.edu; Tel.: +1-781-736-2559

Received: 21 June 2019; Accepted: 14 August 2019; Published: 18 August 2019

Abstract: Multiple factors involving the methionine salvage pathway (MSP) and polyamine biosynthesis have been found to be involved in cancer cell proliferation, migration, invasion and metastasis. This review summarizes the relationships of the MSP enzyme acireductone dioxygenase (ARD), the *ADI1* gene encoding ARD and other gene products (ADI1GP) with carcinomas and carcinogenesis. ARD exhibits structural and functional differences depending upon the metal bound in the active site. In the penultimate step of the MSP, the Fe^{2+} bound form of ARD catalyzes the on-pathway oxidation of acireductone leading to methionine, whereas Ni^{2+} bound ARD catalyzes an off-pathway reaction producing methylthiopropionate and carbon monoxide, a biological signaling molecule and anti-apoptotic. The relationship between ADI1GP, MSP and polyamine synthesis are discussed, along with possible role(s) of metal in modulating the cellular behavior of ADI1GP and its interactions with other cellular components.

Keywords: *ADI1*; nickel-dependent enzyme; methionine salvage pathway; methionine; S-adenosylmethionine (SAM); methylthioadenosine (MTA); enolase phosphatase 1 (ENOPH1); polyamine; matrix metalloproteinase MT1 (MT1-MMP)

---

1. Introduction

Acireductone dioxygenase (ARD) is a metalloenzyme of the cupin superfamily that is ubiquitous among aerobic cellular organisms. It has been identified in bacteria, plants, fungi and animals. Those ARDs that have been structurally characterized all exhibit the standard cupin fold, a double-stranded β-helix domain fringed by three pseudosymmetrically arranged α-helices (Figure 1). In all organisms in which it has been characterized, ARD functions primarily in the methionine salvage pathway (MSP). As the name implies, ARD catalyzes the oxidative cleavage of the penultimate intermediate in the pathway (1-(thiomethyl)-3-keto-4,5-dihydroxy-pent-4-ene, acireductone in Figure 2) to formate and the ketoacid precursor of methionine, MTOB (4-methylthio-2-oxobutyrate).

ARD was first identified in the methionine salvage pathway of the bacterium *Klebsiella*, as part of an effort to better understand mechanisms of methionine metabolism [1]. It had been known for some time that, unlike the cell lines from which they are derived, many carcinomas are either strictly or partially dependent on an external supply of methionine (Met) for survival (Table 1). However, it was (and is yet) unclear whether this is due to an inability to recover Met from S-adenosylmethionine (SAM or AdoMet) or simply Met usage in excess of what normal cellular processes require that cause this dependence. The efforts of the Abeles group in the late 1980s and 1990s were aimed at a clearer understanding of the methionine recycle and salvage pathways. While methionine recycling involves S-methylation of homocysteine (a by-product of activated methyl transfers from SAM), methionine salvage retains only the S–CH_3 group of the original methionine, with all of the other carbon atoms of methionine originating from the ribose moiety of methylthioadenosine (MTA). MTA is produced as the result of ethylene biosynthesis in plants and polyamines in other organisms (Figure 2).

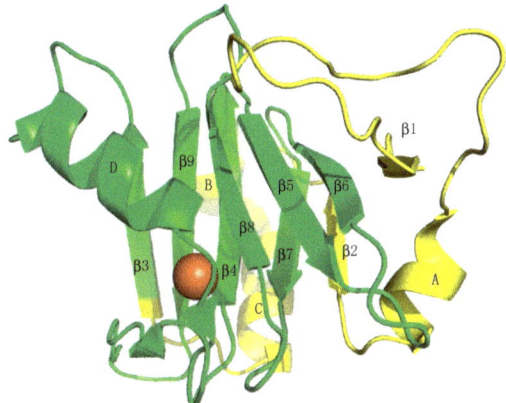

Figure 1. Crystal structure 4QGN [2] of HsARD (ADI1GP), showing arrangements of secondary structures referred to in the text. Residues in yellow are truncated in SipL (submergence-induced-protein-like). Bound metal is shown as an orange sphere. Secondary structural features correspond to residue numbers as follows: β1, A4-Y6; helix A, G27-R33; β2, L37-K40; helix B, K45-D49; helix C, P50-R59; β3, W63-I69; helix D, N76-E86; β4, E94-G101; β5, G103-R108; β6, W114-M119; β7, D123-L127; β8, H133-V137; β9, T143-F149. C-terminal residues 163–178 are not shown; while helical in the crystal, they are disordered in solution (T. Pochapsky and X. Liu, unpublished results).

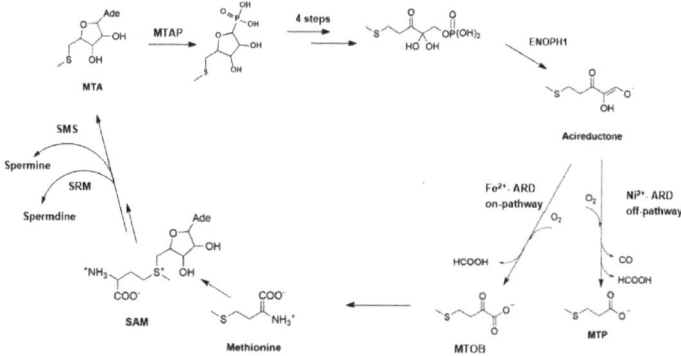

Figure 2. The methionine salvage pathway.

A remarkable discovery by that group was that the enzyme eventually named ARD could be isolated from *Klebsiella pneumoniae* (later reclassified as *K. oxytoca*) in two chromatographically separable forms with different enzymatic activities. While the polypeptides of both isoforms were identical, it was found that ARD to which Fe^{2+} was bound catalyzed the on-MSP reaction resulting in methionine biosynthesis, but if Ni^{2+} was bound, an off-pathway reaction occurs, leading to the formation of formate, methylthiopropionate and carbon monoxide (CO) [1,3]. Subsequent research showed that the same four residues bound the metal in either case, and the structural differences between the two enzymes are the result of subtle differences in metal–ligand bond lengths, which trigger the formation of secondary structural features in the Ni-bound form that are absent in Fe–ARD [4]. The metal-dependent dual functionality of ARD has been confirmed in vitro for mammalian ARDs from mouse (*Mm*ARD) and humans (*Hs*ARD) [5,6].

The metal binding motif of ARD is 3-His 1-Glu, unlike the 2-His 1-Glu/Asp scheme used by α-ketoglutarate (KG)-dependent dioxygenases (also members of the cupin superfamily). The 2-His 1-Glu/Asp ligands in the KG-dependent dioxygenases are located within the wide end of the cupin β-barrel, where they are relatively immobilized. The extra histidine ligand in ARD is located in a turn on the edge of the barrel, adjacent to a loop that appears to regulate metal-dependent conformational differences between the Fe- and Ni-bound enzymes via changes in hydrogen bonding patterns [4]. While the conformational differences between FeARD and NiARD have only been confirmed for the *Klebsiella* enzyme, the fact that the *Mm*ARD and *Hs*ARD enzymes are also separable chromatographically in their Fe- and Ni-bound forms, as well as their differential thermal stabilities, strongly support the likelihood of a metal-dependent conformational shift in these enzymes as well. Whether the conformational shift is responsible for or merely coincident with the different activities of the ARD isozymes is not known. These and related questions have been thoroughly reviewed recently, and interested readers are referred to that review [7]. Rather, we will look at the intriguing and (often) confusing links between the ARD gene product and human health, in particular, cancer and carcinogenesis.

Table 1. Observed methionine dependence of various cancer cell lines.

Citation	Type of Cancer	Cell Lines	Met Dependence Ccomplete, ++, Partial, +, Independent, −
Najim et al. (2009) [8]	Central nervous system	Daoy (medulloblastoma)	++
		D-54 (glioma)	++
Kokkinakis et al. (2001) [9]	Brain	D-54	++
		SWB77 (glioblastoma)	++
		Daoy	++
Willmann et al. (2015) [10]	Breast cancer	epithelial cell line MCF-10A	++
Kano et al. (1982) [11]	Leukemia	Raji (Burkitt)	++
		BALL (B-cell)	++
		TALL (T-cell)	++
		MOLT-3 (T-cell)	−
		MOLT 4B (T-cell)	++
		HL60 (promyelocytic)	++
		K562 (chronic myelogenous leukemia in blastic crisis)	++
Poirson-Bichat et al. (2000) [12]	Colon, lung, glioma	TC71-MA (colon)	++
		SCLC6 (small cell lung)	++
		SNB19 (glioma)	++
Lu et al. (2000) [13]	Prostate	LNCaP (lymph node metastasis)	−
		PC-3 (distant metastasis)	++
		DU 145(distant metastasis)	+
Guo et al. (1993) [14]	Prostate, lung, fibrosarcoma	PC3	++
		SKLU-I (lung carcinoma)	++
		HT 1080 (fibrosarcoma)	++
Mecham et al. (1983) [15]	Bladder, breast, cervical, colon, kidney, lung, prostate, fibrosarcoma, osteogenic sarcoma, glioblastoma, neuroblastoma	EJ (bladder)	++
		J82 (bladder)	++
		SK-BR-2-II (breast)	++
		MCF-7 (breast)	++
		HeLa (cervical)	++
		SK-CO-1 (colon)	++
		A498 (kidney)	++
		A2182 (lung)	++
		PC-3	++
		8387 (fibrosarcoma)	++
		HT-1080	++
		HOS (osteogenic sarcoma)	++
		Human neurological tumors	++
		A172 (glioblastoma)	++
		SK-N-SH (neuroblastoma)	++

2. *ADI1* and ADI1GP

*Hs*ARD is encoded by the *ADI1* gene, which is located on chromosome 2 at locus 2p25.3 and is comprised of four exons. Given that different forms of *ADI1* gene product may have different enzymatic and regulatory effects in vivo, we will refer to them collectively as ADI1GP. The full-length ADI1GP is 179 amino acids, with metal ligands provided by His88, His90, Glu94 and His133 (Figure 3). A partial ADI1GP, discovered prior to identification of ARD, was termed SipL (submergence-induced-protein-like) based on homology with a submergence-induced rice plant gene (*Sip*) that was subsequently shown to be an ARD (OsARD) [6,16–18]. SipL is a truncated ADI1GP, missing 63 residues from the N-terminus. Given that residue 64 of the ADI1GP is a Met, and a portion of the second exon is missing in the SipL sequence, it is likely that SipL is the result of a truncated ribosomal translation rather than an alternate mRNA splicing. The deletion removes strands β1 and β2 from the β-helix, as well as helix A (see Figure 1) but retains all four metal binding ligands. Both SipL and the full-length ADI1GP were found to be involved in human hepatitis C virus replication, permitting viral infection of otherwise resistant tissues [19–21].

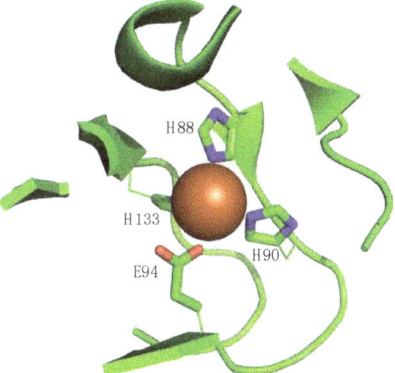

Figure 3. Metal ligation scheme in *Hs*ARD (PDB entry 4QGN). Protein-based ligands are His88, His90, Glu94 and His133. The metal is reported to be Fe^{3+}, although how this was determined is not reported.

Multiple tissues show high levels of ADI1GP based on Western blot analysis, including liver, kidney, prostate, thyroid, and skeletal muscle. Lower expression levels are seen in the adrenal gland, trachea, spinal cord and stomach. Only trace amounts of *ADI1* mRNA have been observed in heart, brain, mammary gland, lymph node, bone marrow, placenta, bladder and leukocytes [22]. ADI1GP can be localized both in the cytosolic and nuclear compartments [23,24]. Multiple carcinoma are found to have *ADI1* and ADI1GP dependences (Table 2).

2.1. A Role for ADI1GP in Hepatitis C Virus (HCV) Infection

The first hints that ADI1GP might play important roles in disease progression were found even before the discovery of the enzyme, in the work by Yeh et al. on the role of SipL in hepatitis C infection (vide supra). The hepatitis C virus (HCV) is a blood-borne (+)-single-stranded RNA virus with a genome 9.6 kb in length. HCV is the major cause of chronic hepatitis worldwide with severe complications, including cirrhosis, hepatic failure and hepatocellular carcinoma. HCV entry into the cell begins with the HCV E2 coat protein specifically binding to four transmembrane domains of tetraspanin CD81 located on the hepatocyte surface. However, this interaction alone is not sufficient to cause infection, other cofactors appear to be required [25,26]. Yeh et al. injected mouse hepatoma cells, Hepa1-6-CD81-SipL infected with HCV-positive serum into a mouse model, and detected HCV RNA by days 2–6 via RT-PCR. Immunofluorescence assays of resected tumor tissue identified SipL as a hepatic factor supporting HCV infection and replication, in combination with CD81 and other cofactors in an otherwise non-permissive cell line [19]. Further studies on mouse hepatomas verified that co-expressed human CD81 and SipL in Hepa1-6cells are permissive for HCV infection and replication. As noted above, SipL was identified to be a truncated version of ADI1GP. The group suggested the function of SipL in HCV infection was to bind with membrane-type 1 matrix metalloproteinase (MT1-MMP), thereby facilitating cell entry of HCV, albeit with low efficiency [20]. That group later showed that expression of ADI1GP alone can lead to a small amount of cell entry and replication of HCV, although the efficiency of cell entry and replication was enhanced significantly by co-expression with CD81. In addition, their replicon transfection experiments indicated ADI1GP expression did not increase replication efficiency. Hence, the group proposed expression of human ADI1GP could increase cell uptake of HCV, but not replication [21].

A recent study indicated that an interaction between MT1-MMP (matrix metalloproteinase MT1) and ADI1GP was correlated with *lower* HCV RNA levels in Huh7.5 cells, with the MT1-MMP–ADI1GP interaction decreasing HCV cell entry. However, when ADI1GP was overexpressed, the inhibitory effect was reversed. These researchers proposed that interaction between ADI1GP and MT1-MMP would draw MT1-MMP away from interacting with CD81 conducive to HCV entry, thereby reversing the inhibitory effect [23].

2.2. ADI1 in Hepatocellular Carcinoma (HCC)

Given the links between HCV infection and liver cancer, it would be reasonable to suspect that *ADI1* gene expression might be correlated with liver neoplasms. A recent paper by Chu et al. found a negative correlation between ADI1GP expression and hepatocellular carcinoma (HCC) cell proliferation [24]. Western blot results indicated a significant reduction of ADI1GP in tumor tissues versus normal liver tissue in a group of 161 patients. The group performed short hairpin *ADI1*-mediated knockdown in human hepatocellular carcinoma cell lines J7 and Huh7, showing that depletion of ADI1GP markedly enhanced cell proliferation. On the other hand, overexpression of ADI1GP resulted in decreased cell proliferation. Terminal deoxynucleotidyl transferase dUTP nick-end labeling (TUNEL) assays indicated that AD1GP overexpression lead to a large increase in the rate of apoptosis. An mRNA analysis of the human hepatocellular carcinoma GSE14520 dataset from the NCBI Gene Expression Omnibus and the human hepatocellular carcinoma dataset from the Cancer Genome Atlas program showed *ADI1* mRNA level reductions in cancerous tissue with more substantial down-regulation of *ADI1* in later stages of HCC progression, a correlation supported by immunohistochemistry analysis [24].

This group also claimed evidence that metal binding modulates ADI1GP effects on cell proliferation [24]. Over-expression of the ADI1GP mutant E94A in J7 and Huh7 cells showed a significant decrease of cell proliferation, while the H133A mutation did not change cell growth rates relative to untransformed cells, with a similar result observed on HCC tissue xenografts. While the authors suggested that these results implied a role for the on-pathway MSP function of *Hs*ARD in HCC growth repression, it is not clear why the removal of one ligand (E94) should be different from another (H133). Mutations of any of the four ligands in *Klebsiella oxytoca* ARD resulted in complete loss of metal binding and enzymatic function [27].

The same group investigated altered gene expression levels due to ADI1GP overexpression. They found that caveolin-1 (CAV1) was consistently down-regulated both at the protein (Western blot) and mRNA levels, with the same effect observed in both tumor and normal tissues. Caveolins are a class of oligomeric proteins involved in caveolae formation [28]. CAV1 is involved in lipid transport, membrane trafficking and signal transduction [29]. Interestingly, CAV1 has also been implicated in oncogenic cell transformation, tumorigenesis, and metastasis [28,30]. Additional mutation studies suggested that only the functional on-MSP ADI1GP would significantly down-regulate CAV1 expression, with CAV1 a downstream effector in ADI1-mediated repression. Due to the strong positive correlation between ADI1GP expression levels and SAM concentrations in the cell, the relation between CAV1 protein level and SAM levels were tested. The group found CAV1 expression decreases as SAM levels increase, with similar amounts of ADI1GP present, suggesting that ADI1GP inhibits *CAV1* expression via the MSP. Computational analysis of the NCBI database suggest ADI1GP levels to be negatively correlated with CAV1 levels in HCC patients [24].

Given that SAM plays an important role in cellular methylation, it was proposed that increased levels of ADI1GP would increase SAM concentrations and result in *CAV1* gene methylation to suppress transcription. They proposed a regulatory mechanism where in non-cancerous HCC cells, high ADI1GP levels would generate a large amount of SAM, modulating genome-wide methylation (with 15% in gene promoter regions), resulting in tumor suppression. However, in cancerous hepatocytes, with less ADI1GP and less SAM available, alterations of genome methylation patterns might promote cell proliferation. While an intriguing possibility, the study does not exclude the potential participation of other factors in determining gene methylation patterns [24].

2.3. ADI1GP Induces Apoptosis in Prostate Cancer Cell Lines

Further evidence that ADI1GP may be important in preventing or regulating carcinogenesis is that elevated *ADI1* expression level led to an increase in apoptosis in prostate cancer cell lines. Oram et al. observed down-regulation of *ADI1* in high-grade prostate tumor cell lines. Epithelial and stromal cells are the two major components of prostate. Non-cancerous prostatic hyperplasia tissue showed ADI1 mRNA expression in epithelial cells and little or no expression in stromal cells. Gleason grade 3 prostate tumor tissues had less ADI1GP than the benign specimen. A previous study using rat prostate epithelial cells and LNCaP epithelial cell lines expressed the ADI1GP ortholog ALP1 [22,31].

By introducing the synthetic androgen miboleron (Mib) in epithelial LNCaP Cells and inhibitor CHX, the researchers have demonstrated that *ADI1* mRNA expression was regulated by Mib, and proposed that Mib directly induced *ADI1* expression. Androgen has previously found to be regulating ortholog *APL1* mRNA levels in rat [22,31].

In order to examine how metal binding to ADI1GP influences apoptosis and growth inhibition of the prostate cell lines, mutations of conserved metal-ligating residues were made. The group examined the localization of ADI1GP mutants in LNCaP and PC3 cells using fluorescence microscopy. Since no ADI1GP expression in stromal cell lines would be observed, the group studied PC3 to illustrate the apoptosis-promoting function of ADI1GP. For all the mutations, ranging from single to quadruple, on ADI1GP, PC3 cell lines all showed approximately the same apoptosis rates as wild-type ADI1GP. In addition, the same ADI1GP mutants and WT in LNCaP cell lines resulted in decreased colony formation relative to controls, consistent with the results for LNCaP cells in their mRNA and tissue studies. In summary, these results suggest that the apoptosis-inducing effects of ADI1GP on stromal cells appear to be independent of metal binding.

On the other hand, apoptosis could be induced in pancreatic carcinoma, breast tumor, and HCC cell lines by supplementing growth media with the on-pathway MSP product of ADI1GP oxidation, MTOB [32–34]. Furthermore, other upstream metabolites (e.g., MTA, SAM) also induce apoptosis [24,32,34,35]. Ornithine decarboxylase 1 (ODC), an enzyme from the polyamine biosynthesis pathway, also inhibits tumor growth [22,36]. As such, further investigations are needed to confirm that enhanced apoptosis and tumor inhibition with elevated ADI1GP expression levels are truly independent of metal binding.

2.4. ADI1GP Regulation of Membrane-Type 1 Matrix Metalloproteinase (MT1-MMP)

Perhaps the most intriguing and direct link between ADI1GP and cancer is the observation that ADI1GP appears to suppress the metastasis-promoting activity of MT1-MMP [31,37,38]. MT1-MMP, also called MMP-14, functions in the pericellular space on the cell surface, where it is involved in the degradation of extracellular matrix involved in cellular functions such as migration, proliferation, and the regulation of cell morphology [26,39]. Due to its high expression level in cancerous tissue, MT1-MMP is believed to have a significant role in tumor metastasis via degradation of the extracellular matrix, freeing tumor cells to migrate. MT1-MMP activity is regulated by a transmembrane tail projecting into the cytoplasm [26,39,40]. Uekita and coworkers reported that ADI1GP down-regulates cell migration and invasion promoted by MT1-MMP. By comparing the interaction of ADI1GP with a FLAG-tagged MT1-MMP against that with an MT1-MMP mutant lacking the regulatory cytoplasmic tail, these researchers found that wild-type MT1-MMP formed a complex with the ADI1GP (which they called MTCBP-1) at the cytoplasmic tail of MT1-MMP. Further investigation showed ADI1GP to co-localize with MT1-MMP at the plasma membrane, and that ADI1GP was only recruited into the membrane fraction in the presence of the cytoplasmic tail. They showed ADI1GP co-expression specifically inhibited MT1-MMP-promoted cell migration but not other types of migration. That group also found that ADI1GP significantly reduces tissue invasion caused by MT1-MMP. They concluded that ADI1GP binding to the cytoplasmic tail of MT1-MMP regulates the activity of the enzyme towards the intercellular matrix, and reduced ADI1GP levels in tumor cell lines compared with the non-transformed fibroblasts would provide an advantage to the tumor cells for migration and proliferation [17,37,41].

Table 2. *ADI1* and ADI1GP dependences of cancer cell lines and tissue.

Types of Cancer	Tissue/Cell Type	ADI1GP Expression	Metastatic/Apoptotic	Mechanism
Hepatocellular carcinoma [24]	Tissue, J7 and Huh7	Down regulated by CAV1, overexpressed	No information	Not proposed
Prostate cancer [22,31]	Tissue, LNCaP and PC3	Observed in epithelial cells, little/no expression in stromal cells. Expression may be Mib regulated, Gleason gr. 3 prostate tumor tissue < ADI1GP than benign tissue.	Apoptotic	Associated with metal binding
Pancreatic ductal adenocarcinoma [42]	DanG, BxPC3, and Panc-1	Overexpressed	Metastatic	ADI1GP MT1-MMP interaction

2.5. ADI1GP/MT1-MMP Interaction Restricts Metastasis of Pancreatic Ductal Adenocarcinoma (PDAC)

ADI1GP has also been found to restrict tumor metastasis by disrupting the interactions between MT1-MMP and F-actin in pancreatic ductal adenocarcinoma (PDAC) [42]. Metastatic PDAC tumors are extremely aggressive and invasive, actively remodeling the actin-rich invadopodia that protrude into the extracellular matrix, facilitating invasion of nearby tissues [43–45]. Qiang et al. suggested that ADI1GP may serve as an endogenous antimetastatic factor in PDAC. They showed that decreasing ADI1GP enhances invasive migration and increases the rate of extracellular matrix degradation in PDAC cell lines DanG, BxPC3, and Panc-1. Conversely, ADI1GP overexpression slows invasion, suppresses extracellular matrix degradation and reduces invadopodia counts in PDAC cells. Using fluorescence microscopy with WT and mutant MT1-MMP lacking the cytoplasmic tail, they showed that ADI1GP binds the cytoplasmic tail of MT1-MMP directly, inhibiting the invasive properties of PDAC. In addition, they found ADI1GP in the invadopodia disrupted the interactions between MT1-MMP and F-actin. ADI1GP was thus shown to be an intrinsic inhibitor of stromal remodeling through a direct interaction with MT1-MMP. The localization of ADI1GP to invadopodia significantly reduced the capacity of PDAC cells to invade and metastasize into peripheral tissues [42].

3. Polyamines, MSP and Cancer

Elevated polyamine (spermidine, spermine and putrescine) levels have long been known to be associated with tumors and tumorogenesis [46]. Polyamine biosynthesis and concentrations are tightly regulated under normal conditions, given their relationship with cell replication and proliferation. Increased polyamine levels are associated with malignant transformation, increased cell proliferation and preservation of neoplastic phenotypes [46–49].

The polyamine precursor S-adenosylmethionine (SAM) plays multiple important roles in the cell. SAM is the principle methyl donor required for methylation of nucleic acids, phospholipids, histones, biogenic amines, and proteins [50,51]. Low SAM levels are associated with chronic liver diseases [52–54]. Hepatocellular SAM concentrations were found to affect oxidative stress, mitochondrial function, hepatocellular apoptosis, as well as malignant transformation [50]. Gene methylation has been proposed for tumor suppression, suggesting that SAM might be used for tumor suppression through its methylation function [24]. The processing of SAM metabolites depends upon its use in SAM-supported methylation reactions, which result in homocysteine production via the transsulfuration pathway, which produces S-adenosylhomocystine, followed by hydrolysis to form homocysteine, which then progresses via intermediate cystathionine to cysteine [55].

The biosynthesis of polyamines makes use of the amino acid substituent on decarboxyated SAM, with spermidine synthase (SRM) yielding spermidine and spermine synthase (SMS) producing spermine. In either case, methylthioadenosine (MTA), the first committed intermediate of the MSP, is a byproduct (Figure 2). MTA has been shown to regulate gene expression, inhibit protein methylation, prevent cell proliferation, and regulate apoptosis. MTA also had a role in tumor development, cancer cell invasion and lymphocyte activation [35]. As such, MTA levels are also tightly regulated under normal conditions, regulation that depends heavily on a functional MSP, in which MTA is cleaved and phosphorylated to form 5′-methylthioribose-1-phosphate (MTR-1-P) catalyzed by 5′-methythioadenosine phosphorylase (MTAP). In normal cells, MTA is rapidly metabolized by MTAP. Studies have shown MTAP expression induces a significant reduction in intracellular polyamine levels as well as putrescine to total polyamine ratio changes [56]. Many malignant cell lines lack MTAP activity, and MTAP-deficient cells were found to secrete MTA [35,57,58], In addition to SRM and SMS, ornithine decarboxylase 1 (ODC) also serve a key role in polyamines biosynthesis reacting with ornithine to produce putrescine. MTA inhibits SRM, ODC and strongly inhibits SMS [35,59]. ODC is also linked to tumor progression. Therefore, a substantial reduction in ODC activity can inhibit tumor growth [36]. The products of the methionine salvage pathway negatively regulate ODC. The inhibition of ODC by MTA can be partially mediated by its metabolite MTOB from the MSP in yeast and tumor cells [35,46,56].

Although not well investigated in human cancer cell lines and tissues, the metal dependence of ARD is clearly important in regulating MSP function [6,7,18]. The on-MSP form of ARD sustains the high polyamine synthesis rate essential for cell proliferation and provides methionine for SAM and protein production in normal cells. High polyamine synthesis rates lead to higher MTA concentrations, in turn, requiring upregulation of MTAP to maintain proper MTA levels. Furthermore, functional MTA production suggests proper ODC regulation for tumor suppression.

A recent study has shown that acireductone-generating enzyme ENOPH1 (enolase-phosphatase 1) is also associated with cell cycle regulation. Knockdown of ENOPH1 suppressed cell proliferation and migration in malignant glioma and promoted ADI1GP translocation from the nucleus to cytoplasm, leading to an indirect decrease in MT1-MMP activity [60]. Zhang et al. analyzed cultured brain microvascular endothelial cells in rat under oxygen–glucose deprivation (OGD)-induced ENOPH1 upregulation. However, knockdown ENOPH1 had no effect on OGD-induced *ADI1* upregulation. On the other hand, ENOPH1 was found to regulate OGD-induced ADI1GP translocation from the nucleus to the cytoplasm. ENOPH1 knockdown increased the permeability of the endothelial monolayer to ADI1GP. It was hypothesized that reduced ADI1GP concentrations in the cytoplasm would dysregulate MT1-MMP located on the cell membrane [61].

4. Carbon Monoxide as an Anti-Apoptotic Signal: A Potential off-MSP Role for ADI1GP in Carcinogenesis

With the exception of the potentiation of HCV infection, nearly all of the reports summarized above suggest that ADI1GP plays a protective role in carcinogenesis and tumor metastasis, or at least enables other directly protective processes to occur. However, if metals other than Fe(II) are bound in the active site of *Hs*ARD (e.g., Ni(II), Mn(II), Co(II)), the off-MSP products, 3-(methylthio)propionate and carbon monoxide (CO), are the dominant products of acireductone deoxygenation [5,6]. In this regard, CO is of particular interest. While toxic at high levels, CO is now generally accepted to be a cellular gasotransmitter at low concentrations, modulating inflammatory and apoptotic responses to cell damage [62–64]. In particular, CO is thought to prevent mitochondrial membrane permeabilization, the first step in cytochrome-*c*-initiated apoptosis [62].

This raises the intriguing possibility that off-MSP production of CO by Ni- or Mn-bound *Hs*ARD might be cytoprotective for carcinomas, while also rationalizing the methionine dependence of many cancer cell lines. In this regard, the Ni-bound ARD is of particular interest. The Ni-bound HsARD is the most thermostable of all metal-bound forms of the enzyme, with a midpoint for denaturation curves at 61 °C, with the Fe-bound form at 52 °C [6]. Furthermore, removal of Fe(II) from the ARD active site occurs under mild conditions, while Ni can only be removed by denaturation [3]. As such, nickel binding to ADI1GP would be both a thermodynamic and kinetic sink, permanently preventing expressed HsARD from binding iron and catalyzing the on-MSP chemistry required for methionine salvage. While Mn^{2+}-bound HsARD is less thermostable than the Fe-bound enzyme (denaturation midpoint at 46 °C), the relative abundance of manganese relative to nickel in the cell means it may also represent a reasonable pathway for off-pathway CO production. To date, the only confirmed source of endogenous CO is heme oxygenase-1 (HO-1), which oxidatively breaks down hemin to CO and bilirubin, and endogenous CO production via an off-MSP route has not been detected in normal mammalian tissues. We are currently testing this possibility in transformed tissues, particularly those from Met-dependent cell lines.

Another consideration arises in light of the significant conformational differences that are observed between Fe- and Ni-bound bacterial ARD (and inferred from the differential chromatographic behavior and thermal stabilities of the corresponding HsARDs) [4,6]. As virtually all of the experimental data obtained thus far regarding the role(s) of ADI1GP in carcinogenesis, apoptosis and metastasis have been obtained in cellular cultures, it is not known whether any of the roles of ADI1GP discussed here are dependent upon which metal is present in the active site (or indeed if any metal is present at all). A case can be plausibly made that, for example, regulatory binding of ADI1GP to MMP-MT1 could be modulated by the bound metal, particularly if the conformational differences between metal ARD isozymes involves the MMP-MT1 binding site, which is still unknown.

5. Conclusions

Although *ADI1* gene expression and ADI1GP in multiple type of cells and tissues have been shown to correlate (often negatively) with cancer migration, metastasis and apoptosis, it is not clear whether these effects are due directly to ADI1GP interactions with other cellular components, or indirectly, as a result of enzymatic function, either on- or off-MSP. SAM, MTA, ODC, L-methionine, MTOB, polyamines, ENOPH1 and ADI1GP have all been implicated in regulation of processes such as cell division, migration, metastasis and apoptosis. It is also possible that it is a combination of factors that link ADI1GP to cancer.

Probably the most important unknown is the role of metal binding to ADI1GP in any of these observations. None of the cell or tissue studies have identified what, if any, metal is bound to ADI1GP. The prostate cancer study [22] indicated that while mutation of the four metal ligands in ADI1GP destroys enzymatic activity, it does not affect pro-apoptotic and colony inhibition. Both the HCC and PDAC studies [24,42] showed the ADI1GP binds to the cytoplasmic tail of MT1-MMP, suggesting that this function, at least, is independent of metal binding. However, for the bacterial enzyme, at least, the Fe-bound and apo-ARDs forms are isostructural [4], leaving open the question of whether Ni or Mn binding could interfere with the ADI1GP–MT1-MMP interaction.

The following questions remain to be investigated:

1. Does the metal-dependent on/off MSP divergence occur in human cell lines? If so, under what conditions? Is off-MSP chemistry observed only under certain circumstances (e.g., in transformed cells)?
2. Is the methionine dependence of many cancer cell lines the result of off-MSP chemistry?
3. Do different metals bound to the ADI1GP act as switches for regulation of cell processes such as division, migration, invasion, metastasis and apoptosis? Is it possible to distinguish normal cells from tumor cells based on the metal content of their ADI1GP?
4. Does binding of metals other than Fe^{2+} change the interaction between ADI1 and MT1-MMP?

5. A systems biology approach will be necessary to place all of these possibilities into a complex cellular context, but answering some of these questions will help to define the role(s) of ADI1GP and, particularly, *Hs*ARD in human health.

Author Contributions: X.L. wrote the manuscript; T.C.P. edited and corrected it.

Funding: T.C.P. acknowledges support from NIH grant R01-GM130997.

Conflicts of Interest: The authors declare no conflict of interest.

References

1. Dai, Y.; Wensink, O.C.; Abeles, R.H. One Protein, Two Enzymes. *J. Biol. Chem.* **1999**, *274*, 1193–1195. [CrossRef] [PubMed]
2. Milaczewska, A.; Kot, E.; Amaya, J.A.; Makris, T.M.; Zajac, M.; Korecki, J.; Chumakov, A.; Trzewik, B.; Kedracka-Krok, S.; Minor, W.; et al. On the Structure and Reaction Mechanism of Human Acireductone Dioxygenase. *Chem. A Eur. J.* **2018**, *24*, 5225–5237. [CrossRef] [PubMed]
3. Dai, Y.; Pochapsky, T.C.; Abeles, R.H. Mechanistic Studies of Two Dioxygenases in the Methionine Salvage Pathway of Klebsiella Pneumoniae. *Biochemistry* **2001**, *40*, 6379–6387. [CrossRef] [PubMed]
4. Ju, T.T.; Goldsmith, R.B.; Chai, S.C.; Maroney, M.J.; Pochapsky, S.S.; Pochapsky, T.C. One Protein, Two Enzymes Revisited: A Structural Entropy Switch Interconverts the Two Isoforms of Acireductone Dioxygenase. *J. Mol. Biol.* **2006**, *363*, 523–534. [CrossRef] [PubMed]
5. Deshpande, A.R.; Wagenpfeil, K.; Pochapsky, T.C.; Petsko, G.A.; Ringe, D. Metal-Dependent Function of a Mammalian Acireductone Dioxygenase. *Biochemistry* **2016**, *55*, 1398–1407. [CrossRef] [PubMed]
6. Deshpande, A.R.; Pochapsky, T.C.; Petsko, G.A.; Ringe, D. Dual Chemistry Catalyzed by Human Acireductone Dioxygenase. *Protein Eng. Des. Sel.* **2017**, *30*, 109–206. [CrossRef] [PubMed]
7. Deshpande, A.R.; Pochapsky, T.C.; Ringe, D. The Metal Drives the Chemistry: Dual Functions of Acireductone Dioxygenase. *Chem. Rev.* **2017**, *117*, 10474–10501. [CrossRef] [PubMed]
8. Najim, N.; Podmore, I.D.; McGown, A.; Estlin, E.J. Biochemical Changes and Cytotoxicity Associated with Methionine Depletion in Paediatric Central Nervous System Tumour Cell Lines. *Anticancer Res.* **2009**, *29*, 2971–2976. [PubMed]
9. Kokkinakis, D.M.; Hoffman, R.M.; Frenkel, E.P.; Wick, J.B.; Han, Q.; Xu, M.; Tan, Y.; Schold, S.C. Synergy between Methionine Stress and Chemotherapy in the Treatment of Brain Tumor Xenografts in Athymic Mice. *Cancer Res.* **2001**, *61*, 4017–4023.
10. Willmann, L.; Erbes, T.; Halbach, S.; Brummer, T.; Jäger, M.; Hirschfeld, M.; Fehm, T.; Neubauer, H.; Stickeler, E.; Kammerer, B. Exometabolom Analysis of Breast Cancer Cell Lines: Metabolic Signature. *Sci. Rep.* **2015**, *5*, 13374. [CrossRef]
11. Kano, Y.; Sakamoto, S.; Kasahara, T.; Kusumoto, K.; Hida, K.; Suda, K.; Ozawa, K.; Miura, Y.; Takaku, F. Methionine Dependency of Cell Growth in Normal and Malignant Hematopoietic Cells. *Cancer Res.* **1982**, *42*, 3090–3092. [PubMed]
12. Poirson-Bichat, F.; Gonçalves, R.A.B.; Miccoli, L.; Dutrillaux, B.; Poupon, M.F. Methionine Depletion Enhances the Antitumoral Efficacy of Cytotoxic Agents in Drug-Resistant Human Tumor Xenografts. *Clin. Cancer Res.* **2000**, *6*, 643–653. [PubMed]
13. Lu, S.; Epner, D.E. Molecular Mechanisms of Cell Cycle Block by Methionine Restriction in Human Prostate Cancer Cells. *Nutr. Cancer* **2000**, *38*, 123–130. [CrossRef] [PubMed]
14. Guo, H.-Y.; Herrera, H.; Groce, A.; Hoffman, R.M. Expression of the Biochemical Defect of Methionine Dependence in Fresh Patient Tumors in Primary Histoculture. *Cancer Res.* **1993**, *53*, 2479–2483. [PubMed]
15. Mecham, J.O.; Rowitch, D.; Wallace, C.D.; Stern, P.H.; Hoffman, R.M. The Metabolic Defect of Methionine Dependence Occurs Frequently in Human Tumor Cell Lines. *Biochem. Biophys. Res. Commun.* **1983**, *117*, 429–434. [CrossRef]
16. Sauter, M.; Lorbiecke, R.; OuYang, B.; Pochapsky, T.C.; Rzewuski, G. The Immediate-Early Ethylene Response Gene Osard1 Encodes an Acireductone Dioxygenase Involved in Recycling of the Ethylene Precursor S-Adenosylmethionine. *Plant J.* **2005**, *44*, 718–729. [CrossRef]

17. Hirano, W.; Gotoh, I.; Uekita, T.; Seiki, M. Membrane-Type 1 Matrix Metalloproteinase Cytoplasmic Tail Binding Protein-1 (Mtcbp-1) Acts as an Eukaryotic Aci-Reductone Dioxygenase (Ard) in the Methionine Salvage Pathway. *Genes Cells* **2005**, *10*, 565–574. [CrossRef]
18. Deshpande, A.R.; Wagenpfeil, K.; Pochapsky, T.C.; Petsko, G.; Ringe, D. Metal Drives Chemistry: Dual-Function of Acireductone Dioxygenase Enzymes. *FASEB J.* **2017**, *31*, 1. [CrossRef]
19. Yeh, C.-T.; Lai, H.-Y.; Chen, T.-C.; Chu, C.-M.; Liaw, Y.-F. Identification of a Hepatic Factor Capable of Supporting Hepatitis C Virus Replication in a Nonpermissive Cell Line. *J. Virol.* **2001**, *75*, 11017–11024. [CrossRef]
20. Yeh, C.-T.; Lai, H.-Y.; Yeh, Y.-J.; Cheng, J.-C. Hepatitis C Virus Infection in Mouse Hepatoma Cells Co-Expressing Human Cd81 and Sip-L. *Biochem. Biophys. Res. Commun.* **2008**, *372*, 157–161. [CrossRef]
21. Cheng, J.C.; Yeh, Y.J.; Pai, L.M.; Chang, M.L.; Yeh, C.T. 293 Cells over-Expressing Human ADI1 and Cd81 Are Permissive for Serum-Derived Hepatitis C Virus Infection. *J. Med. Virol.* **2009**, *81*, 1560–1568. [CrossRef]
22. Oram, S.W.; Ai, J.; Pagani, G.M.; Hitchens, M.R.; Stern, J.A.; Eggener, S.; Pins, M.; Xiao, W.; Cai, X.; Haleem, R.; et al. Expression and Function of the Human Androgen-Responsive Gene ADI1 in Prostate Cancer. *Neoplasia* **2007**, *9*, 643–651. [CrossRef]
23. Chang, M.-L.; Huang, Y.-H.; Cheng, J.-C.; Yeh, C.-T. Interaction between Hepatic Membrane Type 1 Matrix Metalloproteinase and Acireductone Dioxygenase 1 Regulates Hepatitis C Virus Infection. *J. Viral Hepat.* **2016**, *23*, 256–266. [CrossRef]
24. Chu, Y.D.; Lai, H.Y.; Pai, L.M.; Huang, Y.H.; Lin, Y.H.; Liang, K.H.; Yeh, C.T. The Methionine Salvage Pathway-Involving ADI1 Inhibits Hepatoma Growth by Epigenetically Altering Genes Expression Via Elevating S-Adenosylmethionine. *Cell Death Dis.* **2019**, *10*, 240. [CrossRef]
25. Miyamori, H.; Takino, T.; Kobayashi, Y.; Tokai, H.; Itoh, Y.; Seiki, M.; Sato, H. Claudin Promotes Activation of Pro-Matrix Metalloproteinase-2 Mediated by Membrane-Type Matrix Metalloproteinases. *J. Biol. Chem.* **2001**, *276*, 28204–28211. [CrossRef]
26. Seiki, M. The Cell Surface: The Stage for Matrix Metalloproteinase Regulation of Migration. *Curr. Opin. Cell Biol.* **2002**, *14*, 624–632. [CrossRef]
27. Chai, S.C.; Ju, T.T.; Dang, M.; Goldsmith, R.B.; Maroney, M.J.; Pochapsky, T.C. Characterization of Metal Binding in the Active Sites of Acireductone Dioxygenase Isoforms from Klebsiella Atcc 8724. *Biochemistry* **2008**, *47*, 2428–2438. [CrossRef]
28. Williams, T.M.; Lisanti, M.P. Caveolin-1 in Oncogenic Transformation, Cancer, and Metastasis. *Am. J. Physiol. Cell Physiol.* **2005**, *288*, C494–C506. [CrossRef]
29. Liu, P.; Rudick, M.; Anderson, R.G.W. Multiple Functions of Caveolin-1. *J. Biol. Chem.* **2002**, *277*, 41295–41298. [CrossRef]
30. Wiechen, K.; Diatchenko, L.; Agoulnik, A.; Scharff, K.M.; Schober, H.; Arlt, K.; Zhumabayeva, B.; Siebert, P.D.; Dietel, M.; Schäfer, R.; et al. Caveolin-1 Is Down-Regulated in Human Ovarian Carcinoma and Acts as a Candidate Tumor Suppressor Gene. *Am. J. Pathol.* **2001**, *159*, 1635–1643. [CrossRef]
31. Oram, S.; Jiang, F.; Cai, X.Y.; Haleem, R.; Dincer, Z.; Wang, Z. Identification and Characterization of an Androgen-Responsive Gene Encoding an Aci-Reductone Dioxygenase-Like Protein in the Rat Prostate. *Endocrinology* **2004**, *145*, 1933–1942. [CrossRef]
32. Lu, S.C.; Mato, J.M. S-Adenosylmethionine in Cell Growth, Apoptosis and Liver Cancer. *J. Gastroenterol. Hepatol.* **2008**, *23*, S73–S77. [CrossRef]
33. Ansorena, E.; García-Trevijano, E.R.; Martínez-Chantar, M.L.; Huang, Z.; Chen, L.; Mato, J.M.; Iraburu, M.; Lu, S.C.; Avila, M.A. S-Adenosylmethionine and Methylthioadenosine Are Antiapoptotic in Cultured Rat Hepatocytes but Proapoptotic in Human Hepatoma Cells. *Hepatology* **2002**, *35*, 274–280. [CrossRef]
34. Tang, B.; Kadariya, Y.; Murphy, M.E.; Kruger, W.D. The Methionine Salvage Pathway Compound 4-Methylthio-2-Oxobutanate Causes Apoptosis Independent of Down-Regulation of Ornithine Decarboxylase. *Biochem. Pharmacol.* **2006**, *72*, 806–815. [CrossRef]
35. Avila, M.A.; García-Trevijano, E.R.; Lu, S.C.; Corrales, F.J.; Mato, J.M. Methylthioadenosine. *Int. J. Biochem. Cell Biol.* **2004**, *36*, 2125–2130. [CrossRef]
36. Pegg, A.E. Regulation of Ornithine Decarboxylase. *J. Biol. Chem.* **2006**, *281*, 14529–14532. [CrossRef]
37. Uekita, T.; Gotoh, I.; Kinoshita, T.; Itoh, Y.; Sato, H.; Shiomi, T.; Okada, Y.; Seiki, M. Membrane-Type 1 Matrix Metalloproteinase Cytoplasmic Tail-Binding Protein-1 Is a New Member of the Cupin Superfamily—A

Possible Multifunctional Protein Acting as an Invasion Suppressor Down-Regulated in Tumors. *J. Biol. Chem.* **2004**, *279*, 12734–12743. [CrossRef]
38. Pratt, J.; Iddir, M.; Bourgault, S.; Annabi, B. Evidence of Mtcbp-1 Interaction with the Cytoplasmic Domain of Mt1-Mmp: Implications in the Autophagy Cell Index of High-Grade Glioblastoma. *Mol. Carcinog.* **2016**, *55*, 148–160. [CrossRef]
39. Seiki, M.; Koshikawa, N.; Yana, I. Role of Pericellular Proteolysis by Membrane-Type 1 Matrix Metalloproteinase in Cancer Invasion and Angiogenesis. *Cancer Metastasis Rev.* **2003**, *22*, 129–143. [CrossRef]
40. Sato, H.; Takino, T.; Okada, Y.; Cao, J.; Shinagawa, A.; Yamamoto, E.; Seiki, M. A Matrix Metalloproteinase Expressed on the Surface of Invasive Tumour Cells. *Nature* **1994**, *370*, 61. [CrossRef]
41. Gotoh, I.; Uekita, T.; Seiki, M. Regulated Nucleo-Cytoplasmic Shuttling of Human Aci-Reductone Dioxygenase (HADI1) and Its Potential Role in Mrna Processing. *Genes Cells* **2007**, *12*, 105–117. [CrossRef]
42. Qiang, L.; Cao, H.; Chen, J.; Weller, S.G.; Krueger, E.W.; Zhang, L.; Razidlo, G.L.; McNiven, M.A. Pancreatic Tumor Cell Metastasis Is Restricted by Mt1-Mmp Binding Protein Mtcbp-1. *J. Cell Biol.* **2019**, *218*, 317–332. [CrossRef]
43. Ridley, A.J. Life at the Leading Edge. *Cell* **2011**, *145*, 1012–1022. [CrossRef]
44. Linder, S. The Matrix Corroded: Podosomes and Invadopodia in Extracellular Matrix Degradation. *Trends Cell Biol.* **2007**, *17*, 107–117. [CrossRef]
45. McNiven, M.A.; Kim, L.; Krueger, E.W.; Orth, J.D.; Cao, H.; Wong, T.W. Regulated Interactions between Dynamin and the Actin-Binding Protein Cortactin Modulate Cell Shape. *J. Cell Biol.* **2000**, *151*, 187–198. [CrossRef]
46. Bae, D.H.; Lane, D.J.R.; Jansson, P.J.; Richardson, D.R. The Old and New Biochemistry of Polyamines. *Biochim. Biophys. Acta Gen. Subj.* **2018**, *1862*, 2053–2068. [CrossRef]
47. Gerner, E.W.; Meyskens, F.L., Jr. Polyamines and Cancer: Old Molecules, New Understanding. *Nat. Rev. Cancer* **2004**, *4*, 781–792. [CrossRef]
48. Nowotarski, S.L.; Woster, P.M.; Casero, R.A. Polyamines and Cancer: Implications for Chemotherapy and Chemoprevention. *Expert Rev. Mol. Med.* **2013**, *15*, e3. [CrossRef]
49. Cervelli, M.; Pietropaoli, S.; Signore, F.; Amendola, R.; Mariottini, P. Polyamines Metabolism and Breast Cancer: State of the Art and Perspectives. *Breast Cancer Res. Treat.* **2014**, *148*, 233–248. [CrossRef]
50. Anstee, Q.M.; Day, C.P. S-Adenosylmethionine (Same) Therapy in Liver Disease: A Review of Current Evidence and Clinical Utility. *J. Hepatol.* **2012**, *57*, 1097–1109. [CrossRef]
51. Lieber, C.S.; Packer, L. S-Adenosylmethionine: Molecular, Biological, and Clinical Aspects—An Introduction. *Am. J. Clin. Nutr.* **2002**, *76*, 1148S–1150S. [CrossRef]
52. Martínez-Chantar, M.L.; García-Trevijano, E.R.; Latasa, M.U.; Pérez-Mato, I.; del Pino, M.M.S.; Corrales, F.J.; Avila, M.A.; Mato, J.M. Importance of a Deficiency in S-Adenosyl-L-Methionine Synthesis in the Pathogenesis of Liver Injury. *Am. J. Clin. Nutr.* **2002**, *76*, 1177S–1182S. [CrossRef]
53. Mato, J.M.; Lu, S.C. Role of S-Adenosyl-L-Methionine in Liver Health and Injury. *Hepatology* **2007**, *45*, 1306–1312. [CrossRef]
54. Lieber, C.S. S-Adenosyl-L-Methionine: Its Role in the Treatment of Liver Disorders. *Am. J. Clin. Nutr.* **2002**, *76*, 1183S–1187S. [CrossRef]
55. Finkelstein, J.D. The Metabolism of Homocysteine: Pathways and Regulation. *Eur. J. Pediatr.* **1998**, *157*, S40–S44. [CrossRef]
56. Subhi, A.L.; Diegelman, P.; Porter, C.W.; Tang, B.; Lu, Z.J.; Markham, G.D.; Kruger, W.D. Methylthioadenosine Phosphorylase Regulates Ornithine Decarboxylase by Production of Downstream Metabolites. *J. Biol. Chem.* **2003**, *278*, 49868–49873. [CrossRef]
57. Williams-Ashman, H.G.; Seidenfeld, J.; Galletti, P. Trends in the Biochemical Pharmacology of 5'-Deoxy-5'-Methylthioadenosine. *Biochem. Pharmacol.* **1982**, *31*, 277–288. [CrossRef]
58. Nobori, T.; Takabayashi, K.; Tran, P.; Orvis, L.; Batova, A.; Yu, A.L.; Carson, D.A. Genomic Cloning of Methylthioadenosine Phosphorylase: A Purine Metabolic Enzyme Deficient in Multiple Different Cancers. *Proc. Natl. Acad. Sci. USA* **1996**, *93*, 6203–6208. [CrossRef]
59. Pegg, A.E. Polyamine Metabolism and Its Importance in Neoplastic Growth and a Target for Chemotherapy. *Cancer Res.* **1988**, *48*, 759–774.

60. Su, L.; Yang, K.; Li, S.; Liu, C.; Han, J.G.; Zhang, Y.; Xu, G.Z. Enolase-Phosphatase 1 as a Novel Potential Malignant Glioma Indicator Promotes Cell Proliferation and Migration. *Oncol. Rep.* **2018**, *40*, 2233–2241. [CrossRef]
61. Zhang, Y.; Wang, T.; Yang, K.; Xu, J.; Ren, L.J.; Li, W.P.; Liu, W.L. Cerebral Microvascular Endothelial Cell Apoptosis after Ischemia: Role of Enolase-Phosphatase 1 Activation and Aci-Reductone Dioxygenase 1 Translocation. *Front. Mol. Neurosci.* **2016**, *9*, 79. [CrossRef]
62. Almeida, A.S.; Figueiredo-Pereira, C.; Vieira, H.L.A. Carbon Monoxide and Mitochondria—Modulation of Cell Metabolism, Redox Response and Cell Death. *Front. Physiol.* **2015**, *6*, 33. [CrossRef]
63. Bakhautdin, B.; Das, D.; Mandal, P.; Roychowdhury, S.; Danner, J.; Bush, K.; Pollard, K.; Kaspar, J.W.; Li, W.; Salomon, R.G.; et al. Protective Role of Ho-1 and Carbon Monoxide in Ethanol-Induced Hepatocyte Cell Death and Liver Injury in Mice. *J. Hepatol.* **2014**, *61*, 1029–1037. [CrossRef]
64. Rochette, L.; Cottin, Y.; Zeller, M.; Vergely, C. Carbon Monoxide: Mechanisms of Action and Potential Clinical Implications. *Pharmacol. Ther.* **2013**, *137*, 133–135. [CrossRef]

© 2019 by the authors. Licensee MDPI, Basel, Switzerland. This article is an open access article distributed under the terms and conditions of the Creative Commons Attribution (CC BY) license (http://creativecommons.org/licenses/by/4.0/).

Review

The Role of Non-Coding RNAs Involved in Nickel-Induced Lung Carcinogenic Mechanisms

Yusha Zhu, Qiao Yi Chen, Alex Heng Li and Max Costa *

Department of Environmental Medicine, New York University School of Medicine, New York, NY 10010, USA
* Correspondence: Max.Costa@nyulangone.org

Received: 29 May 2019; Accepted: 25 June 2019; Published: 28 June 2019

Abstract: Nickel is a naturally occurring element found in the Earth's crust and an International Agency for Research on Cancer (IARC)-classified human carcinogen. While low levels found in the natural environment pose a minor concern, the extensive use of nickel in industrial settings such as in the production of stainless steel and various alloys complicate human exposure and health effects. Notably, interactions with nickel macromolecules, primarily through inhalation, have been demonstrated to promote lung cancer. Mechanisms of nickel-carcinogenesis range from oxidative stress, DNA damage, and hypoxia-inducible pathways to epigenetic mechanisms. Recently, non-coding RNAs have drawn increased attention in cancer mechanistic studies. Specifically, nickel has been found to disrupt expression and functions of micro-RNAs and long-non-coding RNAs, resulting in subsequent changes in target gene expression levels, some of which include key cancer genes such as p53, MDM2, c-myc, and AP-1. Non-coding RNAs are also involved in well-studied mechanisms of nickel-induced lung carcinogenesis, such as the hypoxia-inducible factor (HIF) pathway, oxidative stress, DNA damage and repair, DNA hypermethylation, and alterations in tumor suppressors and oncogenes. This review provides a summary of the currently known epigenetic mechanisms involved in nickel-induced lung carcinogenesis, with a particular focus on non-coding RNAs.

Keywords: nickel; ncRNA; miRNA; lncRNA; lung carcinogenesis

---

1. Introduction

1.1. Nickel Overview

Nickel is a commonly occurring metal on Earth, existing as both soluble and insoluble compounds in soil, fumes, and water. Used as a component in many products such as watches, coins, belt buckles, earrings, mobile phones, and various medical devices, it is as ubiquitous in industrial usage as it is in nature [1]. Nickel is a transition metal that normally exists in nature in a +2 oxidative state [2]. While nickel plays fundamental roles for plants, bacteria, archaea, and unicellular eukaryotes, there are no enzymes in the human body that require nickel to function. As a result, it is only biologically significant as a toxicant for humans [3]. Inhalation is the most prominent route of exposure, with oral and dermal routes considered much less important [4]. On a cellular level, insoluble particulate nickel enters cells by phagocytosis, while nickel carbonyl is a lipophilic compound that can pass through the plasma membrane. Ni(II) can be transported into cells through calcium channels and/or divalent cation transporters such as DMT-1 [5]. Nickel is primarily excreted through urine with elimination following first order kinetics without any evidence of dose-dependent excretion. Excretion may also occur in saliva and sweat, which may be more significant in hot environments [6]. Workplace-related nickel exposure tends to be via inhalation of airborne fumes and dusts containing nickel and its associated compounds, while non-workplace-related nickel exposures are often from diet or dermal contact [7]. Dietary exposure is more significant in vegetables (particularly spinach), cocoa and nuts because of their higher nickel content. Chronic nickel exposure can lead to dermatitis, pulmonary fibrosis, and asthma [8,9]. Nickel sensitivity is a well-documented issue that affects millions of people. It is caused by triggering T-lymphocyte-driven delayed-type hypersensitivity reaction by Ni^{2+}, followed by leukocyte infiltration at the site of exposure [1].

1.2. Nickel-Induced Carcinogenesis

The most scrutinized effect of nickel exposure is carcinogenesis. In 1990, the International Agency for Research on Cancer (IARC) classified soluble nickel as a "group 1" human carcinogen [10]. In 1990, a report from the IARC that aimed to identify which chemical forms of nickel caused elevated risks of cancer mortality in occupationally exposed workers found that exposure to high levels of oxidic nickel compounds, exposure to sulfidic nickel in combination with oxidic nickel, and exposure to water-soluble nickel, alone or together with less soluble compounds, led to higher mortality from cancers of the lung and nasal sinuses [11]. While some studies have since suggested that nickel workers processing and refining sulfidic nickel ores have demonstrated an increased pulmonary and sinonasal cancer risk, some studies involving low level nickel exposure did not observe an increased incidence of cancer, suggesting that there may be threshold-like responses in tumor incidence [12–17].

1.3. Non-Coding RNA

While 80% of the human genome can be transcribed into RNA, only 2% of the genes code for proteins [18]. Non-coding RNAs are classified as either housekeeping non-coding RNAs, such as small nuclear RNAs, transfer RNAs, ribosomal RNAs, and small nucleolar RNAs, or regulatory non-coding RNAs, such as microRNAs (miRNAs), piwiRNAs, and long non-coding RNAs (lncRNAs) based on their size.

MicroRNAs are small non-coding RNAs that are 18–25 nucleotides in length and function by targeting complementary mRNAs at the 3'-untranslated region (3'-UTR) to cause degradation, translation inhibition or regulating gene expression at the post-transcriptional level [19]. They are specifically expressed in different human cells. Dysregulation of miRNA is related to various types of cancer and is involved in multiple processes, such as cell proliferation and differentiation, apoptosis, cancer initiation and progression, and cell metastasis [20]. MicroRNAs are now providing important information for cancer diagnosis, treatment, and prognosis. A lot of miRNAs have been identified with different roles in lung carcinogenesis as summarized by Uddin et al. [21], such as miR-21 in apoptosis, miR-31 in cell division, and miR-101, miR-15 and miR-16 in cell proliferation [22–25].

Long non-coding RNAs (lncRNAs) are RNAs longer than 200 nucleotides. Unlike miRNAs, lncRNAs are poorly conserved and regulate gene expression at different levels including histone modification, transcription regulation, post-transcription regulation, translation regulation and pseudogene regulation [26]. They also serve as scaffolds for cellular components and decoys for miRNAs and other coding RNAs [27,28]. lncRNAs have been given more attention recently since it was discovered that they play an essential role in pathological progressions including cancer invasion and metastasis [29–31]. Many lncRNAs have been reported to be associated with non-small cell lung cancer (NSCLC), such as CAR10, MALAT1, HOTAIR and so on [32–34]. Studies have shown that lncRNA was also involved in nickel-induced tumorigenesis [35,36]. A new lncRNA, termed nickel-related gene (NRG1), was the first to be identified as a lncRNA induced in Ni-induced lung cancer but the mechanisms that regulate nickel-induced NRG1 level is not known [37].

The molecular mechanisms by which nickel causes tumorigenesis can be summarized as DNA hypermethylation, oxidative stress and DNA damage, hypoxia-inducible signaling pathways, tumor suppressor genes, oncogene alterations, etc. [37]. In this review, we will discuss the molecular mechanisms involving non-coding RNA, especially miRNA and lncRNA, in nickel-induced lung cancer. Although the number of ncRNAs studied in this area so far is limited, this review reveals a current state of knowledge and provides directions for future research in ncRNAs involved in nickel-induced carcinogenesis.

2. Mechanisms

2.1. DNA Hypermethylation

While only considered weakly mutagenic, nickel is known to promote tumorigenesis through epigenetic mechanisms including DNA methylation, histone acetylation, and miRNAs. To date, DNA methylation is the primary epigenetic event studied in nickel-induced lung cancer, and has been shown to induce various human diseases and cancers through gene inactivation, including fragile X mental retardation [38–43]. Specifically, DNA methyltransferases (DNMTs) act to transfer a methyl group from S-adenosyl methionine (SAM) to the fifth carbon on cytosine. Notably, as is the case in most cancers, nickel is capable of eliciting global hypermethylation, thereby suppressing key tumor suppressor genes and inducing senescence as a part of its carcinogenic mechanism [44]. As a case in point, while 5-azacytidine and nickel have both been shown to inhibit DNA methyltransferase activity, the effects of nickel are only transient, in that after a recovery period following nickel exposure, levels of methyltransferases will recuperate and surge, leading to genome-wide increases in DNA methylation levels [38,45,46]. In fact, nickel-induced DNA hypermethylation has been confirmed both in vitro and in vivo. For example, in human bronchial epithelial (Beas-2B) cells, a commonly-used cell line for studies involving lung cancer, nickel has been found to induce the suppression of E-cadherin [47]. E-cadherin is a prominent tumor suppressor gene and cell-adhesion molecule. Reductions in E-cadherin levels have been evidenced to trigger dysfunctions of the cell–cell adhesion system, leading to cancer cell metastasis and invasion [48]. In addition, nickel has also been shown to induce the silencing of p16 through CpG site hypermethylation and thereby escape cell senescence [44]. Specifically, using p53 heterozygous mice and Wistar rats, treatment with nickel sulfide promoted promoter hypermethylation of p16, a cyclin-dependent kinase inhibitor and tumor suppressor gene [49]. Previous studies have demonstrated that nickel induces site-specific DNA hypermethylation. Specifically, instead of active euchromatin, nickel selectively targets inactive heterochromatin regions such as the long arm of chromosome X in Chinese hamster ovary cells [50]. In detail, the gpt gene was inserted near an active euchromatin region and the telomere in G10 and G12 cells, respectively [51]. The successful silencing of the gpt gene is validated by exhibition of high resistance to 6TG. Interestingly, while nickel was shown to induce silencing of the gpt gene in both G10 and G12 cells, the silencing effect was much more significant in G12 cells, indicating that nickel specifically targets genes in heterochromatic regions and spread heterochromatinization through DNA condensation [39,52]. The heterochromatin's condensed structure dictates that genes in this region have minimal activity, but its overall structure is intrinsic for the maintenance of overall genomic stability and that the organization of heterochromatin is often compromised in cancer [53,54].

Hypermethylation at the promoter regions of non-coding RNAs upon nickel treatment has been reported in several recent studies. One study observed that nickel led to a dose (0–0.5 mM) and time-dependent (0–12 h) reduction of lncRNA maternally expressed gene 3 (MEG3), in both Beas-2B and BEP2D cells (both human bronchial epithelial cells). Ectopic expression of MEG3 in Beas-2B cells inhibited cell transformation by nickel and knockdown of MEG3 resulted in spontaneous cell transformation of Beas-2B cells [35]. MEG3 is a 1.6 kb lncRNA and is expressed in many normal tissues, while loss of MEG3 expression has been seen in tumors and was reported to be related to tumor development and progression [55–57]. Zhou et al. found that the MEG3 promoter region was hypermethylated in normal human bronchial epithelial cells upon nickel exposure (0.5 µM, from 0–12 h), and treatment of cells with the DNA methylation inhibitor 5-aza-2-deoxycytidine (5-Aza) (5 µM, 72 h) removed the methylation status and increased the expression level of MEG3. They also revealed that the hypermethylation effect was mainly mediated by DNMT3b since nickel exposure upregulated DNMT3b specifically, and knockout of DNMT3b restored MEG3 expression in $NiCl_2$-treated Beas-2B cells [35].

Another study found that DNMT1 was upregulated in NiS-transformed 16HBE cells (NSTCs), a human bronchial epithelial cell line [58], along with a significant inhibition of miR-152 expression [59].

MicroRNA-152 is a member of the miR-148/152 family that has sequence complementarity to the 3′-UTR of DNMT1. Other members in the family include miR-148a and miR-148b. Ectopic expression of miR-152 in NSTCs suppressed cell growth, and knockdown of miR-152 increased cell growth of non-treated 16HBE cells. Interestingly, they first found that miR-152 repressed DNMT1 expression by targeting its 3′-UTR, and then discovered that downregulation of miR-152 in NSTCs was mediated through hypermethylation at the promoter region by DNMT1, indicating a double-negative loop between miR-152 and DNMT1. They speculated that NiS exposure firstly induced upregulation of DNMT1 via other mechanisms such as oxidative stress, and then led to hypermethylation of miR-152 at the CpG island promoter region, resulting in its downregulation, which further promoted DNMT1 expression by reduced methylation at the 3′-UTR by miR-152. It was also shown that miR-152 expression suppressed cell growth and a reduction of miR-152 expression lead to a 30% increase in cell proliferation [59].

MicroRNA-203 is another miRNA that was found to be downregulated in NSTCs [60]. Ectopic expression of miR-203 reduced NSTC cell migration and tumor growth in nude mice. It was found that the CpG island and first exon area of miR-203 was hypermethylated, however the mechanisms that mediate the hypermethylation effect was not elucidated in the study. Intriguingly, besides 5-Aza, TSA (Trichostatin A, a histone deacetylase inhibitor)-treated NSTCs also partially restored the expression level of miR-203, suggesting that downregulation of miR-203 is possibly mediated by DNA hypermethylation and/or other epigenetic silencing mechanisms.

2.2. Hypoxia-Inducible Pathway

The hypoxia-inducible signaling pathway is another important route for nickel-induced carcinogenesis, in which the transcription factor hypoxia-inducible-factor-1 (HIF-1) is activated [61]. HIF-1 is a critical regulator of genes that facilitate cell survival and adaptation in hypoxic conditions [62–64]. It is a heterodimeric transcriptional factor that consists of two subunits, HIF-1α and HIF-1β, both of which are required for HIF-1 to function; HIF-1β is integral in HIF-1 heterodimer formation, while HIF-1α is the key regulatory subunit and is responsible for HIF-1 transcriptional activity [65]. HIF-1 overexpression has been observed in various cancers [66]. Studies suggested that nickel can replace iron in HIF prolyl hydroxylases and thus inhibits the association of HIF-1α with Von-Hippel-Lindau (VHL) E3 ubiquitin ligase, which in turn stabilizes the HIF protein [67–69]. It was reported that elevated HIF-1 aided in tumor progression by inducing glycolytic activity and increasing production of lactic acid in tumor cells (Warburg effect) [70]. A collection of hypoxia-inducible genes were also induced by nickel including glycolytic enzymes and glucose transporters, and tumor oncogenes such as NDRG1 (N-myc Downstream-Regulated Gene 1/Cap43), and some other HIF-dependent genes involved in carcinogenesis such as BCL-2 (B-Cell Lymphoma 2) binding protein Nip3, EGLN1 (Egl-9 Family Hypoxia Inducible Factor 1), HIG1 (Hypoxia-Induced Gene 1) and P4H (Prolyl 4-Hydroxylase) [61]. High fidelity DNA repair mechanisms such as base excision repair and nucleotide repair were found to be compromised by the hypoxic conditions induced by nickel [71,72].

MicroRNA-210 is known as a hypoxia-sensitive miRNA and is one of the target miRNAs for HIF-1α. These are termed as hypoxia-regulated microRNAs (HRMs), and include miR-21, 23, 24, 26a, 213, etc. [73]. One study reported that nickel exposure ($NiCl_2$, 1 mM, 4 h) induced overexpression of miR-210 with stabilization of the HIF-1α protein, resulting in a metabolism shift [74]. In fact, HIF-1α has been demonstrated to be directly recruited to the promoter region of miR-210 during hypoxic conditions to induce miRNA expression [75]. Similar with the DNMT1/miR-152 double-negative loop, another study revealed a non-canonical miR-210 targeting site in the HIF-1α 3′-UTR region, suggesting a potential negative feedback loop in HIF-1α/miR-210 signaling [76].

MEG3, as mentioned, is a long non-coding RNA that was found to be downregulated by DNMT3b under nickel exposure. It has been suggested that MEG3 is a regulator of HIF-1α and nickel-induced HIF-1α accumulation independent of iron metabolism through inhibiting MEG3 [35]. Instead of directly targeting HIF-1α at the promoter region, MEG3 regulates HIF-1α at the protein translational level, since alterations in transcription, mRNA stability and protein degradation were not observed in MEG3 ectopically expressed Beas-2B cells compared with vector control cells in the study. It was further investigated that MEG3 induced protein translation by S6 phosphorylation via the Akt/p70S6K/S6 pathway, and they identified the upstream regulator of the pathway, PHLPP1 (PH Domain and Leucine Rich Repeat Protein Phosphatase 1), and c-Jun (a component of the transcription factor activator protein 1), a direct target of MEG3 and a negative regulator for PHLPP1. Together, nickel induces DNMT3b expression and hypermethylation of the MEG3 promoter, reduces MEG3 binding with the c-Jun transcription factor and increases its activity, which negatively regulates PHLPP1 by binding to its promoter, and in turn activates the Akt/p70S6K/S6 pathway, resulting in increased protein translation of HIF-1α.

2.3. Oxidative Stress

Oxidative stress was suggested to be another essential component in nickel-induced carcinogenesis [77]. Oxidative stress is known as the imbalance of oxidants, such as overproduction of reactive oxygen species (ROS) and nitrogen species (NOS), and antioxidants in favor of the oxidants [78,79]. Sustained high levels of ROS can cause oxidative damage to nucleic acids, lipids, and proteins, alter oxidative equilibrium, and regulate cell viability [80]. Oxidative damage to DNA is known to be one of the most important mechanisms in cancers [81].

The carcinogenic effect of nickel is theorized to be caused in part by its role in the generation of reactive oxidative species. Nickel is known to play a role in the generation of R· and RO· radicals in conjunction with glutathione [82]. Oxidative stress-related biomarkers can be observed in nickel electroplating workers exposed to nickel, such as increased lipid peroxidation, and reduced anti-oxidative enzymes as GSH (Glutathione), SOD (Superoxide Dismutase), and oxidative DNA damage marker 8-OH-dG [83]. In vitro, Beas-2B lung epithelial cells exposed to nickel nanoparticles, NiO nanoparticles, and $NiCl_2$ were shown to exhibit chromosomal damage, with NiO being the most potent in causing DNA strand breaks and increasing intracellular ROS [84]. Both water soluble $NiCl_2$ and insoluble Ni_3S_2 have been shown to enhance the formation of intracellular oxidants after 6 h of exposure, and free radicals have been observed in the nucleus when cells have been exposed to Ni_3S_2 for more than 18 h [85]. In addition, nickel and ROS exhibit synergistic inhibition towards both DNA polymerization and ligation and cause protein fragmentation, resulting in impaired DNA repair [86]. Nickel is also shown to induce transcriptional downregulation of homology-dependent DNA double-strand break repair (HDR) and mismatch repair (MMR) pathways [87]. MicroRNA regulation of oxidative stresses has been observed in several diseases including cancers, such as miRNA-34a-5p, miRNA-1915-3p, miRNA-638, and miRNA-150-3p in hepatocellular carcinoma (HCC) [88].

Hypoxic conditions favor the increase of reactive oxygen species and oxidative stress, and the hypoxia sensitive miRNA, miR-210, was found to respond to oxidative stress induced by H_2O_2 in H_9C_2 cardiomyocytes. It indirectly regulates the iron–sulfur cluster assembly protein (ISCU) and proved its activity in mitochondrial electron transport chains and energy metabolism, which are important mechanisms in the progression of ischemic heart disease [89]. Another hypoxia-related non-coding RNA, MEG3, was found to be downregulated in H_2O_2-treated RF/6A cells and played an important role in diabetes-induced microvascular abnormalities. Increased expression of miR-152 was observed in hypertrophic conditions and in H_2O_2-treated HL-1 cells, indicating its role in oxidative stress-related cardiac hypertrophy [90]. However, the exact role of these miRNAs in the nickel-induced oxidative stress pathways in lung carcinogenesis was not clarified and needs further investigation.

2.4. Tumor Suppressor Gene and Proto-Oncogene Alterations

Cell proliferation, differentiation, immortalization and transformation are key steps in cell carcinogenesis. Carcinogenic mechanisms induced by nickel will ultimately lead to activation of proto-oncogenes or inactivation of tumor suppressor genes. p53 is one of the most studied tumor suppressors in various cancers. Although nickel-induced mutations in the p53 gene were observed in several studies, indicating that nickel is a possible mutagen in mammalian cells [91–93], its regulation of p53 activity is not rare. This study proved that both water soluble and insoluble nickel were able to alter the expression of p53 and lead to cell transformation [61]. MEG3 has been described as a tumor suppressor [56], and found to exert its antitumor effect by modulating the activity of p53, Mdm2 (Mouse double minute 2 homolog), Rb (Retinoblastoma) and other cell cycle regulators [94]. MEG3 inhibited NSCLC cell proliferation and apoptosis by MEG3–p53 pathways [95], and it can be increased by chemotherapeutic agents such as paclitaxel (PTX), a first-line chemotherapy drug for NSCLC [96]. It was suggested that MEG3 can regulate p53 activity independent of p21 signaling [97,98], or via regulating the activity of MDM2, the main inhibitor of p53 [99–101].

In fact, lncRNA MEG3 also interferes with other miRNAs and functions as competing endogenous RNAs (ceRNAs) to bind and compete for miRNA target genes [102]. c-Myc is a known oncogene related to cell transformation, immortalization, differentiation, and cell apoptosis. MEG3 was found to compete with miR-27a for PHLPP2 to downregulate c-Myc and thereby reduced bladder cancer invasion [103]. MicroRNA-27a is an inhibitor for PHLPP2 [104], and it was reported to be upregulated in NSCLC and was possibly associated with lung carcinogenesis by targeting the TGF-β (Transforming Growth Factor beta) signaling pathway [105]. Reduction of MEG3 via nickel exposure is likely to attenuate its competing effect with miR-27a and thus suppress PHLPP2 and increase c-Myc and TGF-β signaling, inducing epithelial–mesenchymal transition (EMT) and transformation in lung carcinogenesis.

Other than MEG3, dysregulations of many lncRNAs in lung cancer have been identified and summarized by Khandelwal et al. [32,106], of which some have been shown to be associated with occupational PAH (Polycyclic Aromatic Hydrocarbon) exposures in coke oven workers, such as HOTAIR, TUG1, MALAT1 and GAS5 [107]. Cadmium was also found to induce the expression of lncRNA, ENST00000414355, to modulate DNA damage and repair in 16HBE cells. Other than being an upstream regulator of MEG3 and many other lncRNAs in carcinogenic signaling pathways, they can possibly be downstream of other genes; for example, TUG1 was found to be upregulated by p53 in lung cancer [108], indicating that TUG1 could be a potential lncRNA involved in the MEG3–p53 signaling pathway in nickel-induced lung carcinogenesis.

The target genes for miR-210, MNT, a MYC antagonist [109], and E2F3, a key regulator of the cell cycle [110], were of importance to study in the mechanisms of carcinogenesis. Another study revealed that miR-210 was upregulated in NSCLC human tissue samples, and six potential target genes of miR-210 were identified (IL-6, GNG11, CXCL12, ADRB2, ADCY9, and CHRM2), suggesting its role in NSCLC signaling pathways [111]. The diagnostic value of miR-210 in lung cancer was also indicated in a meta-analysis study by Yang et al. [112].

MicroRNA-203 was hypermethylated following nickel exposure both in vivo and in vitro, and inhibited miR-203 led to dysregulation of its target gene ABL1, contributing to nickel-induced cancers [60]. ABL1 is a known oncogene and plays an important role in cell differentiation, division, adhesion, and stress responses. It was proved to be the miRNA-203 direct target gene along with several other tumor proto-oncogenes, such as ABCE1, E2F3 and p63 [113–117]. An upregulation of miR-222 was observed in both NSTCs and Ni-induced tumor tissues of rats, and it was suggested that miR-222 could be involved in nickel-induced tumorigenesis by targeting tumor suppressor genes CDKN1C (p57) and CDKN1B (p27) [118]. Dysregulation of p27 via miR-222 was reported to be related to increased cancer cell proliferation [119]. Besides lung cancer, CDKN1C and CDKN1B were also found to be regulated by miR-222 in human hepatocellular carcinoma [120]. $NiCl_2$ also induced expression of miR-4417 in Beas-2B cells and A549 cells, which targeted the TAB2 gene and was involved in the mechanism of nickel-induced EMT, lung fibrosis and tumor progression. Ectopic expression of miR-4417 in Beas-2B and A549 cells induced fibronectin, while nickel exposure failed to induce fibronectin in miR-4417 inhibited Beas-2B cells. Although TGF-β was involved in nickel-induced cell EMT and significantly induced miR-4417 in Beas-2B cells, its inhibitor SB525334 only partially reduced miR-4417 level and failed to restore TAB2 level in TGF-β-treated cells, indicating that nickel can induce miR-4417 through pathways independent of TGF-β [121].

3. Conclusions

Nickel is a naturally occurring element that can be found in soils, sediments, water, and air. Humans can be exposed to nickel from different sources and occupational exposure to nickel in nickel refineries and processing plants has been a big concern for workers and was reported to be related to lung cancer. The mechanisms of nickel-induced carcinogenesis that have been studied include hypoxia-inducible factor pathways and the introduction of oxidative stress which generates DNA damage and impairs DNA repair pathways. Nickel also disrupts the integrity of heterochromatin, and subsequently silences tumor suppressor genes, such as p53, and activates c-Myc, resulting in dysregulated cell differentiation, proliferation, transformation, and cancer progression.

A newly found role played by non-coding RNAs has recently been studied in nickel-induced lung carcinogenesis. With increased attention in non-coding RNAs and the development of new approaches for understanding non-coding RNAs, various new mechanisms resulting in cancer initiation and progression have been identified. Several studies have reported that nickel was able to alter the expression of miRNAs, such as miR-210, miR-203, miR-152, miR-4417, and miR-222, and lncRNAs, such as NRG1 and MEG3. These studies suggest important roles for non-coding RNAs in nickel-induced lung carcinogenesis, providing insight into the pathogenesis and mechanisms of the disease.

However, only a small amount of non-coding RNAs related to nickel exposure have been characterized so far and their functions and mechanisms in carcinogenesis remain to be elucidated. The mechanisms by which ncRNAs were dysregulated upon nickel exposure were mainly through the increased expression of DNMTs followed by hypermethylation at the promoter region of the ncRNAs, resulting in altered expression levels of their target genes which are involved in cell cycle regulation, proliferation, apoptosis, EMT, metastasis and many aspects of tumorigenesis. Instead of being upstream regulators, some other ncRNAs were found to be regulated by HIF, TGF-β and p53 during the signaling pathways induced by nickel. Figure 1 provides a schematic diagram of the network that ncRNAs are involved in in nickel-induced carcinogenesis. Future studies are warranted to elucidate the functions of these ncRNAs as potential therapeutic targets, diagnostic markers and prognosis predictions for lung cancer induced by nickel.

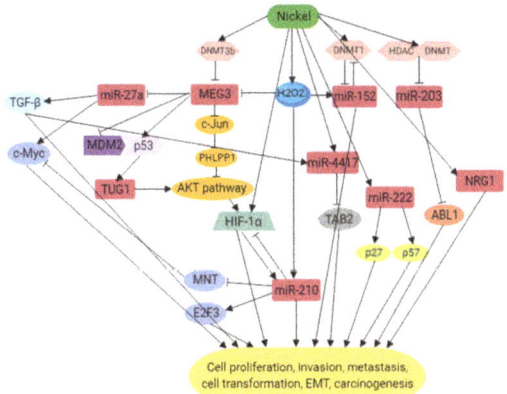

Figure 1. Regulatory networks of non-coding RNAs (ncRNAs) involved in nickel-induced carcinogenesis.

Funding: This research was funded by National Institutes Health (NIH) grants number ES000260, ES022935, ES023174, ES026138.

Conflicts of Interest: The authors declare no conflict of interest.

References

1. Schmidt, M.; Goebeler, M. Nickel allergies: Paying the Toll for innate immunity. *J. Mol. Med.* **2011**, *89*, 961–970. [CrossRef] [PubMed]
2. Cempel, M.; Nikel, G. Nickel: A Review of Its Sources and Environmental Toxicology. *Pol. J. Environ. Stud.* **2006**, *15*, 375–382.
3. Zambelli, B.; Uversky, V.N.; Ciurli, S. Nickel impact on human health: An intrinsic disorder perspective. *Biochim. Biophys. Acta* **2016**, *1864*, 1714–1731. [CrossRef] [PubMed]
4. Coogan, T.P.; Latta, D.M.; Snow, E.T.; Costa, M. Toxicity and carcinogenicity of nickel compounds. *Crit. Rev. Toxicol.* **1989**, *19*, 341–384. [CrossRef] [PubMed]
5. Barceloux, D.G. Nickel. *J. Toxicol. Clin. Toxicol.* **1999**, *37*, 239–258. [CrossRef] [PubMed]
6. Rezuke, W.N.; Knight, J.A.; Sunderman, F.W., Jr. Reference values for nickel concentrations in human tissues and bile. *Am. J. Ind. Med.* **1987**, *11*, 419–426. [CrossRef] [PubMed]
7. Duda-Chodak, A.D.A.; Błaszczyk, U. The impact of nickel on human health. *J. Elementol.* **2008**, *13*, 685–696.
8. Morgan, L.G.; Usher, V. Health problems associated with nickel refining and use. *Ann. Occup. Hyg.* **1994**, *38*, 189–198.
9. Berge, S.R.; Skyberg, K. Radiographic evidence of pulmonary fibrosis and possible etiologic factors at a nickel refinery in Norway. *J. Environ. Monit.* **2003**, *5*, 681–688. [CrossRef]
10. Oller, A.R.; Costa, M.; Oberdorster, G. Carcinogenicity assessment of selected nickel compounds. *Toxicol. Appl. Pharmacol.* **1997**, *143*, 152–166. [CrossRef]
11. Doll, R. *Report of the International Committee on Nickel Carcinogenesis in Man*; Final Report (No. PB-91-109801/XAB); Program Resources, Inc.: Research Triangle Park, NC, USA, 1990; p. 84.
12. Huvinen, M.; Pukkala, E. Cancer incidence among Finnish ferrochromium and stainless steel production workers in 1967–2011: A cohort study. *BMJ Open* **2013**, *3*, e003819. [CrossRef] [PubMed]
13. Grimsrud, T.K.; Andersen, A. Evidence of carcinogenicity in humans of water-soluble nickel salts. *J. Occup. Med. Toxicol.* **2010**, *5*, 7. [CrossRef] [PubMed]
14. Andersen, A.; Berge, S.R.; Engeland, A.; Norseth, T. Exposure to nickel compounds and smoking in relation to incidence of lung and nasal cancer among nickel refinery workers. *Occup. Environ. Med.* **1996**, *53*, 708–713. [CrossRef] [PubMed]
15. Seilkop, S.K.; Oller, A.R. Respiratory cancer risks associated with low-level nickel exposure: An integrated assessment based on animal, epidemiological, and mechanistic data. *Regul. Toxicol. Pharmacol.* **2003**, *37*, 173–190. [CrossRef]
16. Moulin, J.J.; Clavel, T.; Roy, D.; Dananche, B.; Marquis, N.; Fevotte, J.; Fontana, J.M. Risk of lung cancer in workers producing stainless steel and metallic alloys. *Int. Arch. Occup. Environ. Health* **2000**, *73*, 171–180. [CrossRef]
17. Anttila, A.; Pukkala, E.; Aitio, A.; Rantanen, T.; Karjalainen, S. Update of cancer incidence among workers at a copper/nickel smelter and nickel refinery. *Int. Arch. Occup. Environ. Health* **1998**, *71*, 245–250. [CrossRef]
18. Carninci, P.; Kasukawa, T.; Katayama, S.; Gough, J.; Frith, M.C.; Maeda, N.; Oyama, R.; Ravasi, T.; Lenhard, B.; Wells, C.; et al. The transcriptional landscape of the mammalian genome. *Science* **2005**, *309*, 1559–1563.
19. Zhao, Y.; Srivastava, D. A developmental view of microRNA function. *Trends Biochem. Sci.* **2007**, *32*, 189–197. [CrossRef]
20. Di Leva, G.; Garofalo, M.; Croce, C.M. MicroRNAs in cancer. *Annu. Rev. Pathol.* **2014**, *9*, 287–314. [CrossRef]
21. Uddin, A.; Chakraborty, S. Role of miRNAs in lung cancer. *J. Cell. Physiol.* **2018**. [CrossRef]
22. Ofir, M.; Hacohen, D.; Ginsberg, D. MiR-15 and miR-16 are direct transcriptional targets of E2F1 that limit E2F-induced proliferation by targeting cyclin E. *Mol. Cancer Res.* **2011**, *9*, 440–447. [CrossRef]
23. Hatley, M.E.; Patrick, D.M.; Garcia, M.R.; Richardson, J.A.; Bassel-Duby, R.; van Rooij, E.; Olson, E.N. Modulation of K-Ras-dependent lung tumorigenesis by MicroRNA-21. *Cancer Cell* **2010**, *18*, 282–293. [CrossRef] [PubMed]
24. Liu, X.; Sempere, L.F.; Ouyang, H.; Memoli, V.A.; Andrew, A.S.; Luo, Y.; Demidenko, E.; Korc, M.; Shi, W.; Preis, M.; et al. MicroRNA-31 functions as an oncogenic microRNA in mouse and human lung cancer cells by repressing specific tumor suppressors. *J. Clin. Investig.* **2010**, *120*, 1298–1309. [CrossRef] [PubMed]
25. Luo, L.; Zhang, T.; Liu, H.; Lv, T.; Yuan, D.; Yao, Y.; Lv, Y.; Song, Y. MiR-101 and Mcl-1 in non-small-cell lung cancer: Expression profile and clinical significance. *Med. Oncol.* **2012**, *29*, 1681–1686. [CrossRef]

26. Wapinski, O.; Chang, H.Y. Long noncoding RNAs and human disease. *Trends Cell Biol.* **2011**, *21*, 354–361. [CrossRef] [PubMed]
27. Poliseno, L.; Salmena, L.; Zhang, J.; Carver, B.; Haveman, W.J.; Pandolfi, P.P. A coding-independent function of gene and pseudogene mRNAs regulates tumour biology. *Nature* **2010**, *465*, 1033–1038. [CrossRef] [PubMed]
28. Ling, H.; Fabbri, M.; Calin, G.A. MicroRNAs and other non-coding RNAs as targets for anticancer drug development. *Nat. Rev. Drug Discov.* **2013**, *12*, 847–865. [CrossRef]
29. Sullenger, B.A.; Nair, S. From the RNA world to the clinic. *Science* **2016**, *352*, 1417–1420. [CrossRef]
30. Schmitt, A.M.; Chang, H.Y. Long Noncoding RNAs in Cancer Pathways. *Cancer Cell* **2016**, *29*, 452–463. [CrossRef]
31. Schmitz, S.U.; Grote, P.; Herrmann, B.G. Mechanisms of long noncoding RNA function in development and disease. *Cell. Mol. Life Sci.* **2016**, *73*, 2491–2509. [CrossRef]
32. Wei, M.M.; Zhou, G.B. Long Non-coding RNAs and Their Roles in Non-small-cell Lung Cancer. *Genom. Proteom. Bioinform.* **2016**, *14*, 280–288. [CrossRef] [PubMed]
33. Shen, L.; Chen, L.; Wang, Y.; Jiang, X.; Xia, H.; Zhuang, Z. Long noncoding RNA MALAT1 promotes brain metastasis by inducing epithelial-mesenchymal transition in lung cancer. *J. Neurooncol.* **2015**, *121*, 101–108. [CrossRef] [PubMed]
34. Loewen, G.; Jayawickramarajah, J.; Zhuo, Y.; Shan, B. Functions of lncRNA HOTAIR in lung cancer. *J. Hematol. Oncol.* **2014**, *7*, 90. [CrossRef] [PubMed]
35. Zhou, C.; Huang, C.; Wang, J.; Huang, H.; Li, J.; Xie, Q.; Liu, Y.; Zhu, J.; Li, Y.; Zhang, D.; et al. LncRNA MEG3 downregulation mediated by DNMT3b contributes to nickel malignant transformation of human bronchial epithelial cells via modulating PHLPP1 transcription and HIF-1α translation. *Oncogene* **2017**, *36*, 3878–3889. [CrossRef] [PubMed]
36. Zhang, J.; Zhou, Y.; Wu, Y.; Ma, L.; Fan, Y.; Kang, X.; Shi, H.; Zhang, J. Isolation and characterization of a novel noncoding RNA from nickel-induced lung cancer. *Biol. Trace. Elem. Res.* **2012**, *150*, 258–263. [CrossRef] [PubMed]
37. Cameron, K.S.; Buchner, V.; Tchounwou, P.B. Exploring the molecular mechanisms of nickel-induced genotoxicity and carcinogenicity: A literature review. *Rev. Environ. Health* **2011**, *26*, 81–92. [CrossRef] [PubMed]
38. Sakai, T.; Toguchida, J.; Ohtani, N.; Yandell, D.W.; Rapaport, J.M.; Dryja, T.P. Allele-specific hypermethylation of the retinoblastoma tumor-suppressor gene. *Am. J. Hum. Genet.* **1991**, *48*, 880–888. [PubMed]
39. Lee, Y.W.; Klein, C.B.; Kargacin, B.; Salnikow, K.; Kitahara, J.; Dowjat, K.; Zhitkovich, A.; Christie, N.T.; Costa, M. Carcinogenic nickel silences gene expression by chromatin condensation and DNA methylation: A new model for epigenetic carcinogens. *Mol. Cell. Biol.* **1995**, *15*, 2547–2557. [CrossRef] [PubMed]
40. Ohtani-Fujita, N.; Fujita, T.; Aoike, A.; Osifchin, N.E.; Robbins, P.D.; Sakai, T. CpG methylation inactivates the promoter activity of the human retinoblastoma tumor-suppressor gene. *Oncogene* **1993**, *8*, 1063–1067. [PubMed]
41. Sutcliffe, J.S.; Nelson, D.L.; Zhang, F.; Pieretti, M.; Caskey, C.T.; Saxe, D.; Warren, S.T. DNA methylation represses FMR-1 transcription in fragile X syndrome. *Hum. Mol. Genet.* **1992**, *1*, 397–400. [CrossRef] [PubMed]
42. Greger, V.; Debus, N.; Lohmann, D.; Hopping, W.; Passarge, E.; Horsthemke, B. Frequency and parental origin of hypermethylated RB1 alleles in retinoblastoma. *Hum. Genet.* **1994**, *94*, 491–496. [CrossRef] [PubMed]
43. Hansen, R.S.; Gartler, S.M.; Scott, C.R.; Chen, S.H.; Laird, C.D. Methylation analysis of CGG sites in the CpG island of the human FMR1 gene. *Hum. Mol. Genet.* **1992**, *1*, 571–578. [CrossRef] [PubMed]
44. Yasaei, H.; Gilham, E.; Pickles, J.C.; Roberts, T.P.; O'Donovan, M.; Newbold, R.F. Carcinogen-specific mutational and epigenetic alterations in INK4A, INK4B and p53 tumour-suppressor genes drive induced senescence bypass in normal diploid mammalian cells. *Oncogene* **2013**, *32*, 171–179. [CrossRef] [PubMed]
45. Lee, Y.W.; Broday, L.; Costa, M. Effects of nickel on DNA methyltransferase activity and genomic DNA methylation levels. *Mutagen. Res.* **1998**, *415*, 213–218. [CrossRef]
46. Zingg, J.M.; Jones, P.A. Genetic and epigenetic aspects of DNA methylation on genome expression, evolution, mutation and carcinogenesis. *Carcinogenesis* **1997**, *18*, 869–882. [CrossRef] [PubMed]
47. Wu, C.H.; Tang, S.C.; Wang, P.H.; Lee, H.; Ko, J.L. Nickel-induced epithelial-mesenchymal transition by reactive oxygen species generation and E-cadherin promoter hypermethylation. *J. Biol. Chem.* **2012**, *287*, 25292–25302. [CrossRef] [PubMed]

48. Pecina-Slaus, N. Tumor suppressor gene E-cadherin and its role in normal and malignant cells. *Cancer Cell Int.* **2003**, *3*, 17. [CrossRef] [PubMed]
49. Sato, F.; Ono, T.; Kawahara, A.; Kawaguchi, T.; Tanaka, H.; Shimamatsu, K.; Kakuma, T.; Akiba, J.; Umeno, H.; Yano, H. Prognostic impact of p16 and PD-L1 expression in patients with oropharyngeal squamous cell carcinoma receiving a definitive treatment. *J. Clin. Pathol.* **2019**. [CrossRef] [PubMed]
50. Chen, Q.Y.; DesMarais, T.; Costa, M. Metals and Mechanisms of Carcinogenesis. *Annu. Rev. Pharmacol. Toxicol.* **2019**, *59*, 537–554. [CrossRef]
51. Lee, Y.W.; Pons, C.; Tummolo, D.M.; Klein, C.B.; Rossman, T.G.; Christie, N.T. Mutagenicity of soluble and insoluble nickel compounds at the gpt locus in G12 Chinese hamster cells. *Environ. Mol. Mutagen.* **1993**, *21*, 365–371. [CrossRef] [PubMed]
52. Ellen, T.P.; Kluz, T.; Harder, M.E.; Xiong, J.; Costa, M. Heterochromatinization as a potential mechanism of nickel-induced carcinogenesis. *Biochemistry* **2009**, *48*, 4626–4632. [CrossRef] [PubMed]
53. Carone, D.M.; Lawrence, J.B. Heterochromatin instability in cancer: From the Barr body to satellites and the nuclear periphery. *Semin. Cancer Biol.* **2013**, *23*, 99–108. [CrossRef] [PubMed]
54. Morgan, M.A.; Shilatifard, A. Chromatin signatures of cancer. *Genes Dev.* **2015**, *29*, 238–249. [CrossRef] [PubMed]
55. Miyoshi, N.; Wagatsuma, H.; Wakana, S.; Shiroishi, T.; Nomura, M.; Aisaka, K.; Kohda, T.; Surani, M.A.; Kaneko-Ishino, T.; Ishino, F. Identification of an imprinted gene, Meg3/Gtl2 and its human homologue MEG3, first mapped on mouse distal chromosome 12 and human chromosome 14q. *Genes Cells* **2000**, *5*, 211–220. [CrossRef] [PubMed]
56. Zhou, Y.; Zhang, X.; Klibanski, A. MEG3 noncoding RNA: A tumor suppressor. *J. Mol. Endocrinol.* **2012**, *48*, R45–R53. [CrossRef] [PubMed]
57. Benetatos, L.; Vartholomatos, G.; Hatzimichael, E. MEG3 imprinted gene contribution in tumorigenesis. *Int. J. Cancer* **2011**, *129*, 773–779. [CrossRef] [PubMed]
58. Ji, W.; Yang, L.; Yu, L.; Yuan, J.; Hu, D.; Zhang, W.; Yang, J.; Pang, Y.; Li, W.; Lu, J.; et al. Epigenetic silencing of O6-methylguanine DNA methyltransferase gene in NiS-transformed cells. *Carcinogenesis* **2008**, *29*, 1267–1275. [CrossRef] [PubMed]
59. Ji, W.; Yang, L.; Yuan, J.; Yang, L.; Zhang, M.; Qi, D.; Duan, X.; Xuan, A.; Zhang, W.; Lu, J.; et al. MicroRNA-152 targets DNA methyltransferase 1 in NiS-transformed cells via a feedback mechanism. *Carcinogenesis* **2013**, *34*, 446–453. [CrossRef]
60. Zhang, J.; Zhou, Y.; Wu, Y.J.; Li, M.J.; Wang, R.J.; Huang, S.Q.; Gao, R.R.; Ma, L.; Shi, H.J.; Zhang, J. Hyper-methylated miR-203 dysregulates ABL1 and contributes to the nickel-induced tumorigenesis. *Toxicol. Lett.* **2013**, *223*, 42–51. [CrossRef]
61. Salnikow, K.; Davidson, T.; Zhang, Q.; Chen, L.C.; Su, W.; Costa, M. The involvement of hypoxia-inducible transcription factor-1-dependent pathway in nickel carcinogenesis. *Cancer Res* **2003**, *63*, 3524–3530.
62. Wang, G.L.; Jiang, B.H.; Rue, E.A.; Semenza, G.L. Hypoxia-inducible factor 1 is a basic-helix-loop-helix-PAS heterodimer regulated by cellular O_2 tension. *Proc. Natl. Acad. Sci. USA* **1995**, *92*, 5510–5514. [CrossRef] [PubMed]
63. Semenza, G.L. Hypoxia-inducible factor 1: Master regulator of O_2 homeostasis. *Curr. Opin. Genet. Dev.* **1998**, *8*, 588–594. [CrossRef]
64. Costa, M.; Davidson, T.L.; Chen, H.; Ke, Q.; Zhang, P.; Yan, Y.; Huang, C.; Kluz, T. Nickel carcinogenesis: Epigenetics and hypoxia signaling. *Mutagen. Res.* **2005**, *592*, 79–88. [CrossRef] [PubMed]
65. Lee, J.W.; Bae, S.H.; Jeong, J.W.; Kim, S.H.; Kim, K.W. Hypoxia-inducible factor (HIF-1)α: Its protein stability and biological functions. *Exp. Mol. Med.* **2004**, *36*, 1–12. [CrossRef] [PubMed]
66. Ke, Q.; Costa, M. Hypoxia-inducible factor-1 (HIF-1). *Mol. Pharmacol.* **2006**, *70*, 1469–1480. [CrossRef] [PubMed]
67. Li, Q.; Chen, H.; Huang, X.; Costa, M. Effects of 12 metal ions on iron regulatory protein 1 (IRP-1) and hypoxia-inducible factor-1 alpha (HIF-1α) and HIF-regulated genes. *Toxicol. Appl. Pharmacol.* **2006**, *213*, 245–255. [CrossRef] [PubMed]
68. Davidson, T.L.; Chen, H.; Di Toro, D.M.; D'Angelo, G.; Costa, M. Soluble nickel inhibits HIF-prolyl-hydroxylases creating persistent hypoxic signaling in A549 cells. *Mol. Carcinog.* **2006**, *45*, 479–489. [CrossRef]

69. Davidson, T.; Chen, H.; Garrick, M.D.; D'Angelo, G.; Costa, M. Soluble nickel interferes with cellular iron homeostasis. *Mol. Cell. Biochem.* **2005**, *279*, 157–162. [CrossRef]
70. Salnikow, K.; Blagosklonny, M.V.; Ryan, H.; Johnson, R.; Costa, M. Carcinogenic Nickel Induces Genes Involved with Hypoxic Stress. *Cancer Res.* **2000**, *60*, 38–41.
71. Rezvani, H.R.; Mahfouf, W.; Ali, N.; Chemin, C.; Ged, C.; Kim, A.L.; de Verneuil, H.; Taieb, A.; Bickers, D.R.; Mazurier, F. Hypoxia-inducible factor-1alpha regulates the expression of nucleotide excision repair proteins in keratinocytes. *Nucleic Acids Res.* **2010**, *38*, 797–809. [CrossRef]
72. Chan, N.; Ali, M.; McCallum, G.P.; Kumareswaran, R.; Koritzinsky, M.; Wouters, B.G.; Wells, P.G.; Gallinger, S.; Bristow, R.G. Hypoxia provokes base excision repair changes and a repair-deficient, mutator phenotype in colorectal cancer cells. *Mol. Cancer Res.* **2014**, *12*, 1407–1415. [CrossRef] [PubMed]
73. Kulshreshtha, R.; Ferracin, M.; Negrini, M.; Calin, G.A.; Davuluri, R.V.; Ivan, M. Regulation of microRNA expression: The hypoxic component. *Cell Cycle* **2007**, *6*, 1426–1431. [CrossRef] [PubMed]
74. He, M.; Lu, Y.; Xu, S.; Mao, L.; Zhang, L.; Duan, W.; Liu, C.; Pi, H.; Zhang, Y.; Zhong, M.; et al. MiRNA-210 modulates a nickel-induced cellular energy metabolism shift by repressing the iron-sulfur cluster assembly proteins ISCU1/2 in Neuro-2a cells. *Cell Death Dis.* **2014**, *5*, e1090. [CrossRef] [PubMed]
75. Kulshreshtha, R.; Ferracin, M.; Wojcik, S.E.; Garzon, R.; Alder, H.; Agosto-Perez, F.J.; Davuluri, R.; Liu, C.G.; Croce, C.M.; Negrini, M.; et al. A microRNA signature of hypoxia. *Mol. Cell. Biol.* **2007**, *27*, 1859–1867. [CrossRef] [PubMed]
76. Wang, H.; Flach, H.; Onizawa, M.; Wei, L.; McManus, M.T.; Weiss, A. Negative regulation of Hif1a expression and TH17 differentiation by the hypoxia-regulated microRNA miR-210. *Nat. Immunol.* **2014**, *15*, 393–401. [CrossRef] [PubMed]
77. Valko, M.; Rhodes, C.J.; Moncol, J.; Izakovic, M.; Mazur, M. Free radicals, metals and antioxidants in oxidative stress-induced cancer. *Chem. Biol. Interact.* **2006**, *160*, 1–40. [CrossRef] [PubMed]
78. Pero, R.W.; Roush, G.C.; Markowitz, M.M.; Miller, D.G. Oxidative stress, DNA repair, and cancer susceptibility. *Cancer Detect. Prev.* **1990**, *14*, 555–561.
79. Sies, H. Oxidative stress: Oxidants and antioxidants. *Exp. Physiol.* **1997**, *82*, 291–295. [CrossRef]
80. Halliwell, B. Reactive oxygen species in living systems: Source, biochemistry, and role in human disease. *Am. J. Med.* **1991**, *91*, 14s–22s. [CrossRef]
81. Lee, J.D.; Cai, Q.; Shu, X.O.; Nechuta, S.J. The Role of Biomarkers of Oxidative Stress in Breast Cancer Risk and Prognosis: A Systematic Review of the Epidemiologic Literature. *J. Womens Health* **2017**, *26*, 467–482. [CrossRef]
82. Shi, X.; Dalal, N.S.; Kasprzak, K.S. Generation of free radicals from lipid hydroperoxides by Ni^{2+} in the presence of oligopeptides. *Arch. Biochem. Biophys.* **1992**, *299*, 154–162. [CrossRef]
83. Tsao, Y.C.; Gu, P.W.; Liu, S.H.; Tzeng, I.S.; Chen, J.Y.; Luo, J.J. Nickel exposure and plasma levels of biomarkers for assessing oxidative stress in nickel electroplating workers. *Biomarkers* **2017**, *22*, 455–460. [CrossRef] [PubMed]
84. Di Bucchianico, S.; Gliga, A.R.; Åkerlund, E.; Skoglund, S.; Wallinder, I.O.; Fadeel, B.; Karlsson, H.L. Calcium-dependent cyto- and genotoxicity of nickel metal and nickel oxide nanoparticles in human lung cells. *Part. Fibre Toxicol.* **2018**, *15*, 32. [CrossRef] [PubMed]
85. Huang, X.; Klein, C.B.; Costa, M. Crystalline Ni_3S_2 specifically enhances the formation of oxidants in the nuclei of CHO cells as detected by dichlorofluorescein. *Carcinogenesis* **1994**, *15*, 545–548. [CrossRef] [PubMed]
86. Lynn, S.; Yew, F.H.; Chen, K.S.; Jan, K.Y. Reactive oxygen species are involved in nickel inhibition of DNA repair. *Environ. Mol. Mutagen.* **1997**, *29*, 208–216. [CrossRef]
87. Scanlon, S.E.; Scanlon, C.D.; Hegan, D.C.; Sulkowski, P.L.; Glazer, P.M. Nickel induces transcriptional down-regulation of DNA repair pathways in tumorigenic and non-tumorigenic lung cells. *Carcinogenesis* **2017**, *38*, 627–637. [CrossRef]
88. Wan, Y.; Cui, R.; Gu, J.; Zhang, X.; Xiang, X.; Liu, C.; Qu, K.; Lin, T. Identification of Four Oxidative Stress-Responsive MicroRNAs, miR-34a-5p, miR-1915-3p, miR-638, and miR-150-3p, in Hepatocellular Carcinoma. *Oxid. Med. Cell. Longev.* **2017**, *2017*, 5189138. [CrossRef]
89. Sun, W.; Zhao, L.; Song, X.; Zhang, J.; Xing, Y.; Liu, N.; Yan, Y.; Li, Z.; Lu, Y.; Wu, J.; et al. MicroRNA-210 Modulates the Cellular Energy Metabolism Shift During H_2O_2-Induced Oxidative Stress by Repressing ISCU in H9c2 Cardiomyocytes. *Cell. Physiol. Biochem.* **2017**, *43*, 383–394. [CrossRef]

90. Ali, T.; Mushtaq, I.; Maryam, S.; Farhan, A.; Saba, K.; Jan, M.I.; Sultan, A.; Anees, M.; Duygu, B.; Hamera, S.; et al. Interplay of N acetyl cysteine and melatonin in regulating oxidative stress-induced cardiac hypertrophic factors and microRNAs. *Arch. Biochem. Biophys.* **2019**, *661*, 56–65. [CrossRef]
91. Clemens, F.; Verma, R.; Ramnath, J.; Landolph, J.R. Amplification of the Ect2 proto-oncogene and over-expression of Ect2 mRNA and protein in nickel compound and methylcholanthrene-transformed 10T1/2 mouse fibroblast cell lines. *Toxicol. Appl. Pharmacol.* **2005**, *206*, 138–149. [CrossRef]
92. Chiocca, S.M.; Sterner, D.A.; Biggart, N.W.; Murphy, E.C., Jr. Nickel mutagenesis: Alteration of the MuSVts110 thermosensitive splicing phenotype by a nickel-induced duplication of the 3′ splice site. *Mol. Carcinog.* **1991**, *4*, 61–71. [CrossRef] [PubMed]
93. Maehle, L.; Metcalf, R.A.; Ryberg, D.; Bennett, W.P.; Harris, C.C.; Haugen, A. Altered p53 gene structure and expression in human epithelial cells after exposure to nickel. *Cancer Res.* **1992**, *52*, 218–221. [PubMed]
94. Shi, Y.; Lv, C.; Shi, L.; Tu, G. MEG3 inhibits proliferation and invasion and promotes apoptosis of human osteosarcoma cells. *Oncol. Lett.* **2018**, *15*, 1917–1923. [CrossRef] [PubMed]
95. Lu, K.H.; Li, W.; Liu, X.H.; Sun, M.; Zhang, M.L.; Wu, W.Q.; Xie, W.P.; Hou, Y.Y. Long non-coding RNA MEG3 inhibits NSCLC cells proliferation and induces apoptosis by affecting p53 expression. *BMC Cancer* **2013**, *13*, 461. [CrossRef] [PubMed]
96. Xu, J.; Su, C.; Zhao, F.; Tao, J.; Hu, D.; Shi, A.; Pan, J.; Zhang, Y. Paclitaxel promotes lung cancer cell apoptosis via MEG3-P53 pathway activation. *Biochem. Biophys. Res. Commun.* **2018**, *504*, 123–128. [CrossRef] [PubMed]
97. Zhang, X.; Zhou, Y.; Mehta, K.R.; Danila, D.C.; Scolavino, S.; Johnson, S.R.; Klibanski, A. A pituitary-derived MEG3 isoform functions as a growth suppressor in tumor cells. *J. Clin. Endocrinol. Metab.* **2003**, *88*, 5119–5126. [CrossRef] [PubMed]
98. Zhou, Y.; Zhong, Y.; Wang, Y.; Zhang, X.; Batista, D.L.; Gejman, R.; Ansell, P.J.; Zhao, J.; Weng, C.; Klibanski, A. Activation of p53 by MEG3 non-coding RNA. *J. Biol. Chem.* **2007**, *282*, 24731–24742. [CrossRef] [PubMed]
99. Liu, L.X.; Deng, W.; Zhou, X.T.; Chen, R.P.; Xiang, M.Q.; Guo, Y.T.; Pu, Z.J.; Li, R.; Wang, G.F.; Wu, L.F. The mechanism of adenosine-mediated activation of lncRNA MEG3 and its antitumor effects in human hepatoma cells. *Int. J. Oncol.* **2016**, *48*, 421–429. [CrossRef]
100. Lv, D.; Sun, R.; Yu, Q.; Zhang, X. The long non-coding RNA maternally expressed gene 3 activates p53 and is downregulated in esophageal squamous cell cancer. *Tumour. Biol.* **2016**, *37*, 16259–16267. [CrossRef]
101. Liu, J.; Wan, L.; Lu, K.; Sun, M.; Pan, X.; Zhang, P.; Lu, B.; Liu, G.; Wang, Z. The Long Noncoding RNA MEG3 Contributes to Cisplatin Resistance of Human Lung Adenocarcinoma. *PLoS ONE* **2015**, *10*, e0114586. [CrossRef]
102. Tay, Y.; Rinn, J.; Pandolfi, P.P. The multilayered complexity of ceRNA crosstalk and competition. *Nature* **2014**, *505*, 344–352. [CrossRef] [PubMed]
103. Huang, C.; Liao, X.; Jin, H.; Xie, F.; Zheng, F.; Li, J.; Zhou, C.; Jiang, G.; Wu, X.R.; Huang, C. MEG3, as a Competing Endogenous RNA, Binds with miR-27a to Promote PHLPP2 Protein Translation and Impairs Bladder Cancer Invasion. *Mol. Ther. Nucleic Acids* **2019**, *16*, 51–62. [CrossRef] [PubMed]
104. Ding, L.; Zhang, S.; Xu, M.; Zhang, R.; Sui, P.; Yang, Q. MicroRNA-27a contributes to the malignant behavior of gastric cancer cells by directly targeting PH domain and leucine-rich repeat protein phosphatase 2. *J. Exp. Clin. Cancer Res.* **2017**, *36*, 45. [CrossRef] [PubMed]
105. Chae, D.K.; Ban, E.; Yoo, Y.S.; Kim, E.E.; Baik, J.H.; Song, E.J. MIR-27a regulates the TGF-β signaling pathway by targeting SMAD2 and SMAD4 in lung cancer. *Mol. Carcinog.* **2017**, *56*, 1992–1998. [CrossRef]
106. Khandelwal, A.; Bacolla, A.; Vasquez, K.M.; Jain, A. Long non-coding RNA: A new paradigm for lung cancer. *Mol. Carcinog.* **2015**, *54*, 1235–1251. [CrossRef]
107. Gao, C.; He, Z.; Li, J.; Li, X.; Bai, Q.; Zhang, Z.; Zhang, X.; Wang, S.; Xiao, X.; Wang, F.; et al. Specific long non-coding RNAs response to occupational PAHs exposure in coke oven workers. *Toxicol. Rep.* **2016**, *3*, 160–166. [CrossRef]
108. Zhang, E.B.; Yin, D.D.; Sun, M.; Kong, R.; Liu, X.H.; You, L.H.; Han, L.; Xia, R.; Wang, K.M.; Yang, J.S.; et al. P53-regulated long non-coding RNA TUG1 affects cell proliferation in human non-small cell lung cancer, partly through epigenetically regulating HOXB7 expression. *Cell Death Dis.* **2014**, *5*, e1243. [CrossRef]
109. Zhang, Z.; Sun, H.; Dai, H.; Walsh, R.M.; Imakura, M.; Schelter, J.; Burchard, J.; Dai, X.; Chang, A.N.; Diaz, R.L.; et al. MicroRNA miR-210 modulates cellular response to hypoxia through the MYC antagonist MNT. *Cell Cycle* **2009**, *8*, 2756–2768. [CrossRef]

110. Giannakakis, A.; Sandaltzopoulos, R.; Greshock, J.; Liang, S.; Huang, J.; Hasegawa, K.; Li, C.; O'Brien-Jenkins, A.; Katsaros, D.; Weber, B.L.; et al. miR-210 links hypoxia with cell cycle regulation and is deleted in human epithelial ovarian cancer. *Cancer Biol. Ther.* **2008**, *7*, 255–264. [CrossRef]
111. He, R.Q.; Cen, W.L.; Cen, J.M.; Cen, W.N.; Li, J.Y.; Li, M.W.; Gan, T.Q.; Hu, X.H.; Chen, G. Clinical Significance of miR-210 and its Prospective Signaling Pathways in Non-Small Cell Lung Cancer: Evidence from Gene Expression Omnibus and the Cancer Genome Atlas Data Mining with 2763 Samples and Validation via Real-Time Quantitative PCR. *Cell. Physiol. Biochem.* **2018**, *46*, 925–952. [CrossRef]
112. Yang, J.S.; Li, B.J.; Lu, H.W.; Chen, Y.; Lu, C.; Zhu, R.X.; Liu, S.H.; Yi, Q.T.; Li, J.; Song, C.H. Serum miR-152, miR-148a, miR-148b, and miR-21 as novel biomarkers in non-small cell lung cancer screening. *Tumour. Biol.* **2015**, *36*, 3035–3042. [CrossRef] [PubMed]
113. Bueno, M.J.; Perez de Castro, I.; Gomez de Cedron, M.; Santos, J.; Calin, G.A.; Cigudosa, J.C.; Croce, C.M.; Fernandez-Piqueras, J.; Malumbres, M. Genetic and epigenetic silencing of microRNA-203 enhances ABL1 and BCR-ABL1 oncogene expression. *Cancer Cell* **2008**, *13*, 496–506. [CrossRef] [PubMed]
114. Craig, V.J.; Cogliatti, S.B.; Rehrauer, H.; Wundisch, T.; Muller, A. Epigenetic silencing of microRNA-203 dysregulates ABL1 expression and drives Helicobacter-associated gastric lymphomagenesis. *Cancer Res.* **2011**, *71*, 3616–3624. [CrossRef] [PubMed]
115. Furuta, M.; Kozaki, K.I.; Tanaka, S.; Arii, S.; Imoto, I.; Inazawa, J. miR-124 and miR-203 are epigenetically silenced tumor-suppressive microRNAs in hepatocellular carcinoma. *Carcinogenesis* **2010**, *31*, 766–776. [CrossRef] [PubMed]
116. Noguchi, S.; Mori, T.; Otsuka, Y.; Yamada, N.; Yasui, Y.; Iwasaki, J.; Kumazaki, M.; Maruo, K.; Akao, Y. Anti-oncogenic microRNA-203 induces senescence by targeting E2F3 protein in human melanoma cells. *J. Biol. Chem.* **2012**, *287*, 11769–11777. [CrossRef]
117. Yi, R.; Poy, M.N.; Stoffel, M.; Fuchs, E. A skin microRNA promotes differentiation by repressing 'stemness'. *Nature* **2008**, *452*, 225–229. [CrossRef]
118. Zhang, J.; Zhou, Y.; Ma, L.; Huang, S.; Wang, R.; Gao, R.; Wu, Y.; Shi, H.; Zhang, J. The alteration of miR-222 and its target genes in nickel-induced tumor. *Biol. Trace Elem. Res.* **2013**, *152*, 267–274. [CrossRef]
119. Le Sage, C.; Nagel, R.; Egan, D.A.; Schrier, M.; Mesman, E.; Mangiola, A.; Anile, C.; Maira, G.; Mercatelli, N.; Ciafre, S.A.; et al. Regulation of the p27(Kip1) tumor suppressor by miR-221 and miR-222 promotes cancer cell proliferation. *EMBO J.* **2007**, *26*, 3699–3708. [CrossRef]
120. Fornari, F.; Gramantieri, L.; Ferracin, M.; Veronese, A.; Sabbioni, S.; Calin, G.A.; Grazi, G.L.; Giovannini, C.; Croce, C.M.; Bolondi, L.; et al. MiR-221 controls CDKN1C/p57 and CDKN1B/p27 expression in human hepatocellular carcinoma. *Oncogene* **2008**, *27*, 5651–5661. [CrossRef]
121. Wu, C.H.; Hsiao, Y.M.; Yeh, K.T.; Tsou, T.C.; Chen, C.Y.; Wu, M.F.; Ko, J.L. Upregulation of microRNA-4417 and Its Target Genes Contribute to Nickel Chloride-promoted Lung Epithelial Cell Fibrogenesis and Tumorigenesis. *Sci. Rep.* **2017**, *7*, 15320. [CrossRef]

© 2019 by the authors. Licensee MDPI, Basel, Switzerland. This article is an open access article distributed under the terms and conditions of the Creative Commons Attribution (CC BY) license (http://creativecommons.org/licenses/by/4.0/).

Review

Theoretical Studies of Nickel-Dependent Enzymes

Per E. M. Siegbahn [1,*], Shi-Lu Chen [2] and Rong-Zhen Liao [3]

1. Department of Organic Chemistry, Arrhenius Laboratory, Stockholm University, SE-10691 Stockholm, Sweden
2. School of Chemistry and Chemical Engineering, Beijing Institute of Technology, Beijing 100081, China
3. School of Chemistry and Chemical Engineering, Huazhong University of Science and Technology, Wuhan 430074, China
* Correspondence: per.siegbahn@su.se

Received: 10 June 2019; Accepted: 23 July 2019; Published: 29 July 2019

Abstract: The advancements of quantum chemical methods and computer power allow detailed mechanistic investigations of metalloenzymes. In particular, both quantum chemical cluster and combined QM/MM approaches have been used, which have been proven to successfully complement experimental studies. This review starts with a brief introduction of nickel-dependent enzymes and then summarizes theoretical studies on the reaction mechanisms of these enzymes, including NiFe hydrogenase, methyl-coenzyme M reductase, nickel CO dehydrogenase, acetyl CoA synthase, acireductone dioxygenase, quercetin 2,4-dioxygenase, urease, lactate racemase, and superoxide dismutase.

Keywords: nickel enzymes; reaction mechanism; quantum chemical calculations

1. Introduction

Nature harnesses transition metals to catalyze many different types of biological reactions that are crucial to life. Although about one-third of the enzymes are metalloenzymes, the use of nickel as a cofactor is not very common [1–9]. It was in 1975 that Zerner discovered the first nickel-dependent enzyme, namely urease [10]. Until now, ten different types of nickel enzymes (Table 1) have been reported with diverse biological functions. Among the known nickel enzymes, four of them are involved in the processing of anaerobic gases, namely, H_2, CO, CO_2, and CH_4, which have been proposed to be crucial to the origin of life [7]. Ni–Fe hydrogenase mediates the reversible two-electron reduction of protons to H_2 [11]. Methyl-coenzyme M reductase catalyzes the reversible transformation of methyl-coenzyme M plus coenzyme B into CH_4 and heterodisulfide CoM–S–S–CoB [12]. CO dehydrogenase catalyzes the reversible oxidation of CO to CO_2 using water as the oxygen source [13]. Acetyl-CoA synthase interacts tightly with CO dehydrogenase and uses the CO generated by CO dehydrogenase, CoA-SH, and a methyl group from a corrinoid/FeS protein to synthesize acetyl-CoA [14]. Acireductone dioxygenase [15], and quercetin 2,4-dioxygenase [16] are both nickel-dependent dioxygenases, and produce CO as one of the products. Ni-dependent superoxide dismutase targets the toxic reactive oxygen species superoxide and catalyzes its conversion to O_2 and H_2O_2 [17]. All these enzymes belong to the oxidoreductase class, while the remaining three enzymes are involved in the hydrolysis of urea (urease) [10], the isomerization of methylglyoxal to lactate (glyoxylase I) [18], and the racemization of lactate (lactate racemase, or LarA) [19].

Understanding the reaction mechanisms of nickel-dependent enzymes is of fundamental and practical importance. A particularly interesting question is why these enzymes select nickel as a catalytic center. One important advantage of using nickel has been suggested to be related to its flexible coordination geometry [8]. For redox reactions, the ligand environment has been demonstrated to play a crucial role in tuning the redox potentials of nickel, ranging from +0.89 V in superoxide dismutase to −0.60 V in Methyl-coenzyme M reductase and CO dehydrogenase [6]. Sulfur-donor ligands, such as sulfides, thiolates of cysteine residues, are commonly used to fulfill this purpose. In addition, the Lewis acidic character of nickel has also been suggested to be important for urease, glyoxalase I, acireductone dioxygenase, and quercetin 2,4-dioxygenase [8]. Importantly, these enzymes preferentially adopt O/N-donor ligands. Another important question is how the enzyme catalyzes the reaction. To understand the mechanistic details, especially the structures of transition states and intermediates for all elementary steps, quantum chemical calculations have proven to be useful to complement experimental work.

Table 1. Reactions catalyzed by Ni-dependent enzymes.

Enzyme Name	Reaction
Hydrogenase	$2H^+ + 2e^- \rightleftharpoons H_2$
Methyl-coenzyme M reductase	$CoM-S-CH_3 + H-S-CoB \rightleftharpoons CH_4 + CoM-S-S-CoB$
CO dehydrogenase	$CO + H_2O \rightleftharpoons CO_2 + 2H^+ + 2e^-$
Acetyl-CoA synthase	$CH_3-Co(III)-FeSP + CO + CoA-SH \rightleftharpoons Co(I)-FeSP + CoA-S-C(O)-CH_3$
Acireductone dioxygenase	(see figure)
Quercetin 2,4-dioxygenase	(see figure)
Superoxide dismutase	$2O_2^- + 2H^+ \rightarrow O_2 + H_2O_2$
Urease	(see figure)
Glyoxylase I	(see figure)
Lactate racemase	(see figure)

Two popular but different approaches have been developed in the modeling of metalloenzymes. The first one is termed the quantum chemical cluster approach, developed by Siegbahn and Himo et al. [20–26]. More than 30 years of experience has demonstrated that a relatively small model of the active site, from about 30 atoms initially to more than 200 atoms nowadays, is capable of describing all important mechanistic features of the catalysis. The alternative approach is called the combined quantum mechanics/molecular mechanics (QM/MM) method, which was first proposed by Warshel and Levitt in 1976 [27]. It has been shown by Liao and Thiel that with proper selection and an increase of the number of atoms in the QM region, both approaches gave similar results and the same conclusion [28,29].

In the present review, we summarize the mechanistic studies on nickel enzymes by quantum chemical calculations.

2. NiFe Hydrogenase

In nature, hydrogenases are the enzymes designed to form dihydrogen from protons and electrons in a reversible process:

$$2H^+ + 2e^- \rightleftharpoons H_2 \quad (1)$$

The most common forms of hydrogenases are NiFe- and FeFe-hydrogenases. The NiFe-enzymes are primarily used for hydrogen oxidation, and the FeFe-enzymes for proton reduction. NiFe-hydrogenase is probably the Ni-enzyme that has been most intensively studied by theoretical methods. The active site [30] is shown in Figure 1. Apart from the four cysteines, there are three additional ligands on iron, two cyanides, and one carbon monoxide, which are very unusual in biological systems, since they are potentially poisonous. A complicated machinery involving many enzymes is, therefore, used to construct this complex.

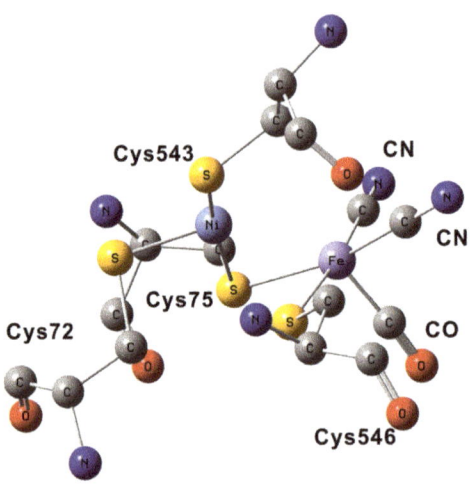

Figure 1. The X-ray structure of the active site of NiFe hydrogenase (Protein Data Bank (PDB) code: 1YRQ) [30].

The theoretical work prior to 2007 has been reviewed [31,32]. A variety of different models had been used, ranging from a minimal one, with only the direct ligands of the NiFe complex, to larger ones, with some charged second shell residues. All studies agreed on the general mechanism described in Figure 2. It contains three intermediate states observed experimentally, Ni_a-C^*, Ni_a-S and Ni_a-SR. Ni_a-C^* is the electron paramagnetic resonance (EPR)-active resting state, which has a bridging hydride between the metals. The EPR-silent Ni_a-SR state is reached by reduction of Ni_a-C^*, while Ni_a-S, which is also EPR-silent, is obtained by oxidation. To complete the cycle, Ni_a-SR can be reached by adding dihydrogen to Ni_a-S. Several other states have also been observed experimentally under varying conditions. The TS for H–H cleavage was generally described as heterolytic, leading to a bridging hydride and a protonation of a cysteine, usually Cys543. This mechanism could be regarded as being in consensus among the theoretical groups [31,32], and had strong support from experiments. For the energy profile, it could be concluded that entropy and dispersion played important roles [31], even though the dispersion effect could not be routinely included in the calculations, as it can be now.

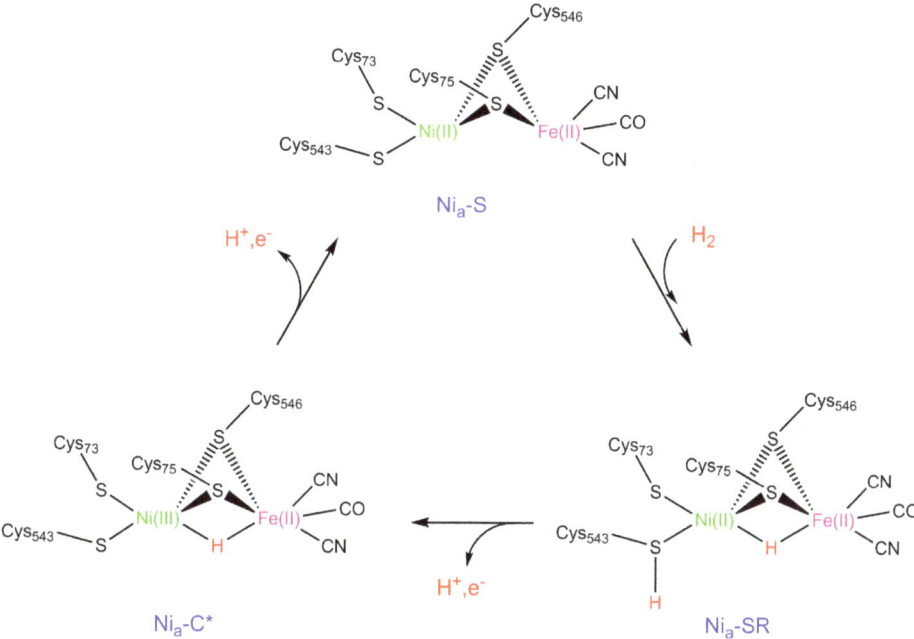

Figure 2. Schematic drawing of the catalytic cycles for NiFe-hydrogenases, suggested by experiments. Adapted from [31], Copyright © 2007, American Chemical Society.

In the years following the early period before 2007, some new possibilities and aspects were investigated. This development has also been briefly reviewed [26,33]. First, from experiments, it was suggested that a variant of the mechanism in Figure 2 might be operative, leading to a mechanism in Figure 3 [34]. The main difference to the mechanism of Figure 2 described above was that Ni_a–S does not participate in the active cycle. Instead, catalysis starts only after an initial reaction between Ni_a–S and dihydrogen, where Ni_a–C* is created. Ni_a–C* is then only in equilibrium with Ni_a–SR during turnover. Some support for the alternative mechanism was given in one study [35]. An interesting state in the context of that mechanism was a Ni(I)-state (Ni_a–R*), appearing at the end of the single step heterolytic mechanism. A TS for oxidative addition was found leading to two hydrides, one bridging and one bound to nickel (Ni_a–X*).

Another development was that the interest switched over to the oxidized states. The reason for this interest is the possibility to generate fuel by combining water oxidation with H_2 formation. A major problem in this context is that most hydrogenases are inactivated with O_2 present. Two main oxidized states of the NiFe cofactor have been identified. In Ni–B, there is a bridging hydroxide between Ni(III) and Fe(II). It is inactive for H_2 cleavage, but is relatively easy to reactivate. The Ni–A state is more difficult to reactivate. At the highest resolution (1.1 Å) it was initially concluded that Ni–A is a peroxide with one oxygen bridging between the metals and the other oxygen on nickel [30,36]. The early modeling studies were reviewed in 2007 [31]. While there was no problem obtaining the Ni–B structure with a bridging hydroxide [37,38], the modeling of Ni–A was more problematic. A structure similar to the one found experimentally could rather easily be obtained, but the energy was too high. A detailed analysis of the density of the experimental structure showed that it was likely to be a mixture of states, and the structure was probably also over reduced [39]. Calculations on the mechanism for the reactivation of the initially suggested structure for Ni–A were made [37]. It was suggested that the full reactivation of the enzyme requires two additional electrons and protons and this is the reason that it was slow. A reactivation mechanism for an oxidized cysteine ligand was also investigated. In 2015, a re-analysis of the experimental structure for Ni–A was performed, and it was concluded that Ni–A instead has a bridging hydroxide and an oxidized sulfenated form of Cys75 [40]. DFT calculations were then done for the new structure using larger models than before [41,42], and a revised mechanism for the reactivation was made.

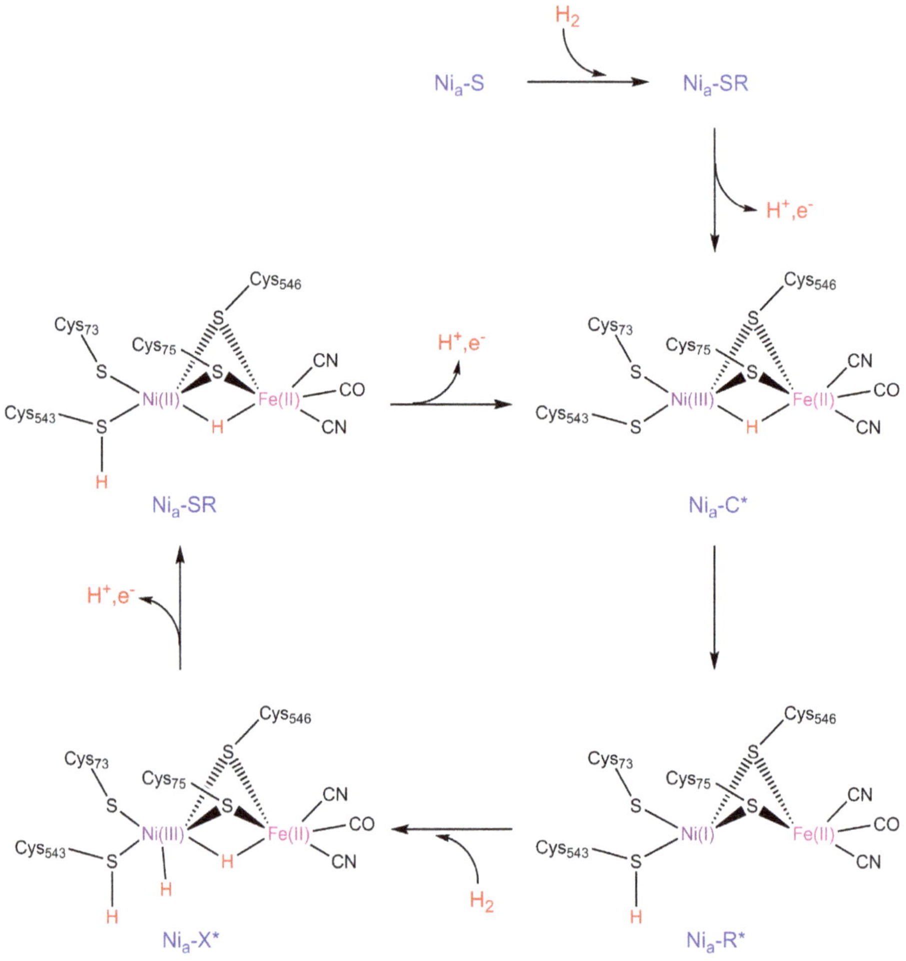

Figure 3. A variant of the mechanism for NiFe-hydrogenases. Adapted from [31], Copyright © 2007, American Chemical Society.

Several oxygen insensitive NiFe hydrogenases have been found. In these enzymes, the Ni–A state is activated by two electrons supplied from the proximal FeS cluster [43]. A few DFT studies have been performed to determine the details of the mechanism of how the electrons are released from the FeS cluster [44–46]. The oxygen tolerant hydrogenases have been found to have specifically designed proximal Fe–S clusters, where a sulfide in a corner is missing and replaced by two cysteines. This allows the Fe–S cluster to deliver two electrons to the NiFe-cluster. Together with two electrons from the active site, the accumulation of the Ni–A state is diminished. DFT calculations were used to determine electronic and structural changes of the Fe–S cluster.

After these initial modeling studies, the interest in the mechanism of NiFe-hydrogenase has decreased. The studies mentioned above gave mechanisms that are quite similar with minor differences, and in good agreement with available experiments [31]. However, in recent years there was a set of studies by Ryde et al. in 2014–2018, where many of the results obtained previously were seriously criticized [47–51]. In particular, the accuracy of the Becke, 3-parameter, Lee–Yang–Parr exchange-correlation functional (B3LYP), used in practically all previous studies, was claimed to be insufficient. Comparisons between B3LYP and more advanced methods of ab initio type seemed to give results, showing that there were large errors using B3LYP. For the bare cofactor without hydrogen, the singlet–triplet splitting (S–T) error was tolerable with a usual error for B3LYP of a few kcal/mol compared to the advanced methods [47]. However, when H_2 was added, the S–T splitting showed large errors for B3LYP [48], with a value of 13.2 kcal/mol compared to high level studies that gave 2–3 kcal/mol [49], but there are severe problems with that comparison. The most important one is that the structures were not optimized, but taken directly from a QM/MM study. This meant that the structures, as used in the comparison, are about 40 kcal/mol higher in energy than the optimized ones. The result most emphasized in the new DFT studies was that Ni_a–R* should not be an intermediate in the reaction sequence [51], as it was in some previous studies [35]. However, there is an interesting experimental result in this context using electro-chemical methods, where the conclusion was that Ni_a–R* was indeed observed as an intermediate [52]. Another more important aspect of the Ryde et al. studies is that they showed that it is actually possible to do calculations with the most accurate methods in quantum chemistry, such as the coupled cluster single-double and perturbative triple method (CCSD(T)) and the density matrix renormalization group method (DMRG). They have not yet been used for the reaction mechanism but it will be interesting to follow that development in the future.

3. Methyl-Coenzyme M Reductase

Methane, acting as the second most important anthropogenic greenhouse gas after CO_2 [53,54], has a significant influence on the global carbon cycling and the climate [53]. With a methane emission annually of around 500–600 Tg, about 69% is biologically released by microbial metabolisms, i.e., methanogenic archaea [55]. Meanwhile, a large amount of methane is oxidatively consumed by aerobic methanotrophic bacteria and anaerobic methane oxidizing archaea (ANME) [56,57]. The anaerobic methane oxidation is usually coupled with the reduction of sulfate (sulfate-reducing δ-proteobacteria) [58,59] iron [60,61], manganese [61], nitrate [62], soluble metal complexes [63], etc. [64], which may critically affect the Proterozoic and Neoproterozoic global climate systems [65] and the breeding of animals [66]. The final step of microbial methane formation and the first step of anaerobic methane oxidation are both performed by the Ni-dependent methyl-coenzyme M reductase (MCR) [56,67], which is capable of reversibly catalyzing the conversion of methyl-coenzyme M (CH_3–SCoM) and coenzyme B (CoB–SH) to a heterodisulfide (CoM–S–S–CoB) and methane using a Ni(I) porphyrinoid cofactor (F_{430}) (Figure 4) [6,9]. It is further found that MCR-like enzymes may constitute a family of alkyl-coenzyme M reductases, which are essential for the microbial anaerobic oxidation of ethane [68], propane [56,68], and butane [69] following the same reversible reaction pattern as MCR, i.e., alkane + CoM–S–S–CoB $\rightleftharpoons$ alkyl–SCoM + CoB–SH. Thus, the MCR mechanism may serve as a template for the family of alkyl-coenzyme M reductases [68], rendering it significant for the understanding of the alkane biosynthesis and the development of biomimetic catalysts for the alkane conversion.

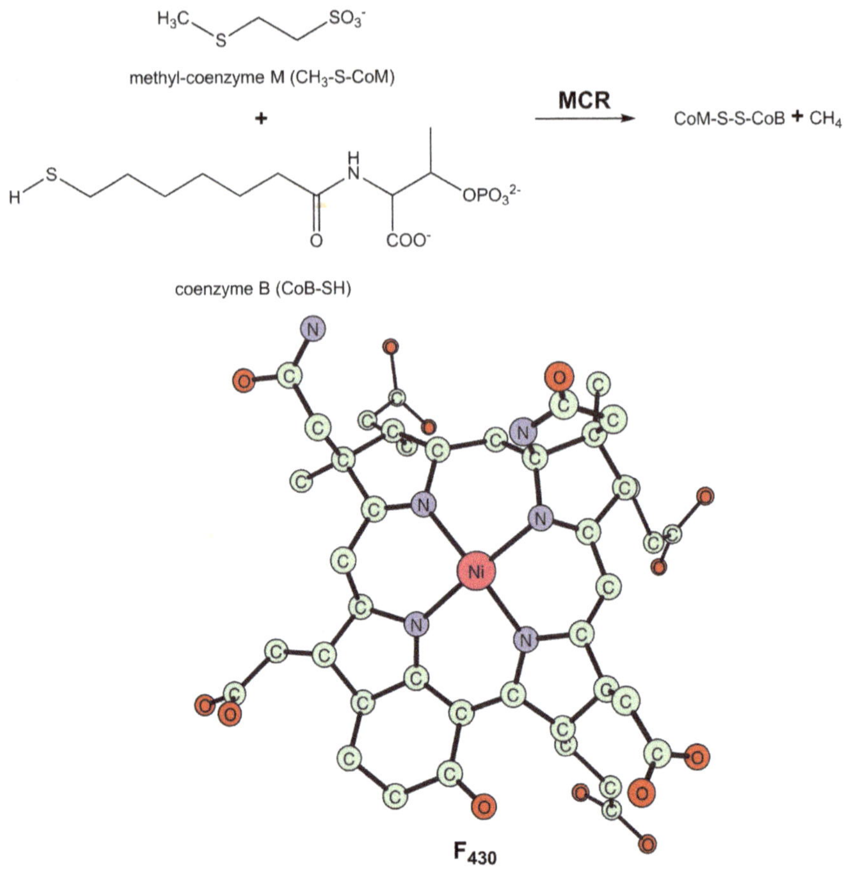

Figure 4. The reaction catalyzed by methyl-coenzyme M reductase (MCR) using a F_{430} cofactor.

Mainly four mechanisms for MCR have been proposed and examined by experiments or theoretical calculations (Figure 5). Accompanying the crystallization of MCR [12], mechanism I (Figure 5a) was hypothesized to include an organometallic Ni(III)-methyl intermediate (referred to as MCR_{Me}) resulting from the Ni-activated C–S bond dissociation of CH_3–SCoM and the proton transfer from CoB–SH to the sulfur of CH_3–SCoM. In mechanism I, the first step of C–S bond dissociation is most likely not rate-limiting, since the following electron transfer was suggested to lead to an unstable CoM–S·H radical. Methane is formed in the next step of S–S bond formation. This mechanism received the strongest support from the direct synthesis of MCR_{Me} from the active MCR (MCR_{red1}) with methyl bromide [70] or iodide [71], although its formation from the native substrate (CH_3–SCoM) has never been found. However, a rough estimation by means of the B3LYP bond dissociation energies (BDEs) predicted a very large endothermicity for the MCR_{Me} formation in the native MCR reaction [72]. A more elaborate B3LYP investigation based on a chemical model built from the crystal structure (Figure 6) showed that the MCR_{Me} formation is dependent on the acidity of the methyl-X leaving group, i.e., the C–X bond dissociation energy [73]. The MCR_{Me} formation is facile with a barrier of a few kcal/mol when X is a halide (I^-, Br^-, or Cl^-), while it has a very high barrier when starting from the native substrate with a stronger C–S bond, which was predicted to have a large endothermicity of 23.5 kcal/mol [73]. This endothermicity was then updated to 21.8 kcal/mol with dispersion and entropy effects added [74], which is sufficient to rule out mechanism I, considering the possible additional barriers in the subsequent steps (proton transfer and electron transfer).

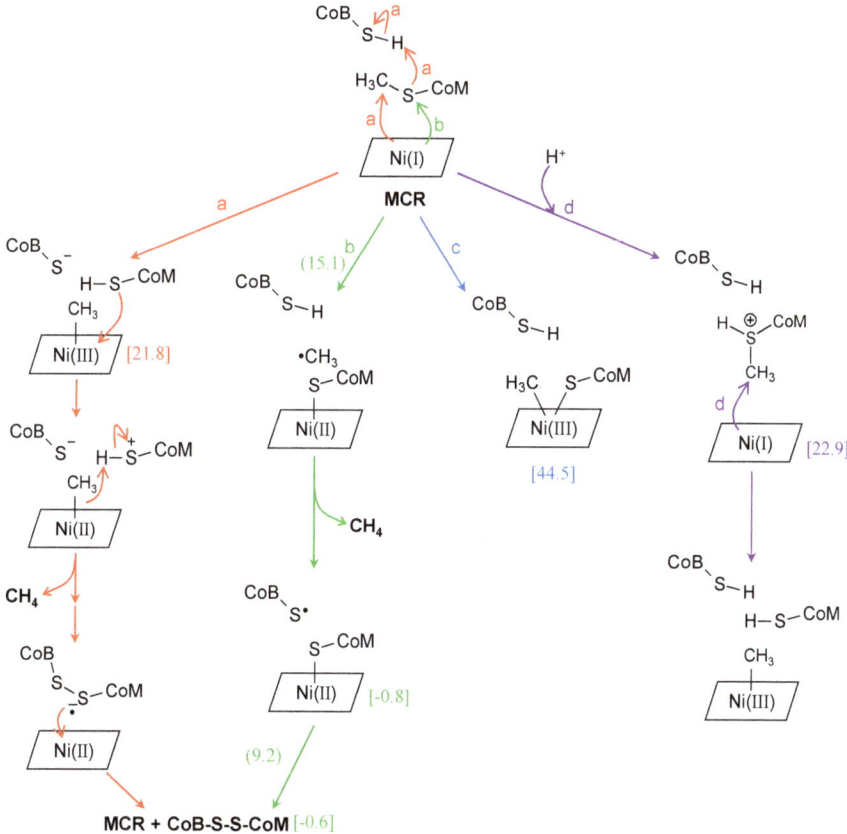

Figure 5. Four proposed mechanisms for the MCR reaction. (**a**) Mechanism I, including a methyl-Ni(III)F$_{430}$ intermediate. (**b**) Mechanism II, involving a methyl radical and a CoM–S–Ni(II)F$_{430}$ intermediate. (**c**) Mechanism III, including a ternary intermediate with side-on methyl and CoM–S$^−$ coordinating to Ni. (**d**) Mechanism IV, involving a proton binding at the sulfur of CH$_3$–SCoM. The energies for minima and transition states are given in brackets and parentheses, respectively. The energies for mechanisms I and II are adapted from [74], copyright © WILEY-VCH Verlag GmbH & Co. KGaA, Weinheim, while the ones for mechanisms III and IV are adapted from [75], with permission from the PCCP Owner Societies.

A more detailed DFT cluster modeling of mechanism II (Figure 5b) was then performed [72,76]. Mechanism II proceeds through a C–S bond dissociation forming Ni(II)–SCoM and a methyl radical, the immediate quenching of the methyl radical by CoB–SH leading to methane and a CoB–S· radical. This is followed by a rebound mechanism between Ni(II)–SCoM and CoB–S· to form a CoM-S-S-CoB and a regeneration of Ni(I) F$_{430}$. This mechanism has the following characteristics that are compatible with most of the mechanistic experiments: (i) The first step of C–S bond dissociation is rate-limiting with a barrier of 19.5 kcal/mol, which is supported by a substantial carbon kinetic isotope effects (^{12}CH$_3$–SCoM/^{13}CH$_3$–SCoM) of 1.04 ± 0.01 [77]. (ii) The methyl radical is captured rapidly by CoB–SH, so that methane formation happens almost simultaneously with the C–S bond dissociation and the whole MCR reaction can be regarded as two chemical steps via two transition states. (iii) An inversion of configuration at the methyl carbon takes place via a trigonal planar configuration during the methane formation; step (iii) combined with step (ii) is consistent with the finding using an isotopically chiral form of ethyl-coenzyme M [78] and a large α-secondary kinetic H/D isotope effect ($k_H/k_D = 1.19 \pm 0.01$), which indicates a geometric change of the methyl group from tetrahedral to trigonal planar at the highest transition state [77]. However, the energetics obtained in 2002 and 2003 [72,76] are not accurate enough to give a quantitatively correct description of the MCR mechanism, mainly because of the use of methanol as the model of Tyr333 and Tyr367, the lack of dispersion effects, and the omission of entropy effects in the second step of the S–S bond formation [74]. This leads to an excessively high barrier of ~26 kcal/mol for the reverse reaction (i.e., the anaerobic oxidation of methane) and an excessively large barrier difference of 10.0 kcal/mol between the two transition states of C–S bond dissociation and S–S bond formation [72,76]. This process faced additional challenges posed by two new isotope exchange experiments. These new experiments proved the feasibility of the reverse oxidation of methane [79] and showed a small barrier difference between the two chemical steps [80]. With dispersion and more accurate entropy effects included, a new DFT investigation with methylphenols as tyrosine models gave more reasonable barriers of 15.1 and 21.1 kcal/mol for the forward and reverse reactions, respectively, and a smaller barrier difference of 3.6 kcal/mol for the two transition states [74], which demonstrated that mechanism II is still acceptable and in agreement with mechanistic experiments at that time. Interestingly, an ingenious spectroscopic experiment using a shorter CoB–SH with one methylene less in order to elongate the distance of CoB–SH to the reacting core, lowering the MCR reaction rate, leads to a direct observation of the Ni(II)–SCoM intermediate [81]. This provided the clearest experimental evidence for mechanism II and made mechanism II a consensus MCR mechanism [82].

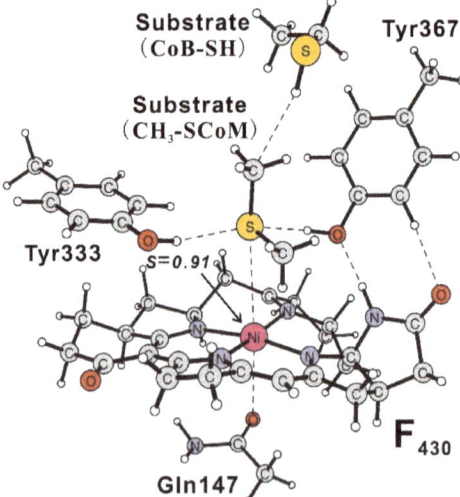

Figure 6. Model used in the calculations of the methyl-coenzyme M reductase (MCR). Adapted from [74], copyright © WILEY-VCH Verlag GmbH & Co. KGaA, Weinheim.

An oxidative addition mechanism (i.e., Mechanism III in Figure 5c) proposed by DFT calculations [83] and experiments [80,84] involves a side-on C–S (for CH_3–SCoM) [84] or C–H (for the reverse methane oxidation) [80] coordination to the Ni ion, forming a ternary intermediate with both methyl and CoM–S$^-$/hydride coordinating to the metal. However, from a DFT study based on the same cluster model that was used in the previous calculations of mechanism II, the formation of a ternary intermediate was predicted to have a very large endothermicity of 44.5 kcal/mol starting from CH_3–SCoM, and also a very high barrier starting from methane [75]. This can safely rule out mechanism

III. Another proposal (mechanism IV in Figure 5d) suggested a protonation at the sulfur position of CH_3–SCoM, which should facilitate the C–S bond activation and the subsequent methyl–Ni(III)F_{430} formation. Nevertheless, the DFT calculations showed that the substrate sulfur has a much smaller proton affinity than the F_{430} cofactor, in particular the carbonyl oxygen in F_{430}, which means that a proton entering the MCR active site should prefer binding at F_{430} instead of CH_3–SCoM, and that an additional energy of ~23 kcal/mol is required to move the proton from F_{430} to the substrate sulfur before the C–S bond activation [75]. This makes mechanism IV infeasible.

In summary, mechanism II (Figure 5b) is finally the most acceptable one for the MCR enzyme after two decades of computational and experimental studies. It fits reasonably well with several structural characteristics of the MCR active site. For example, a deeply-buried active site and the strict arrangement of F_{430}, CH_3–SCoM, and CoB–SH from bottom to top [12,85], ensuring the quick quenching of the methyl radical. In fact, the absence of CoB–SH cannot initiate the C–S bond cleavage and a CoB–SH analogue with one methylene less makes the MCR reaction much slower [86], which may be explained by the fast rebound between Ni(II)–SCoM and the unstable methyl radical. Even if the methyl radical escapes from CoB–SH, it cannot easily cause irreparable damage to the MCR active site, since the latter is hydrophobic and four active-site residues (His257, Arg271, Gln400, and Cys452) have been uncommonly methylated [12,85]. With this, MCR presents a perfect example of a mechanistic proposal by quantum chemical modeling finally leading to the convergence between theory and experiment. The understanding of the MCR mechanism could lead to new ways of producing methane/alkanes as fuels [87], of utilizing methane/alkanes as chemical feedstocks [82,87], and of breaking the C–S bond [88].

4. Nickel CO Dehydrogenase and Acetyl CoA Synthase

Carbon dioxide fixation is an extremely important process in nature, with implications on present day discussions on the greenhouse effect. The dominant enzymes under anaerobic conditions are Ni–CODH and Acetyl CoA synthase, which both have an active site containing Ni. Structures of the enzymes were first determined in 2002–2003 [14,89,90]. In the latest structure [13], obtained at a high resolution of 1.1 Å, an intermediate with bound CO_2 was found. The Ni–CODH and Acetyl CoA synthase enzymes are contained in one complex, and are connected by a 140 Å long channel in which CO is transported [91,92]. In Ni–CODH, CO can be reversibly transformed to CO_2. The reduction potential used for oxidation of CO is −0.3 V, while for reduction of CO_2 it has to be decreased to −0.6 V. Experimentally, the rate for reduction is 10 s^{-1} at pH 7, and at pH 8 the oxidation rate is as high as 3.9×10^4.

The active site of Ni–CODH has an unusual $NiFe_3S_4$ complex, termed the C-cluster, connected to another iron termed Fe_u. The experimental X-ray structure with a bound CO_2 [13], termed C_{red1}, has carbon bound to nickel and one of the oxygens bound to Fe_u, as shown in Figure 7. There are two important hydrogen bonds to CO_2, with one being from the positively charged Lys563, and the other one from a His93–Asp219 couple, which is most probably protonated. The spin state is a doublet [93] indicating an oxidation state of Ni(II)Fe_3(III)Fe(II) for the $NiFe_4$ cluster. Theoretical studies have supported this assignment [94–96]. After adding two electrons, one C–O bond is cleaved and the state termed C_{red2} is reached, which could be either Ni(I)Fe_4(II) or Ni(0)Fe_3(III)Fe(II). In the X-ray structure of C_{red2}, the CO, resulting from the C–O cleavage of CO_2, has disappeared from the cluster leaving an empty site on nickel [13].

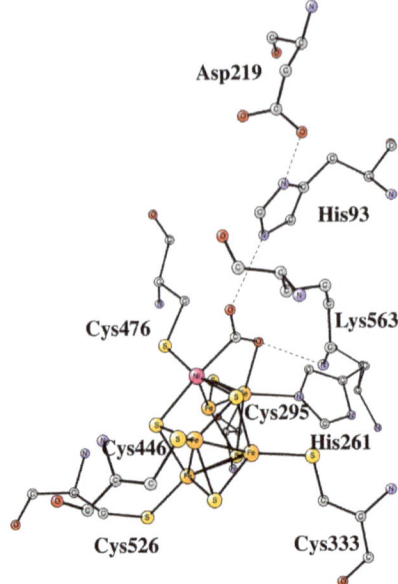

Figure 7. The X-ray structure of the active site of Ni–CODH with bound CO_2 [13].

The first computational study of the mechanism of Ni–CODH was made by Amara et al. [96]. To explain the X-ray structures, they suggested an unusual mechanism, where in C_{red2} a hydride occupies the empty site on nickel. The hydride would then have a role as an electron reservoir in C_{red2}. This would mean that nickel always stays in the Ni(II) state. A bound hydride would not be seen in the X-ray structure and might explain the observation of small amounts of H_2 in the experiments. In the CO_2 reduction direction from C_{red2}, CO_2 was suggested to insert into the Ni–hydride bond. The protein matrix should prevent formate formation. Mössbauer parameter calculations, and an orbital and a charge distribution analysis, were also performed to support their mechanism.

Quite recently, the energetics for the full catalytic cycle of Ni–CODH were computed for the first time (Figure 8) [97]. Energy diagrams for both the reduction of CO_2, using a redox potential of −0.6 V, and oxidation of CO, using a redox potential of −0.3 V, were calculated. In line with experiments, the oxidation is faster with a barrier difference of 4.8 kcal/mol, compared to experiments of 6.2 kcal/mol [92]. The rate-limiting step in both directions is the cleavage (formation) of the C–O bond of CO_2. The strong hydrogen bond from Lys563 and the proton transfer from the His93–Asp219 couple are very important. To obtain accurate energetics, it was found that one of the sulfides in the NiFeS cluster should be protonated. A water molecule in the active site was also found to be reasonably important.

An important difference between the computed and experimental structures found was that CO was bound to nickel in the reduced structure, in contrast to the experimental structure. To explain this result, it was suggested that the X-ray structure was over-reduced. The calculations for the over-reduced structure showed that CO is initially exchanged with a hydride. The hydride and the water molecule then form a hydrogen molecule, suggested to explain the presence of small amounts of H_2.

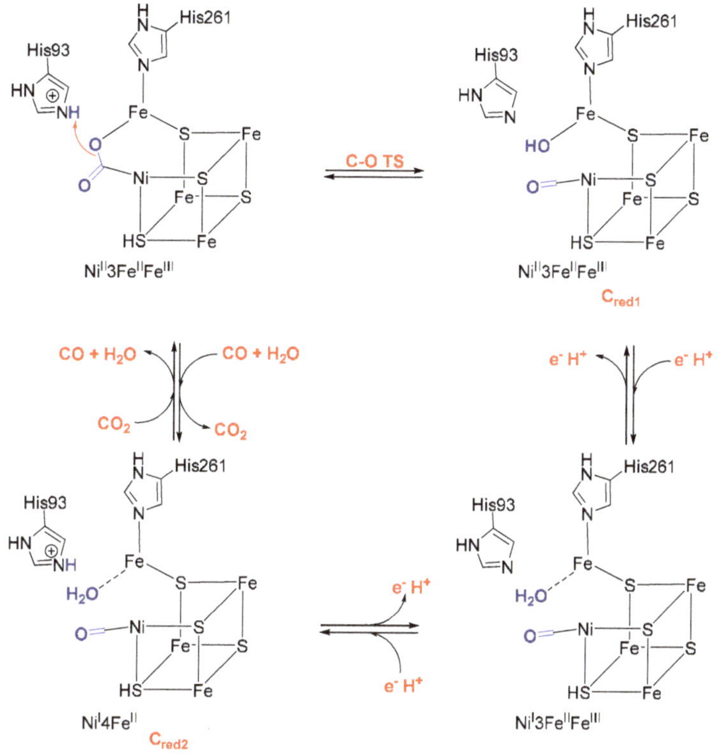

Figure 8. Mechanism of nickel CO dehydrogenase suggested on the basis of density functional calculations. Adapted from [97], Copyright © 2019, American Chemical Society.

As mentioned above, acetyl CoA synthase (ACS) is located in the same complex as Ni–CODH, at the other end of the long channel through which CO is transported. The active catalyst in ACS, termed the A-cluster, is a nickel dimer connected by two sulfide bridges to an Fe_4S_4 cluster (Figure 9). The nickel closest to the FeS cluster is termed Ni_p and the other one Ni_u. Ni_p is, apart from the sulfide bridges to the FeS cluster, connected by two bridging cysteines to Ni_u. Ni_u is furthermore ligated to two backbone nitrogens of two connected cysteines. Ni_p obtains a square planar coordination with a Ni(II) oxidation state before the substrates enter. Ni_p is five-coordinated and has been suggested to be the active metal in the dimer. There are three substrates, CO, CH_3, and coenzyme A (CoA). CH_3 is donated to the cluster from a cobalamine. The order of the entering substrates is debated. There are reasons to believe that the order might matter, since the protein has at least two different conformations, of which only one—the open conformation—allows CO to be transported all the way to ACS. In the closed conformation the transport of CO is blocked.

There has been one theoretical study of the mechanism of ACS [98]. A cluster QM approach was used. A problem in the modeling is that no X-ray structures with bound substrates are available. Possible intermediates in the catalytic cycle were studied, including their oxidation states. Ni(0) had earlier been suggested as a possible state of Ni_p, but the study shows that this is unlikely. The zero-valent state is stabilized only if the FeS cluster is reduced to a state with $S = 1/2$ ($[Fe_4S_4]^{1+}$). As CO binds, $Ni_p(0)$ is still favored. If the FeS cluster is oxidized, $Ni_p(I)$ is instead favored. It is suggested that CO and CH_3 both bind to Ni_p in a mononuclear mechanism on Ni_p. There are major remaining issues in the mechanism, mostly because of the limited structural information. In particular, the transfer of the methyl from cobalamine to the nickel dimer would be very interesting to study in the future, but also many questions remain regarding the final stages of CoA synthase.

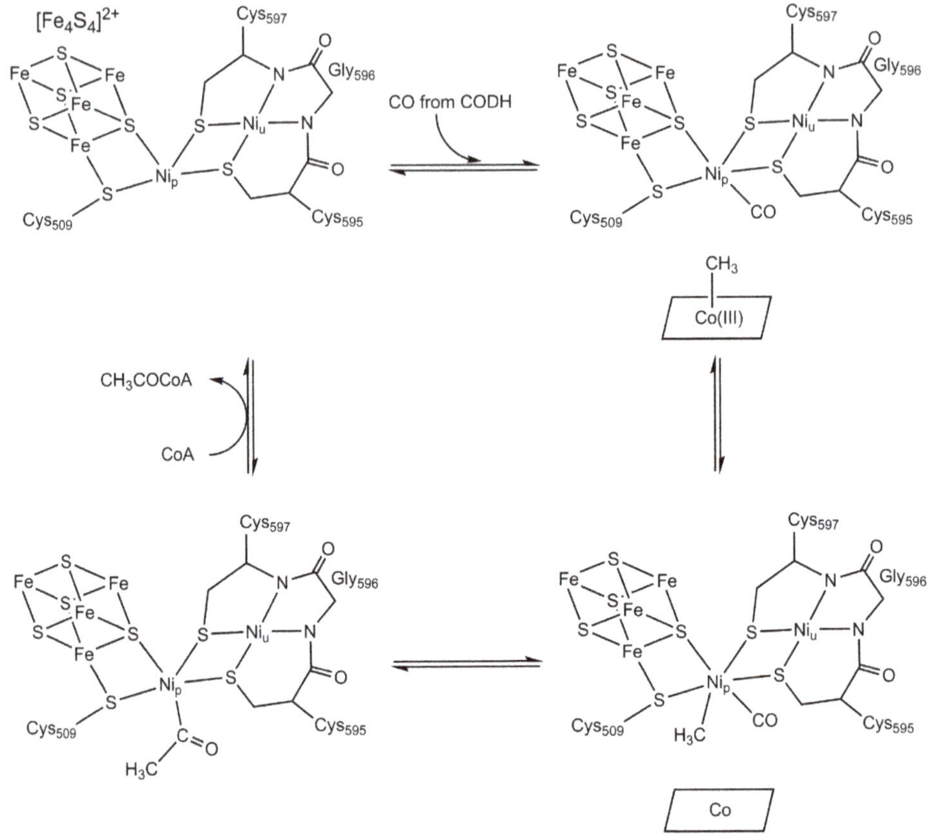

Figure 9. Mechanism of Acetyl CoA suggested on the basis of quantum mechanics (QM) cluster calculations. Adapted from [98], Copyright © 2005, American Chemical Society.

A theoretical study of the Ni-dimer in ACS with emphasis on pK_a values and redox potentials has also been performed [99]. Based on these computed values, the initial part of a mechanism for ACS was suggested. The low reduction potential computed indicates that the reduced form is protonated. Before or during methylation of the A-cluster, the proton could be detached by a base. A proton coupled electron transfer (PCET) mechanism should be operational in the methylation step. Furthermore, a short early note on the ACS mechanism by another group has been published [100]. Their calculations suggest that Ni_p is better described by the oxidation state Ni(I) than Ni(0).

5. Acireductone Dioxygenase

Acireductone dioxygenases (ARDs) are a family of metalloenzymes that are capable of using different transition metal ions (Ni^{2+} and Fe^{2+}) to catalyze the dioxygenation of acireductone [101,102]. Fe^{2+}-ARD and Ni^{2+}-ARD share exactly the same protein sequence. However, Fe^{2+}-ARD catalyzes the formation of methylthioketobutyrate and formate [101], whereas Ni^{2+}-ARD oxidizes acireductone into three products, namely formate, 3-methylthiopropionic acid, along with carbon monoxide [102]. It has been previously suggested that the substrate binding mode for Fe^{2+}-ARD is a five-membered ring, while there is a six-membered ring for Ni^{2+}-ARD (Scheme 1) [103], and this should be the main reason for the formation of different products.

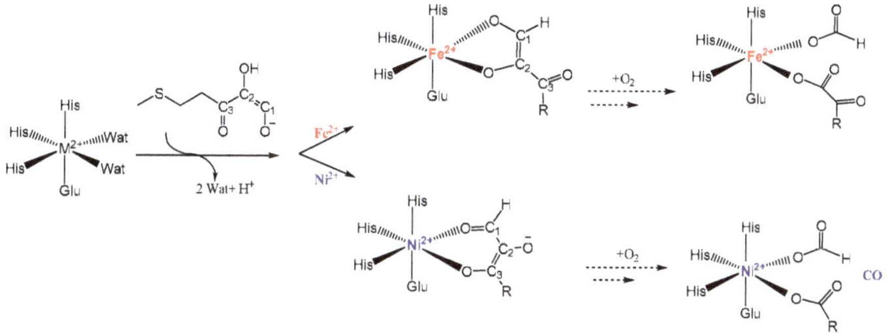

Scheme 1. The previously suggested mechanisms for ARDs. Adapted from [103], Copyright © 2001, American Chemical Society.

Sparta et al. performed discrete molecular dynamics simulations (QM/DMD) to investigate the hypothesis of the different substrate binding modes in Fe^{2+}-ARD and Ni^{2+}-ARD [104]. The QM/DMD simulations suggested that both Fe^{2+}-ARD and Ni^{2+}-ARD exhibit a six-membered ring coordination, with the O1 and O3 atoms binding to the central metal ion (Scheme 2, React). A second-shell residue Arg154 forms a hydrogen bond to the O2 atom of the substrate, which stabilizes this coordination mode. One snapshot was then selected for DFT calculations in order to understand the mechanisms of Fe^{2+}-ARD and Ni^{2+}-ARD. The energy barrier for the first dioxygen attack was not considered. The starting complex is a dioxygen adduct, with a peroxy bridging C1 and C3 of the substrate (reaction, Scheme 2). For Ni^{2+}-ARD, the reaction takes place in the triplet state. After a concerted cleavage of the O–O, C1–C2, and C2–C3 bonds, carbon monoxide and two carboxylates are generated (production, Scheme 2). The energy barrier is only 3.7 kcal/mol.

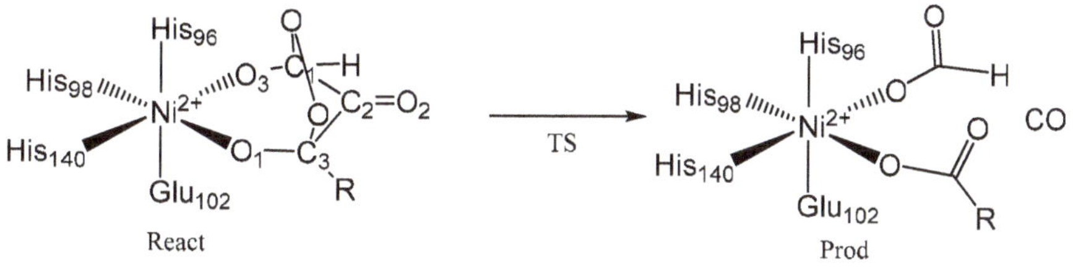

Scheme 2. Reaction mechanism for Ni-acireductone dioxygenase supported by the results of a DFT study. Adapted from [104], Copyright © 2013 Elsevier Ltd.

Recently, based on the structure of human acireductone dioxygenase, Borowski et al. have performed QM/MM calculations to revisit the reaction mechanisms of ARDs [105]. For the Ni^{2+}-ARD, the calculations suggested that the mechanism involves four major steps (Scheme 3). A proton is first transferred from the substrate to His88 when the substrate enters into the active site, and His88 dissociates from the nickel ion. Dioxygen binds to the metal coupled with one electron transfer from the substrate to the dioxygen moiety to generate a Ni^{2+}-superoxide intermediate (complex 1), which is quite similar to quercetin 2,3-dioxygenases [106,107]. The ground state of complex **1** is a quintet state (high-spin nickel(II) ions, one electron on the dioxygen, one electron on the substrate), while the following reaction is in the triplet state (two unpaired electrons on the nickel (II) ion). Hence, the energy of the triplet complex **1** was taken as the reference. After the distal oxygen atom attacks the C2 atom, a peroxo intermediate (complex 2) is formed. This step was found to be barrier-less. The next step is the migration of the C2-bound oxygen atom to the C3 atom (complex 3), which is ready for the following formation of the O–O bridge between C1 and C3 (complex 4). The final concerted cleavages of the O–O, C1–C2, and C2–C3 bonds were found to be the rate-limiting steps (TS45) for the substrate oxidation, with a barrier of only 6.8 kcal/mol. It is more likely that the substrate binding or product release is the rate-limiting step for the whole reaction.

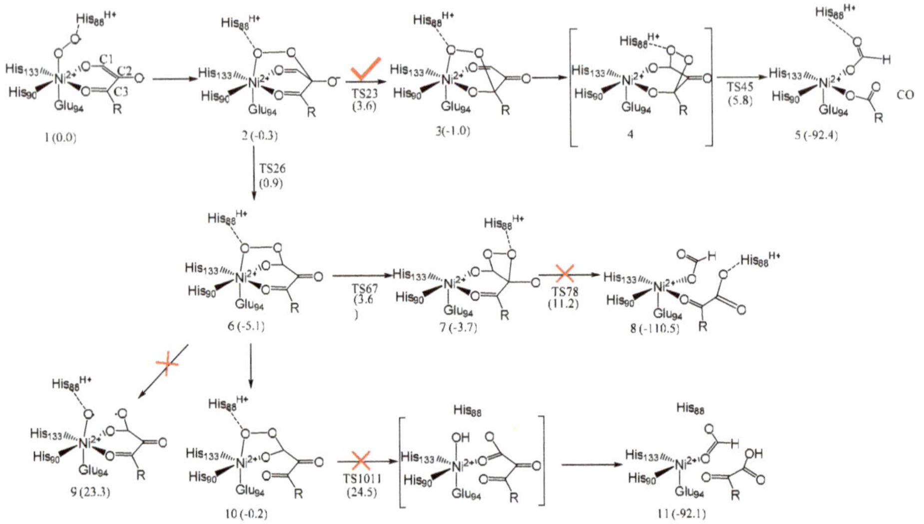

Scheme 3. Reaction mechanisms for Ni-ARD supported by the results of a quantum mechanics/molecular mechanics (QM/MM) study. Adapted from [105], copyright © WILEY-VCH Verlag GmbH & Co. KGaA, Weinheim.

Alternative pathways have also been considered from complex **2**. The C2-bound oxygen atom may attack the C1 atom to form complex **6**. After the formation of a dioxethane intermediate (complex **7**) and the cleavage of the O–O and C1–C2 bonds (TS78), methylthioketobutyrate and formate are generated (complex **8**). However, the barrier of TS78 was found to be 11.2 kcal/mol, higher than that of TS45. Other pathways that pass through a Baeyer-Villiger type rearrangement (**2** → **6** → **10** → **11**) and the homolytic O–O cleavage (**2** → **6** → **9**) were also ruled out due to their higher energy barriers compared with the most favorable mechanism.

6. Quercetin 2,4-Dioxygenase

Quercetin 2,4-dioxygenase from *Streptomyces* sp. strain FLA is a nickel-dependent enzyme catalyzing the oxidative ring-cleavage of quercetin to produce CO and 2-protocatechuoylphloroglucinol carboxylic acid using O_2 as the oxidant [16,108]. The mechanism for the dioxygen activation and substrate oxidation was first investigated by Liu and co-workers [109]. From the 19 ns molecular dynamics simulations, one snapshot was chosen for the following QM/MM calculations. The first-shell ligand Glu76 set to be deprotonated as a proton is transferred from the neutral quercetin substrate. For the reactant complex with the O_2 binding to the Ni ion in an end-on fashion, their calculations showed that the quintet is the ground state, the triplet lies 2.3 kcal/mol higher in energy, while the singlet was found to be much higher in energy. It is likely that the closed-shell singlet was obtained, while the open-shell singlet was not considered. In the triplet state, one electron is transferred from the substrate to the O_2 moiety, leading to the formation of a Ni^{2+}-superoxide-substrate radical species. Importantly, the reaction takes place in the triplet state involving four major steps (Scheme 4). First, the terminal oxygen of the superoxide attacks C2 of the substrate to form the first C–O bond, which has a barrier of only 5.0 kcal/mol. This is followed by a conformational change by a rotation around the newly-formed C2–O bond, and this step turned out to be rate-limiting, with a barrier of 19.9 kcal/mol. Subsequently, the terminal oxygen of the peroxide attacks C4 of the substrate, resulting in a peroxide-bridging intermediate. Finally, simultaneous cleavage of the O–O bond and two C–C bonds produces a CO molecule and the 2-protocatechuoylphloroglucinol carboxylic acid. When the central nickel ion is replaced by an iron ion, the reaction proceeds via a similar pathway in the quintet state. However, the last step becomes rate-limiting, associated with a very high barrier of 29.9 kcal/mol.

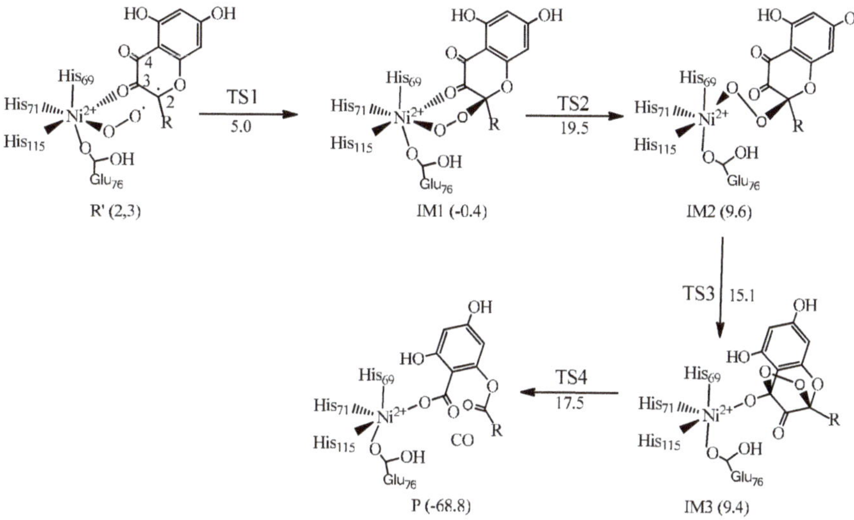

Scheme 4. Reaction mechanism for Ni-QueD suggested by Liu et al. Adapted from [109], with permission from The Royal Society of Chemistry.

Liao and co-workers independently investigated the reaction mechanism of this enzyme almost at the same time [107]. In their QM/MM calculations, the model started directly from the X-ray conformation. Importantly, the first-shell ligand Glu74 was considered to be both neutral and deprotonated. The aim of that study was to elucidate the mechanism and also to rationalize the chemoselectivity, in which only 2,4-dioxygenolytic cleavage takes place, but not 2,3-dioxygenolytic cleavage. They found a similar mechanism as Liu et al. [109] when Glu74 is protonated. However, the last step was found to be rate-limiting with a barrier of 24.8 kcal/mol. Unexpectedly, the 2,3-dioxygenolytic pathway was found to be more favorable, with a barrier of 21.8 kcal/mol. Therefore, the model with a protonated Glu74 residue could not reproduce the observed chemoselectivity. Instead, a model with a deprotonated Glu74 was demonstrated to reproduce the chemoselectivity (Scheme 5). In this case, the total barrier for the 2,4-dioxygenolytic pathway decreases to 17.4 kcal/mol, while it increases to 30.6 kcal/mol for the 2,3-dioxygenolytic pathway. The calculated barrier of 17.4 kcal/mol is in good agreement with the experimental kinetic data [108], which gave a barrier of about 15 kcal/mol.

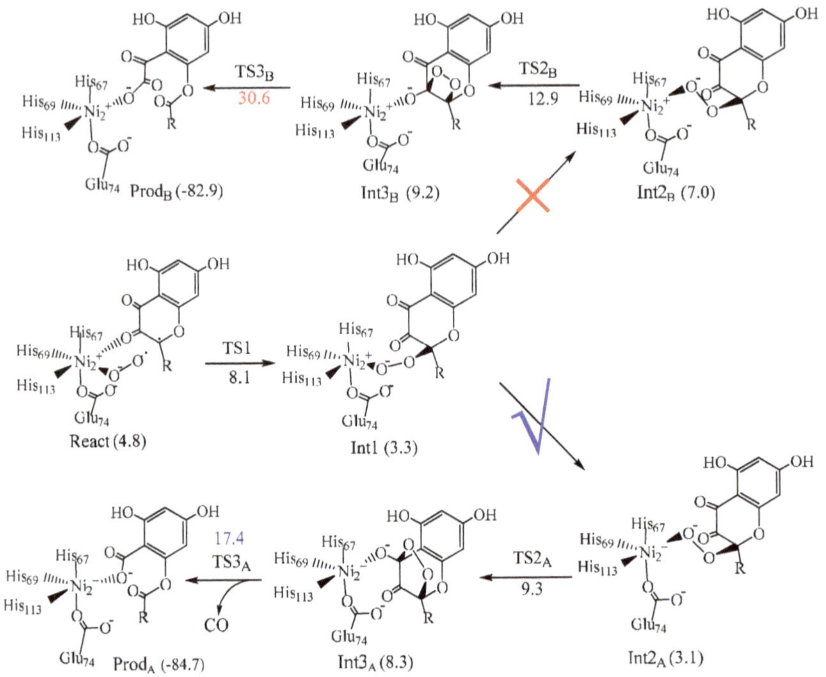

Scheme 5. Reaction mechanism and chemoselectivity for Ni-QueD suggested by Liao et al. Adapted from [107], with permission from the PCCP Owner Societies.

7. Urease

Urease is a nickel-containing enzyme that catalyzes the hydrolysis of urea to yield ammonia and carbon dioxide in the last step of nitrogen mineralization [110]. The uncatalyzed reaction has been suggested by Merz and Estiu [111] to proceed via an elimination pathway. For the enzyme catalyzed reaction, there have been three different mechanistic proposals. First, Karplus et al. suggested a mechanism in which a Nickel-coordinated terminal hydroxide performs the nucleophilic attack, while an adjacent protonated His320 residue delivers a proton to the leaving amino group [110,112]. Second, Benini et al. proposed a mechanism involving a nucleophilic attack of a bridging hydroxide on the carbonyl carbon of urea, followed by the protonation of the leaving amino group by this hydroxide [113]. Third, Lippard et al. suggested an elimination mechanism, similarly to the uncatalyzed reaction [114]. However, this elimination pathway could not explain the experimental observation of carbamate as the first product found by Zerner et al. [115]. A number of computational studies have, therefore, been conducted to elucidate the reaction mechanism of this enzyme [116–123].

The two different hydrolytic mechanisms were investigated by Merz et al. using DFT calculations (Schemes 6 and 7) [120]. For the bridging hydroxide attack mechanism, the calculations suggested that the reaction starts from a Ni^{II}_2 complex with a bridging hydroxide (Scheme 6). In the reactant complex S3, the urea substrate binds in a bi-dentate fashion, where the carbonyl oxygen coordinates to Ni1, while one of the amino groups coordinate to Ni2. Both Ni^{II} ions are in the triplet state, and the complex is, thus, a quintet for the ferromagnetically-coupled state. The bridging hydroxide performs a nucleophilic attack on the carbonyl carbon of the urea substrate via TS1A, leading to the formation of a tetrahedral intermediate (I1A) with the oxyanion stabilized by one of the two nickel ions. This pathway is consistent with the proposal by Ciurli et al., on the basis of docking studies [120]. Then, a proton is delivered from the bridging hydroxide to the amino group of urea (TS2A), facilitated by Asp360. This leads to the production of an ammonia molecule and a carbamate. The second step was found to be rate-limiting, with a barrier of 19.7 kcal/mol. Protonation of the leaving ammonia group by a protonated His320 residue has also been taken into account, and this step is close to isothermic. It should be noted that the role of this histidine residue has very recently been confirmed by modulating its protonation state and conformation of the mobile flap [124]. After the release of the ammonia molecule, a water molecule binds to Ni2, and this water molecule then performs a nucleophilic attack on the carbon atom of the carbamate anion (TS4A), concertedly with a proton transfer from the water molecule to the nearby Asp360 residue. Finally, the protonated Asp360 residue delivers a proton to the bridging oxyanion (TS6A), coupled with the C–O bond cleavage. The di-nickel complex (I7A) with a bridging hydroxide is regenerated. It should be pointed out that this kind of mechanism agrees very well with the recent crystal structure of the complex between urea and fluoride-inhibited urease, in which an unreactive fluoride replaces the bridging hydroxide [125].

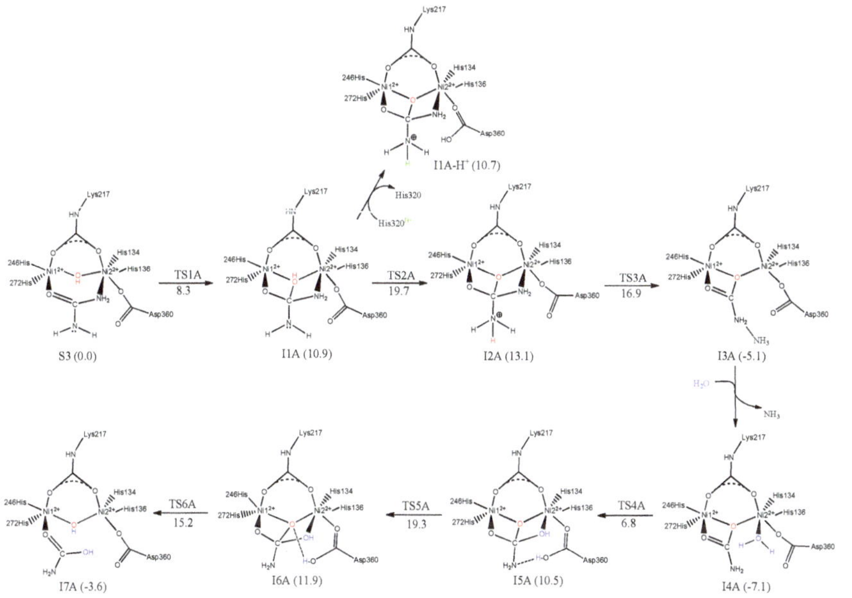

Scheme 6. The bridging hydroxide attack mechanism suggested by Merz et al. on the basis of DFT calculations. Adapted from [120], Copyright © 2003, American Chemical Society. Energies are given in kcal/mol relative to S3.

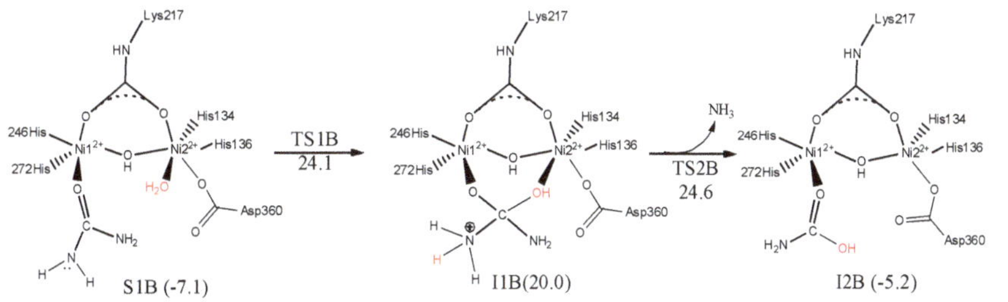

Scheme 7. The nickel bound water nucleophilic attack mechanism considered by Merz et al. on the basis of DFT calculations [120], Copyright © 2003, American Chemical Society. Energies are given in kcal/mol relative to S3 in Scheme 6.

Nucleophilic attack by the nickel bound water molecule was also taken into consideration by Merz et al. [120], and the starting complex was a monodentate urea-bound adduct with the carbonyl oxygen coordinated to Ni1 (S1B, Scheme 7). The calculations suggested that this mechanism involves only two steps (Scheme 7): the Ni2-bound water molecule performs a nucleophilic attack on the urea carbonyl carbon, which is coupled with a proton transfer from the water molecule to the leaving ammonia, assisted by the bridging hydroxide (TS1B). The next step is the C–N bond cleavage with the release of an ammonia molecule. The second step is rate-limiting, with a barrier of 24.6 kcal/mol, which is higher than the one of the bridging attack by a hydroxide.

They have also performed DFT calculations to analyze the elimination mechanism of urease (Scheme 8) [122]. They started from the enzyme without the urea substrate, and the substrate binding was calculated to be endothermic by 5.0 kcal/mol. Then, deprotonation of the amino group of urea by the bridging hydroxide should take place (TS3uw) and a new bridging water molecule is generated (I3uw). In the next step, a proton is transferred from the bridging water to the other amino group of urea (TS4uw), with a barrier of 30.5 kcal/mol. Subsequently, a water molecule enters into the active site and forms hydrogen bonds with the two Ni-bound water molecules, which facilitates the release of NH_4^+ and OCN^-. This mechanism has a substantially higher barrier than those for the two other mechanisms.

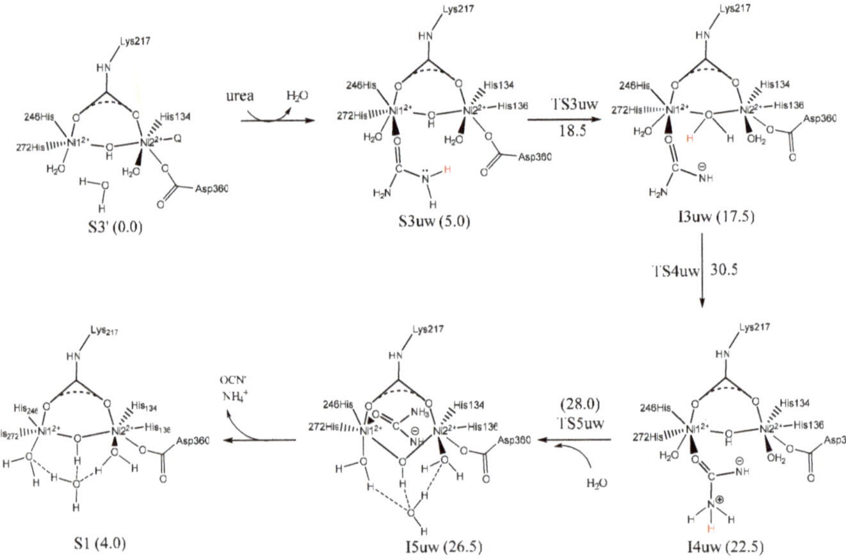

Scheme 8. The elimination mechanism of urea considered by Merz et al. on the basis of DFT calculations. Adapted from [122], Copyright © 2007, American Chemical Society.

8. Lactate Racemase

Lactate racemase (LarA) has been identified to be the enzyme that catalyzes the final step in the racemization between D- and L-lactic acids [126], using a previously unknown Ni cofactor (Ni-PTTMN), pyridinium-3-thioamide-5-thiocarboxylic acid mononucleotide Ni pincer complex [19,127] (Figure 10). Ni-PTTMN is formed by the coordination between a Ni(II) ion and a prosthetic group derived from nicotinic acid (Figure 10). The latter has a thioamide (modified from the nicotinic moiety) covalently bound to Lys184 and a thiocarboxylate, and acts as a tridentate pincer ligand with the two sulfurs and one pyridine carbon ligating to the nickel, where a Ni–C bond is formed. The His200 works as the fourth nickel ligand. The Ni-PTTMN cofactor in LarA serves as a rare example of a pincer complex found in enzymes and its Ni–C bond is the first case of a metal-carbon bond originating from nicotinic acid in biology [128]. Considering the common and widespread NAD$^+$-type cofactors (nicotinamide adenine dinucleotide and its phosphorylated form, NADP$^+$) derived from nicotinic acid, it is surprising that such a complicated Ni-PTTMN cofactor has been biologically evolved with extra consumption of resources. A question is whether it has evolutional advantages compared with NAD$^+$ [128].

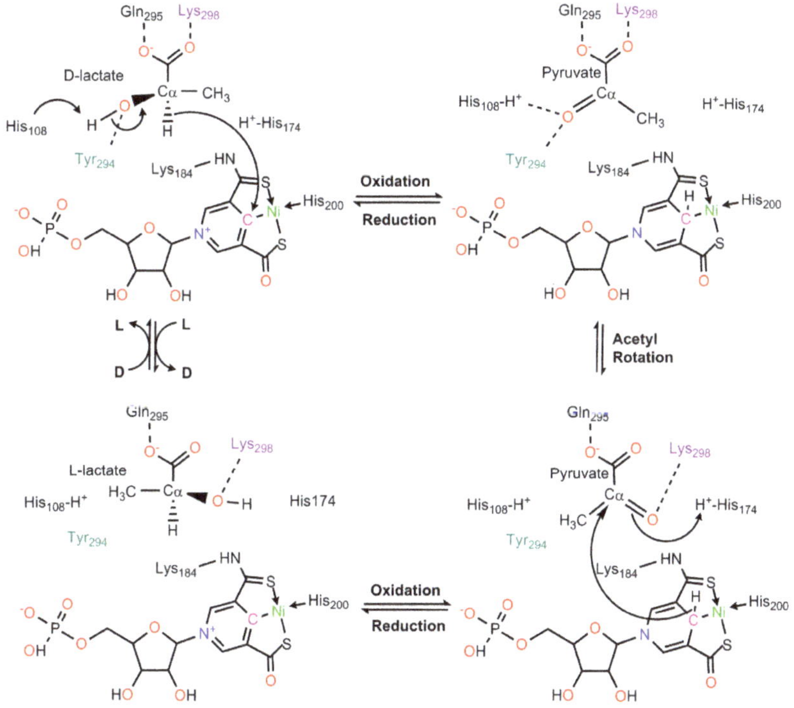

Figure 10. A modified proton-coupled hydride transfer mechanism for lactate racemase (LarA) proposed by Chen et al. Adapted from [129], copyright © WILEY-VCH Verlag GmbH & Co. KGaA, Weinheim. Tyr294 and Lys298 play crucial roles in orienting substrates and stabilizing the negative charge developing at the substrate hydroxyl oxygen in the transition states, thus lowering reaction barriers significantly.

A proton-coupled hydride transfer (PCHT) mechanism using a Ni(II) ion (Figure 10) has been verified by two DFT cluster modeling investigations by Chung et al. [130] and Chen et al. [129]. In this mechanism, a hydride is transferred to the Ni-PTTMN pyridine carbon from the substrate α-carbon, coupled with a proton transfer from the substrate hydroxyl to a histidine, followed by a transfer back to the α-carbon of the resultant pyruvate from the opposite side. This is achieved via a rotation of the pyruvate acetyl moiety. Chung et al. used an active-site model of 139 atoms to give a barrier of ~18 kcal/mol for the hydride transfer at the B3LYP-D3 level [130]. Based on small models (mainly involving cofactors and a proton-acceptor imidazole) and low-polarity solvation effects with a dielectric constant (ε) of 4, the further comparison of Ni-PTTMN with NAD$^+$-like cofactors showed that the Ni-PTTMN cofactor has a higher racemization barrier than NAD$^+$ [130], which makes it interesting to explore the LarA activity in the full active-site environment. In the subsequent DFT calculations by Chen et al. using a 200-atom model with Tyr294 and full side chains of the residues included, a lower rate-limiting barrier of 12 kcal/mol for the hydride transfer was predicted. The catalytic acceleration effect is achieved via the stabilization of the transition states by Tyr294 and Lys298 [129]. Further calculations were performed with various modified Ni-PTTMN cofactors in the

LarA active site. They showed that the barrier of 12 kcal/mol for the native LarA with a Ni-PTTMN cofactor is much lower than those (at least by 19 kcal/mol) for the LarA enzymes with NAD$^+$-like cofactors [129], indicating an enhanced racemization activity of Ni-PTTMN. To further analyze the acceleration effects of Ni-PTTMN, the hydride affinities of the Ni-PTTMN and NAD$^+$-like cofactors were calculated [129]. It was revealed that compared with NAD$^+$-like cofactors, Ni-PTTMN has a hydride affinity less sensitive to the environmental polarity, thus keeping a stronger hydride-addition reactivity in moderately and highly polar surroundings ($\varepsilon \geq 8$), while a weaker one in a low-polarity environment ($\varepsilon < 8$) [129]. This not only explains why a higher barrier for Ni-PTTMN was obtained in the calculations by Chung et al. [130], but also indicates the evolutionary advantage of Ni-PTTMN using an architecture with a Ni pincer, which fits perfectly in the moderately polar active site of LarA, in which a dielectric constant of around 14–18 was estimated by a barrier analysis [130]. Inspired by the Ni-PTTMN-dependent LarA reaction, a series of Ni-PTTMN-like metal pincer complexes (metal = Ni and Pd) were proposed by Yang et al. using DFT calculations to be potential catalysts for lactate racemization with a barrier as low as ~26 kcal/mol, adopting the same PCHT mechanism as LarA [131], which is a step closer to utilizing the novel Ni-PTTMN cofactor and its related chemistry in man-made catalysis.

Another mechanism starting from a Ni(III) state was proposed in a QM/MM study by Shaik et al. [132]. It includes the electron transfer from the lactate carboxylate to Ni(III), the C–C bond dissociation leading to a CO_2 radical anion and an acetaldehyde, the rotation of acetaldehyde, and then a C–C rebound step and the electron transfer back to the carboxylate. In this mechanism, the Ni-PTTMN cofactor works as a reversible electrode to accept and donate back an electron and does not result in covalent bonding to the lactate moiety. However, it is not clear if a Ni(II) Ni-PTTMN cofactor can be oxidized to the Ni(III) state in the LarA active site [133].

9. Superoxide Dismutase

Superoxide dismutases (SOD) catalyze the disproportionation of the toxic superoxide radicals into the less toxic hydrogen peroxide and molecular oxygen,

$$2O_2^{-}\cdot + 2H^+ \rightarrow O_2 + H_2O_2 \quad (2)$$

They have an important protective role, particularly in aerobic metabolism and photosynthesis. There are at least three different types of SOD's, and in one of them there is an active nickel complex.

The metal complex in Ni-SOD has a mononuclear nickel, coordinated to two cysteines and two backbone nitrogens in a square planar configuration (Figure 11). This is very similar to the coordination of Ni_u in acetyl CoA synthase (see above). However, there is one additional, axial, histidine ligand, which stabilizes a Ni(III) oxidation state. The disproportionation reaction occurs in two consecutive half-reactions, each one with one superoxide radical substrate. In the oxidative phase, the first superoxide radical binds and is transformed to dioxygen,

$$Ni(III)\text{-}L + O_2^{-}\cdot + H^+ \rightarrow Ni(II)\text{-}LH + O_2 \quad (3)$$

In the reductive phase, the second superoxide substrate binds and is transformed to hydrogen peroxide,

$$Ni(II)\text{-}LH + O_2^{-}\cdot + H^+ \rightarrow Ni(III)\text{-}L + H_2O_2 \quad (4)$$

Shortly after the first high-resolution X-ray structure appeared in 2004 [17], there were two independent quantum chemical studies of the mechanism [134,135]. Essentially two different mechanisms had been suggested based on experiments. The most popular one was based on the findings on the structure formed by X-ray reduction [17,136]. It was found that upon reduction, the axial histidine lost its coordination to nickel. Therefore, it was suggested that the His became protonated as the proton mediator in the mechanism. The alternative is that one of the cysteines becomes protonated instead. The calculations used essentially minimal cluster models, including only the directly coordinating ligands and the backbone connecting the two cysteines.

The mechanism obtained in the first study [134] is shown in Figure 11. It was suggested that the superoxide substrate enters as an $O_2H\cdot$ radical and binds in the empty axial site, where the cost for the protonation was included in the energetics. Anti-ferromagnetic coupling between nickel and the radical was found to be best. After substrate binding, the proton moves over to Cys2 and O_2 leaves. Nickel has now been reduced to Ni(II). The second substrate radical then enters and binds at the same axial position. The proton on Cys2 is now well suited to move over to the O_2H^- ligand to form H_2O_2, and the disproportionation is completed. The protonation of His1 was found to be much less favorable. The barrier for the first half-reaction was found to be 9.7 kcal/mol and for the second one 11.5 kcal/mol, in good agreement with expectations based on experiments.

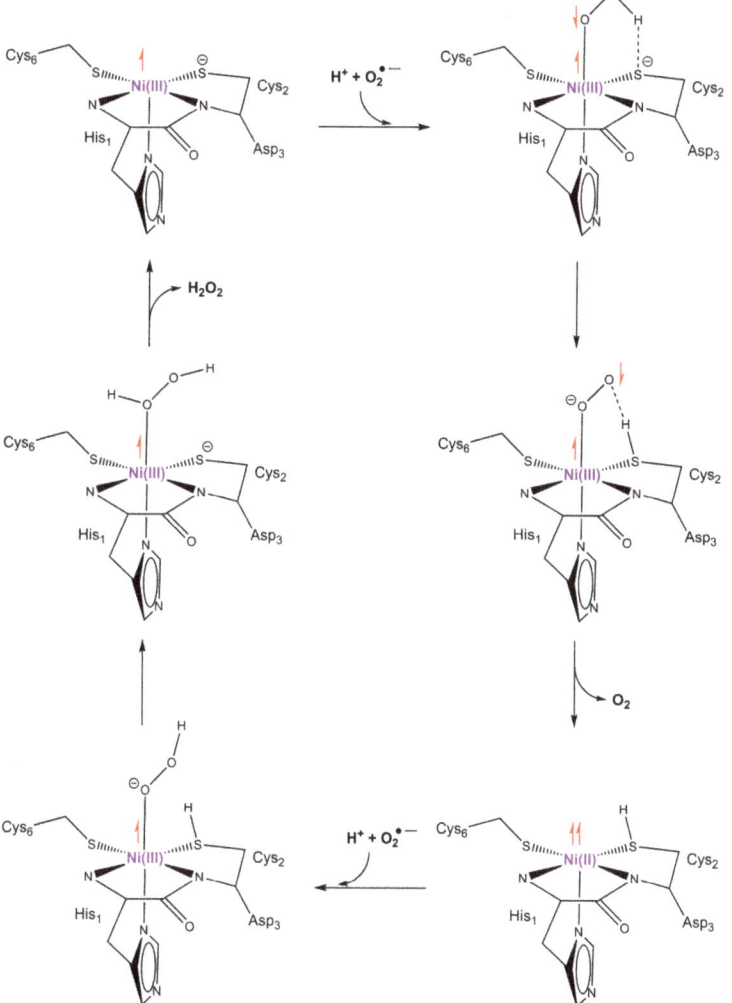

Figure 11. Reaction mechanism for superoxide dismutase (Ni-SOD) suggested by Pelmenschikov et al. Adapted from [134], Copyright © 2006, American Chemical Society.

In the second study, a different mechanism was found for the first half-reaction, which was the only one studied [135]. A few second shell residues were included in the model apart from the directly binding ones. It was suggested that Cys6 started out protonated and donated its proton to the first superoxide. H_2O_2 can then be formed by a donation of a proton from the medium via His1 and Tyr9. A proton donating role of His1 was, therefore, suggested in line with earlier experimental suggestions, based on the X-ray reduced structure [17,136]. Transition states were not determined due to the large size of the model.

Two additional studies can be mentioned. Several years later in 2011, a study was made [137] that reached similar conclusions as the earlier ones. An energy diagram was calculated showing a rate-limiting step with a barrier for one of them of around 19 kcal/mol. In another study on a mimic of SOD, multireference effects were studied with the use of the complete active space self-consistent field method (CASSCF) [138]. It was found that the near-degeneracy effects were small, which was considered surprising at the time. Multi-reference effects were expected, since a radical became bound to a transition metal.

10. Conclusions

In this review, we have presented the progress of theoretical studies on the reaction mechanisms of nickel-dependent enzymes. Both quantum chemical cluster and QM/MM approaches have been successfully used to answer mechanistic questions in these enzymes. Detailed information can be obtained on the structures, energies, electron densities, and spin densities of all relevant intermediates and transition states.

Nickel is redox active and the common oxidation states are Ni^{I}, Ni^{II}, and Ni^{III} in Ni enzymes. Due to its flexible coordination geometry, its redox potential can be tuned by different ligand coordinations with a span from +0.89 V to −0.60 V. Therefore, it can be used for challenging redox reactions, such as for superoxide dismutase, selective reduction of CO_2, and methane formation. In addition, the nickel ion can also function as a Lewis acid to stabilize anionic intermediates and transition states during the reactions.

Author Contributions: Writing—original draft preparation, P.E.M.S., S.-L.C., and R.-Z.L.; writing—review and editing, P.E.M.S., S.-L.C., and R.-Z.L.

Funding: This research was funded by the Swedish Research Council, the Knut and Alice Wallenberg Foundation, the National Natural Science Foundation of China, grant numbers 21673019 and 21873031.

Conflicts of Interest: The authors declare no conflict of interest.

References

1. Walsh, C.T.; Orme-Johnson, W.H. Nickel Enzymes. *Biochemistry* **1987**, *26*, 4901–4906. [CrossRef] [PubMed]
2. Ermler, U.; Grabarse, W.; Shima, S.; Goubeaud, M.; Thauer, R.K. Active sites of transition-metal enzymes with a focus on nickel. *Curr. Opin. Struct. Biol.* **1998**, *8*, 749–758. [CrossRef]
3. Maroney, M.J. Structure/function relationships in nickel metallobiochemistry. *Curr. Opin. Chem. Biol.* **1999**, *3*, 188–199. [CrossRef]
4. Mulrooney, S.B.; Hausinger, R.P. Nickel uptake and utilization by microorganisms. *FEMS Microbiol. Rev.* **2003**, *27*, 239–261. [CrossRef]
5. Li, Y.; Zamble, D.B. Nickel Homeostasis and Nickel Regulation: An Overview. *Chem. Rev.* **2009**, *109*, 4617–4643. [CrossRef]
6. Ragsdale, S.W. Nickel-based Enzyme Systems. *J. Biol. Chem.* **2009**, *284*, 18571–18575. [CrossRef]
7. Fontecilla-Camps, J.C.; Amara, P.; Cavazza, C.; Nicolet, Y.; Volbeda, A. Structure–function relationships of anaerobic gas-processing metalloenzymes. *Nature* **2009**, *460*, 814–822. [CrossRef] [PubMed]
8. Maroney, M.J.; Ciurli, S. Nonredox Nickel Enzymes. *Chem. Rev.* **2014**, *114*, 4206–4228. [CrossRef] [PubMed]
9. Boer, J.L.; Mulrooney, S.B.; Hausinger, R.P. Nickel-dependent metalloenzymes. *Arch. Biochem. Biophys.* **2014**, *544*, 142–152. [CrossRef] [PubMed]

10. Dixon, N.E.; Gazzola, C.; Blakeley, R.L.; Zerner, B. Jack Bean Urease (EC 3.5.1.5). A Metalloenzyme. A Simple Biological Role for Nickel? *J. Am. Chem. Soc.* **1975**, *97*, 4131–4133. [CrossRef]
11. Volbeda, A.; Charon, M.H.; Piras, C.; Hatchikian, E.C.; Frey, M.; Fontecilla-Camps, J.C. Crystal structure of the nickel–iron hydrogenase from Desulfovibrio gigas. *Nature* **1995**, *373*, 580–587. [CrossRef] [PubMed]
12. Ermler, U.; Grabarse, W.; Shima, S.; Goubeaud, M.; Thauer, R.K. Crystal Structure of Methyl-Coenzyme M Reductase: The Key Enzyme of Biological Methane Formation. *Science* **1997**, *278*, 1457–1462. [CrossRef] [PubMed]
13. Jeoung, J.H.; Dobbek, H. Carbon dioxide activation at the Ni, Fe-cluster of anaerobic carbon monoxide dehydrogenase. *Science* **2007**, *318*, 1461–1464. [CrossRef] [PubMed]
14. Doukov, T.; Iverson, T.M.; Seravalli, J.; Ragsdale, S.W.; Drennan, C.L. A Ni–Fe–Cu Center in a Bifunctional Carbon Monoxide Dehydrogenase/Acetyl-CoA Synthase. *Science* **2002**, *298*, 567–572. [CrossRef] [PubMed]
15. Pochapsky, T.C.; Pochapsky, S.S.; Ju, T.; Hoefler, C.; Liang, J. A refined model for the structure of acireductone dioxygenase from Klebsiella ATCC 8724 incorporating residual dipolar couplings. *J. Biomol. NMR.* **2006**, *34*, 117–127. [CrossRef] [PubMed]
16. Jeoung, J.H.; Nianios, D.; Fetzner, S.; Dobbek, H. Quercetin 2,4-Dioxygenase Activates Dioxygen in a Side-On O_2–Ni Complex. *Angew. Chem. Int. Ed.* **2016**, *55*, 3281–3284. [CrossRef]
17. Wuerges, J.; Lee, J.W.; Yim, Y.I.; Yim, H.S.; Kang, S.O.; Carugo, K.D. Crystal structure of nickel-containing superoxide dismutase reveals another type of active site. *Proc. Natl Acad. Sci. USA* **2004**, *101*, 8569–8574. [CrossRef]
18. He, M.M.; Clugston, S.L.; Honek, J.F.; Matthews, B.W. Determination of the Structure of *Escherichia coli* Glyoxalase I Suggests a Structural Basis for Differential Metal Activation. *Biochemistry* **2000**, *39*, 8719–8727. [CrossRef]
19. Desguin, B.; Zhang, T.; Soumillion, P.; Hols, P.; Hu, J.; Hausinger, R.P. A tethered niacin-derived pincer complex with a nickel–carbon bond in lactate racemase. *Science* **2015**, *349*, 66–69. [CrossRef]
20. Siegbahn, P.E.M.; Blomberg, M.R. Transition-metal systems in biochemistry studied by high-accuracy quantumchemical methods. *Chem. Rev.* **2000**, *100*, 421–437. [CrossRef]
21. Himo, F.; Siegbahn, P.E.M. Quantum chemical studies of radical containing enzymes. *Chem. Rev.* **2003**, *103*, 2421–2456. [CrossRef]
22. Siegbahn, P.E.M.; Borowski, T. Modeling enzymatic reactions involving transition metals. *Acc. Chem. Res.* **2006**, *39*, 729–738. [CrossRef]
23. Siegbahn, P.E.M.; Himo, F. Recent developments of the quantum chemical cluster approach for modeling enzyme reactions. *J. Biol. Inorg. Chem.* **2009**, *14*, 643–651. [CrossRef]
24. Siegbahn, P.E.M.; Himo, F. The quantum chemical cluster approach for modeling enzyme reactions. *WIREs Comput. Mol. Sci.* **2011**, *1*, 323–336. [CrossRef]
25. Himo, F. Recent trends in quantum chemical modeling of enzymatic reactions. *J. Am. Chem. Soc.* **2017**, *139*, 6780–6786. [CrossRef]
26. Blomberg, M.R.; Borowski, T.; Himo, F.; Liao, R.Z.; Siegbahn, P.E.M. Quantum chemical studies of mechanisms for metalloenzymes. *Chem. Rev.* **2014**, *114*, 3601–3658. [CrossRef]
27. Warshel, A.; Levitt, M. Theoretical studies of enzymatic reactions dielectric, electrostatic, and steric stabilization of carbonium-ion in reaction of lysozyme. *J. Mol. Biol.* **1976**, *103*, 227–249. [CrossRef]
28. Liao, R.Z.; Thiel, W. Comparison of QM-only and QM/MM models for the mechanism of tungsten-dependent acetylene hydratase. *J. Chem. Theory Comput.* **2012**, *8*, 3793–3803. [CrossRef]
29. Liao, R.Z.; Thiel, W. Convergence in the QM-only and QM/MM modeling of enzymatic reactions: A case study for acetylene hydratase. *J. Comput. Chem.* **2013**, *34*, 2389–2397. [CrossRef]
30. Volbeda, A.; Martin, L.; Cavazza, C.; Matho, M.; Faber, B.; Roseboom, W.; Albracht, S.; Garcin, E.; Rousset, M.; Fontecilla-Camps, J.C. Structural differences between the ready and unready oxidized states of [NiFe] hydrogenases. *J. Biol. Inorg. Chem.* **2005**, *10*, 239–249. [CrossRef]
31. Siegbahn, P.E.M.; Tye, J.W.; Hall, M.B. Computational studies of [NiFe] and [FeFe] hydrogenases. *Chem. Rev.* **2007**, *107*, 4414–4435. [CrossRef]
32. Bruschi, M.; Zampella, G.; Greco, Y.; Bertini, L.; Fantucci, P.; Gioia, D.L. *Computational Inorganic and Bioinorganic Chemistry*; Solomon, E.I., King, B., Scott, R., Eds.; John and Wiley and Sons: Chichester, UK, 2009.
33. Siegbahn, P.E.M.; Blomberg, M.R.A. Quantum Chemical Studies of Proton-Coupled Electron Transfer in Metalloenzymes. *Chem. Rev.* **2010**, *110*, 7040–7061. [CrossRef]

34. Kurkin, S.; George, S.J.; Thorneley, R.N.F.; Albracht, S.P.J. Hydrogen-induced activation of the [NiFe]-hydrogenase from Allochromatium vinosum as studied by stopped-flow infrared spectroscopy. *Biochemistry* **2004**, *43*, 6820–6831. [CrossRef]
35. Nilsson-Lill, S.O.; Siegbahn, P.E.M. An autocatalytic Mechanism for NiFe-Hydrogenase: Reduction to Ni(I) Followed by Oxidative Addition. *Biochemistry* **2009**, *48*, 1056–1066. [CrossRef]
36. Ogata, H.; Hirota, S.; Nakahara, A.; Komori, H.; Shibata, N.; Kato, T.; Kano, K.; Higuchi, Y. Activation process of [NiFe] hydrogenase elucidated by high-resolution X-ray analyses: Conversion of the ready to the unready state. *Structure* **2005**, *13*, 1635–1642. [CrossRef]
37. Siegbahn, P.E.M. Hybrid density functional study of the oxidized states of NiFe-hydrogenase. *C. R. Chim.* **2007**, *10*, 766–774. [CrossRef]
38. Pardo, A.; Lacey, A.L.D.; Fernandez, V.M.; Fan, Y.B.; Hall, M.B. Characterization of the active site of catalytically inactive forms of [NiFe] hydrogenases by density functional theory. *J. Biol. Inorg. Chem.* **2007**, *12*, 751–760. [CrossRef]
39. Söderhjelm, P.; Ryde, U. Combined computational and crystallographic study of the oxidised states of [NiFe] hydrogenase. *Theochem* **2006**, *770*, 108–116. [CrossRef]
40. Volbeda, A.; Martin, L.; Barbier, E.; Gutiérrez-Sanz, O.; De Lacey, A.L.; Liebgott, P.P.; Dementin, S.; Rousset, M.; Fontecilla-Camps, J.C. Crystallographic studies of [NiFe]-hydrogenase mutants: Towards consensus structures for the elusive unready oxidized states. *J. Biol. Inorg. Chem.* **2015**, *20*, 11–22. [CrossRef]
41. Breglia, R.; Greco, C.; Fantucci, P.; De Gioia, L.; Bruschi, M. Theoretical investigation of aerobic and anaerobic oxidative inactivation of the [NiFe]-hydrogenase active site. *Phys. Chem. Chem. Phys.* **2018**, *20*, 1693–1706. [CrossRef]
42. Breglia, R.; Greco, C.; Fantucci, P.; De Gioia, L. Reactivation of the Ready and Unready Oxidized States of [NiFe]-Hydrogenases: Mechanistic Insights from DFT Calculations. *Inorg. Chem.* **2019**, *58*, 279–293. [CrossRef]
43. Goris, T.; Wait, A.F.; Saggu, M.; Fritsch, J.; Heidary, N.; Stein, M.; Zebger, I.; Lendzian, F.; Armstrong, F.A.; Friedrich, B.; et al. A unique iron–sulfur cluster is crucial for oxygen tolerance of a [NiFe]-hydrogenase. *Nat. Chem. Biol.* **2011**, *7*, 310–318. [CrossRef]
44. Mouesca, J.M.; Fontecilla-Camps, J.C.; Amara, P. The Structural Plasticity of the Proximal [4Fe3S] Cluster is Responsible for the O_2 Tolerance of Membrane-Bound [NiFe] Hydrogenases. *Angew. Chem. Int. Ed.* **2013**, *52*, 2002–2006. [CrossRef]
45. Pandelia, M.E.; Bykov, D.; Izsak, R.; Infossi, P.; Giudici-Orticoni, M.T.; Bill, E.; Neese, F.; Lubitz, W. Electronic structure of the unique [4Fe–3S] clusterin O_2-tolerant hydrogenases characterized by ^{57}Fe Mössbauer and EPR spectroscopy. *Proc. Natl Acad. Sci. USA* **2013**, *110*, 483–488. [CrossRef]
46. Pelmenschikov, V.; Kaupp, M. Redox-Dependent Structural Transformations of the [4Fe–3S] Proximal Cluster in O_2-Tolerant Membrane-Bound [NiFe]-Hydrogenase: A DFT Study. *J. Am. Chem. Soc.* **2013**, *135*, 11809–11823. [CrossRef]
47. Delcey, M.G.; Pierloot, K.; Quan, M.P.; Vancoille, S.; Lind, R.; Ryde, U. Accurate calculations of geometries and singlet–triplet energy differences for active-site models of [NiFe] hydrogenase. *Phys. Chem. Chem. Phys.* **2014**, *16*, 7927–7938. [CrossRef]
48. Dong, G.; Quan, M.P.; Hallaert, S.D.; Pierloot, K.; Ryde, U. H_2 binding to the active site of [NiFe] hydrogenase studied by multiconfigurational and coupled-cluster methods. *Phys. Chem. Chem. Phys.* **2017**, *19*, 10590–10601. [CrossRef]
49. Dong, G.; Ryde, U. Protonation states of intermediates in the reaction mechanism of [NiFe] hydrogenase studied by computational methods. *J. Biol. Inorg. Chem.* **2016**, *21*, 383–394. [CrossRef]
50. Dong, G.; Ryde, U.; Jensen, H.J.; Hedegård, E.D. Exploration of H_2 binding to the [NiFe]-hydrogenase active site with multiconfigurational density functional methods. *Phys. Chem. Chem. Phys.* **2018**, *20*, 794–801. [CrossRef]
51. Dong, G.; Quan, M.P.; Pierloot, K.; Ryde, U. Reaction Mechanism of [NiFe] Hydrogenase Studied by Computational Methods. *Inorg. Chem.* **2018**, *57*, 15289–15298. [CrossRef]
52. Murphy, B.J.; Hidalgo, R.; Roessler, M.M.; Evans, R.M.; Ash, P.A.; Myers, W.K.; Vincent, K.A.; Armstrong, F.A. Discovery of Dark pH-Dependent H^+ Migration in a [NiFe]-Hydrogenase and Its Mechanistic Relevance: Mobilizing the Hydrido Ligand of the Ni–C Intermediate. *J. Am. Chem. Soc.* **2015**, *137*, 8484–8489. [CrossRef]
53. Murrell, J.C.; Jetten, M.S.M. The microbial methane cycle. *Environ. Microbiol. Rep.* **2009**, *1*, 279–284. [CrossRef]

54. Hogan, K.B.; Hoffman, J.S.; Thompson, A.M. Methane on the greenhouse agenda. *Nature* **1991**, *354*, 181–182. [CrossRef]
55. Conrad, R. The global methane cycle: Recent advances in understanding the microbial processes involved. *Environ. Microbiol. Rep.* **2009**, *1*, 285–292. [CrossRef]
56. Evans, P.N.; Boyd, J.A.; Leu, A.O.; Woodcroft, B.J.; Parks, D.H.; Hugenholtz, P.; Tyson, G.W. Methane on the greenhouse agenda. *Nat. Rev. Microbiol.* **2019**, *17*, 219–232. [CrossRef]
57. Thauer, R.K.; Kaster, A.K.; Seedorf, H.; Buckel, W.; Hedderich, R. Methanogenic archaea: Ecologically relevant differences in energy conservation. *Nat. Rev. Microbiol.* **2008**, *6*, 579–591. [CrossRef]
58. Nauhaus, K.; Albrecht, M.; Elvert, M.; Boetius, A.; Widdel, F. In vitro cell growth of marine archaeal-bacterial consortia during anaerobic oxidation of methane with sulfate. *Environ. Microbiol.* **2007**, *9*, 187–196. [CrossRef]
59. Nauhaus, K.; Boetius, A.; Krüger, M.; Widdel, F. In vitro demonstration of anaerobic oxidation of methane coupled to sulphate reduction in sediment from a marine gas hydrate area. *Environ. Microbiol.* **2002**, *4*, 296–305. [CrossRef]
60. Ettwig, K.F.; Zhu, B.; Speth, D.; Keltjens, J.T.; Jetten, M.S.M.; Kartal, B. Archaea catalyze iron-dependent anaerobic oxidation of methane. *Proc. Natl Acad. Sci. USA* **2016**, *113*, 12792–12796. [CrossRef]
61. Beal, E.J.; House, C.H.; Orphan, V.J. Manganese-and iron-dependent marine methane oxidation. *Science* **2009**, *325*, 184–187. [CrossRef]
62. Haroon, M.F.; Hu, S.; Shi, Y.; Imelfort, M.; Keller, J.; Hugenholtz, P.; Yuan, Z.; Tyson, G.W. Anaerobic oxidation of methane coupled to nitrate reduction in a novel archaeal lineage. *Nature* **2013**, *500*, 567–570. [CrossRef]
63. Scheller, S.; Yu, H.; Chadwick, G.L.; McGlynn, S.E.; Orphan, V.J. Artificial electron acceptors decouple archaeal methane oxidation from sulfate reduction. *Science* **2016**, *351*, 703–707. [CrossRef]
64. Rotaru, A.E.; Thamdrup, B. A new diet for methane oxidizers. *Science* **2016**, *351*, 658–659. [CrossRef]
65. Shen, B.; Dong, L.; Xiao, S.H.; Lang, X.G.; Huang, K.J.; Peng, Y.B.; Zhou, C.M.; Ke, S.; Liu, P.J. Molar tooth carbonates and benthic methane fluxes in Proterozoic oceans. *Nat. Commun.* **2016**, *7*, 10317. [CrossRef]
66. Planavsky, N.J.; Reinhard, C.T.; Wang, X.; Thomson, D.; McGoldrick, P.; Rainbird, R.H.; Johnson, T.; Fischer, W.W.; Lyons, T.W. Low Mid-Proterozoic atmospheric oxygen levels and the delayed rise of animals. *Science* **2014**, *346*, 635–638. [CrossRef]
67. Thauer, R.K. The Wolfe cycle comes full circle. *Proc. Natl Acad. Sci. USA* **2012**, *109*, 15084–15085. [CrossRef]
68. Chen, S.C.; Musat, N.; Lechtenfeld, O.J.; Paschke, H.; Schmidt, M.; Said, N.; Popp, D.; Calabrese, F.; Stryhanyuk, H.; Jaekel, U.; et al. Anaerobic oxidation of ethane by archaea from a marine hydrocarbon seep. *Nature* **2019**, *568*, 108–111. [CrossRef]
69. Pérez, R.L.; Wegener, G.; Knittel, K.; Widdel, F.; Harding, K.J.; Krukenberg, V.; Meier, D.V.; Richter, M.; Tegetmeyer, H.E.; Riedel, D.; et al. Thermophilic archaea activate butane via alkyl-coenzyme M formation. *Nature* **2016**, *539*, 396–401. [CrossRef]
70. Yang, N.; Reiher, M.; Wang, M.; Harmer, J.; Duin, E.C. Formation of a nickel-methyl species in methyl-coenzyme M reductase, an enzyme catalyzing methane formation. *J. Am. Chem. Soc.* **2007**, *129*, 11028–11029. [CrossRef]
71. Dey, M.; Telser, J.; Kunz, R.C.; Lees, N.S.; Ragsdale, S.W.; Hoffman, B.M. Biochemical and spectroscopic studies of the electronic structure and reactivity of a methyl-Ni species formed on methyl-coenzyme m reductase. *J. Am. Chem. Soc.* **2007**, *129*, 11030–11032. [CrossRef]
72. Pelmenschikov, V.; Blomberg, M.R.A.; Siegbahn, P.E.M.; Crabtree, R.H. A mechanism from quantum chemical studies for methane formation in methanogenesis. *J. Am. Chem. Soc.* **2002**, *124*, 4039–4049. [CrossRef]
73. Chen, S.L.; Pelmenschikov, V.; Blomberg, M.R.A.; Siegbahn, P.E.M. Is There a Ni-Methyl Intermediate in the Mechanism of Methyl-Coenzyme M Reductase? *J. Am. Chem. Soc.* **2009**, *131*, 9912–9913. [CrossRef]
74. Chen, S.L.; Blomberg, M.R.A.; Siegbahn, P.E.M. How is methane formed and oxidized reversibly when catalyzed by Ni-containing methyl-coenzyme M reductase? *Chem. Eur. J.* **2012**, *18*, 6309–6315. [CrossRef]
75. Chen, S.L.; Blomberg, M.R.A.; Siegbahn, P.E.M. An investigation of possible competing mechanisms for Ni-containing methyl-coenzyme M reductase. *Phys. Chem. Chem. Phys.* **2014**, *16*, 14029–14035. [CrossRef]
76. Pelmenschikov, V.; Siegbahn, P.E.M. Catalysis by methyl-coenzyme M reductase: A theoretical study for heterodisulfide product formation. *J. Biol. Inorg. Chem.* **2003**, *8*, 653. [CrossRef]
77. Scheller, S.; Goenrich, M.; Thauer, R.K.; Jaun, B. Methyl-coenzyme M reductase from methanogenic archaea: Isotope effects on label exchange and ethane formation with the homologous substrate ethyl-coenzyme M. *J. Am. Chem. Soc.* **2013**, *135*, 14975–14984. [CrossRef]

78. Ahn, Y.; Krzycki, J.A.; Floss, H.G. Steric course of the reduction of ethyl coenzyme M to ethane catalyzed by methyl coenzyme m reductase from methanosarcina barkeri. *J. Am. Chem. Soc.* **1991**, *113*, 4700–4701. [CrossRef]
79. Scheller, S.; Goenrich, M.; Boecher, R.; Thauer, R.K.; Jaun, B. The key nickel enzyme of methanogenesis catalyses the anaerobic oxidation of methane. *Nature* **2010**, *465*, 606–609. [CrossRef]
80. Scheller, S.; Goenrich, M.; Mayr, S.; Thauer, R.K.; Jaun, B. Intermediates in the Catalytic Cycle of Methyl Coenzyme? M Reductase: Isotope Exchange is Consistent with Formation of a σ-Alkane-Nickel Complex. *Angew. Chem. Int. Ed.* **2010**, *49*, 8112–8115. [CrossRef]
81. Wongnate, T.; Sliwa, D.; Ginovska, B.; Smith, D.; Wolf, M.W.; Lehnert, N.; Raugei, S.; Ragsdale, S.W. The radical mechanism of biological methane synthesis by methyl-coenzyme M reductase. *Science* **2016**, *352*, 953–958. [CrossRef]
82. Lawton, T.J.; Rosenzweig, A.C. Methane-make it or break it. *Science* **2016**, *352*, 892–893. [CrossRef]
83. Duin, E.C.; Mckee, M.L. A new mechanism for methane production from methyl-coenzyme M reductase as derived from density functional calculations. *J. Phys. Chem. B* **2008**, *112*, 2466–2482. [CrossRef]
84. Harmer, J.; Finazzo, C.; Piskorski, R.; Ebner, S.; Duin, E.C.; Goenrich, M.; Thauer, R.K.; Reiher, M.; Schweiger, A.; Hinderberger, D.; et al. A nickel hydride complex in the active site of methyl-coenzyme m reductase: Implications for the catalytic cycle. *J. Am. Chem. Soc.* **2008**, *130*, 10907–10920. [CrossRef]
85. Grabarse, W.; Mahlert, F.; Duin, E.C.; Goubeaud, M.; Shima, S.; Thauer, R.K.; Lamzin, V.; Ermler, U. On the mechanism of biological methane formation: Structural evidence for conformational changes in methyl-coenzyme M reductase upon substrate binding. *J. Mol. Biol.* **2001**, *309*, 315–330. [CrossRef]
86. Horng, Y.C.; Becker, D.F.; Ragsdale, S.W. Mechanistic studies of methane biogenesis by methyl-coenzyme M reductase: Evidence that coenzyme B participates in cleaving the C–S bond of methyl-coenzyme M. *Biochemistry* **2001**, *40*, 12875–12885. [CrossRef]
87. Borman, S. Carbon dioxide hydrogenated to methanol on large scale. *Chem. Eng. News* **2016**, *94*, 7.
88. Shima, S. The Biological Methane-Forming Reaction: Mechanism Confirmed Through Spectroscopic Characterization of a Key Intermediate. *Angew. Chem. Int. Ed.* **2016**, *55*, 13648–13649. [CrossRef]
89. Dobbek, H.; Svetlitchnyi, V.; Gremer, L.; Huber, R.; Meyer, O. Crystal Structure of a Carbon Monoxide Dehydrogenase Reveals a [Ni–4Fe–5S] Cluster. *Science* **2001**, *293*, 1281–1285. [CrossRef]
90. Darnault, C.; Volbeda, A.; Kim, E.J.; Legrand, P.; Vernede, X.; Lindahl, P.A.; Fontecilla-Camps, J.C. Ni–Zn–[Fe$_4$–S$_4$] and Ni–Ni–[Fe$_4$–S$_4$] clusters in closed and open α subunits of acetyl-CoA synthase/carbon monoxide dehydrogenase. *Nat. Struct. Biol.* **2003**, *10*, 271–279. [CrossRef]
91. Appel, A.M.; Bercaw, J.E.; Bocarsly, A.B.; Dobbek, H.; DuBois, D.L.; Dupuis, M.; Ferry, J.G.; Fujita, E.; Hille, R.; Kenis, P.J.A.; et al. Frontiers, Opportunities, and Challenges in Biochemical and Chemical Catalysis of CO_2 Fixation. *Chem. Rev.* **2013**, *113*, 6621–6658. [CrossRef]
92. Can, M.; Armstrong, F.A.; Ragsdale, S.W. Structure, Function, and Mechanism of the Nickel Metalloenzymes, CO Dehydrogenase, and Acetyl-CoA Synthase. *Chem. Rev.* **2014**, *114*, 4149–4174. [CrossRef]
93. Lindahl, P.A.; Munck, E.; Ragsdale, S.W. CO Dehydrogenase from Clostridium thermoaceticum: EPR and electrochemical studies in CO_2 and argon atmospheres. *J. Biol. Chem.* **1990**, *265*, 3873–3879.
94. Cao, Z.-X.; Mo, Y.-R. Computational Characterization of the Elusive C-Cluster of Carbon Monoxide Dehydrogenase. *J. Theor. Comput. Chem.* **2008**, *7*, 473–484. [CrossRef]
95. Xie, H.-J.; Cao, Z.-X. Computational Characterization of Reactive Intermediates of Carbon Monoxide Dehydrogenase. *Chin. J. Struct. Chem.* **2009**, *28*, 1525–1532.
96. Amara, P.; Mouesca, J.-M.; Volbeda, A.; Fontecilla-Camps, J.C. Carbon Monoxide Dehydrogenase Reaction Mechanism: A Likely Case of Abnormal CO_2 Insertion to a Ni–H$^-$ Bond. *Inorg. Chem.* **2011**, *50*, 1868–1878. [CrossRef]
97. Liao, R.-Z.; Siegbahn, P.E.M. Energetics for the mechanism of nickel-containing CO-dehydrogenase. *Inorg. Chem.* **2019**. [CrossRef]
98. Amara, P.; Volbeda, A.; Fontecilla-Camps, J.C.; Field, M.J. A quntum chemical study of the reaction mechanism of Acetyl-Coenzyme A synthase. *J. Am. Chem. Soc.* **2005**, *127*, 2776–2784. [CrossRef]
99. Chmielowska, A.; Lodowski, P.; Jaworska, M. Redox Potentials and Protonation of the A-Cluster from Acetyl-CoA Synthase. A Density Functional Theory Study. *J. Phys. Chem.* **2013**, *117*, 12484–12496. [CrossRef]

100. Schenker, R.P.; Brunold, T.C. Computational studies on the A cluster of acetyl-coenzyme A synthase: Geometric and electronic properties of the NiFeC species and mechanistic implications. *J. Am. Chem. Soc.* **2003**, *125*, 13962–13963. [CrossRef]
101. Dai, Y.; Wensink, P.C.; Abeles, R.H.J. One protein, two enzymes. *Biol. Chem.* **1999**, *274*, 1193–1195. [CrossRef]
102. Wray, J.W.; Abeles, R.H. The Methionine Salvage Pathway in Klebsiella pneumoniae and Rat Liver identification and characterization of two novel dioxygenases. *J. Biol. Chem.* **1995**, *270*, 3147–3153. [CrossRef]
103. Dai, Y.; Pochapsky, T.C.; Abeles, R.H. Mechanistic studies of two dioxygenases in the methionine salvage pathway of Klebsiella pneumonia. *Biochemistry* **2001**, *40*, 6379–6387. [CrossRef]
104. Sparta, M.; Valdez, C.E.; Alexandrova, A.N. Metal-dependent activity of Fe and Ni acireductone dioxygenases: How two electrons reroute the catalytic pathway. *J. Mol. Biol.* **2013**, *425*, 3007–3018. [CrossRef]
105. Miłaczewska, A.; Kot, E.; Amaya, J.A.; Makris, T.M.; Zając, M.; Korecki, J.; Chumakov, A.; Trzewik, B.; Kędracka-Krok, S.; Minor, W.; et al. On the Structure and Reaction Mechanism of Human Acireductone Dioxygenase. *Chem. Eur. J.* **2018**, *24*, 5225–5237. [CrossRef]
106. Siegbahn, P.E.M. Hybrid DFT study of the mechanism of quercetin 2,3-dioxygenase. *Inorg. Chem.* **2004**, *43*, 5944–5953. [CrossRef]
107. Wang, W.J.; Wei, W.J.; Liao, R.Z. Deciphering the chemoselectivity of nickel-dependent quercetin 2,4-dioxygenase. *Phys. Chem. Chem. Phys.* **2018**, *20*, 15784–15794. [CrossRef]
108. Merkens, H.; Kappl, R.; Jakob, R.P.; Schmid, F.X.; Fetzner, S. Quercetinase QueD of Streptomyces sp. FLA, a Monocupin Dioxygenase with a Preference for Nickel and Cobalt. *Biochemistry* **2008**, *47*, 12185–12196. [CrossRef]
109. Li, H.; Wang, X.; Tian, G.; Liu, Y. Insights into the dioxygen activation and catalytic mechanism of the nickel-containing quercetinase. *Catal. Sci. Technol.* **2018**, *8*, 2340–2351. [CrossRef]
110. Karplus, P.A.; Pearson, M.A.; Hausinger, R.P. 70 years of crystalline urease: What have we learned? *Acc. Chem. Res.* **1997**, *30*, 330–337. [CrossRef]
111. Estiu, G.; Merz, K.M. The hydrolysis of urea and the proficiency of urease. *J. Am. Chem. Soc.* **2004**, *126*, 6932–6944. [CrossRef]
112. Pearson, M.A.; Park, I.S.; Schaller, R.A.; Michel, L.O.; Karplus, P.A.; Hausinger, R.P. Kinetic and Structural Characterization of Urease Active Site Variants. *Biochemistry* **2000**, *39*, 8575–8584. [CrossRef]
113. Benini, S.; Rypniewski, W.R.; Wilson, K.S.; Miletti, S.; Ciurli, S.; Mangani, S. A new proposal for urease mechanism based on the crystal structures of the native and inhibited enzyme from Bacillus pasteurii: Why urea hydrolysis costs two nickels. *Structure* **1999**, *7*, 205–216. [CrossRef]
114. Barrios, A.M.; Lippard, S.J. Interaction of urea with a hydroxide-bridged dinuclear nickel center: An alternative model for the mechanism of urease. *J. Am. Chem. Soc.* **2000**, *122*, 9172–9177. [CrossRef]
115. Blakeley, R.L.; Hinds, J.A.; Kunze, H.E.; Webb, E.C.; Zerner, B. Jack bean urease (EC 3.5.1.5) demonstration of a carbamoyl-transfer reaction and inhibition by hydroxamic acids. *Biochemistry* **1969**, *8*, 1991–2000. [CrossRef]
116. Smyj, R.P. A conformational analysis study of a nickel(II) enzyme: Urease. *J. Mol. Struct. THEOCHEM* **1997**, *391*, 207–223. [CrossRef]
117. Zimmer, M. Molecular mechanics evaluation of the proposed mechanisms for the degradation of urea by urease. *J. Biomol. Struct. Dyn.* **2000**, *17*, 787–797. [CrossRef]
118. Zimmer, M. Are classical molecular mechanics calculations still useful in bioinorganic simulations? *Coord. Chem. Rev.* **2009**, *253*, 817–826. [CrossRef]
119. Musiani, F.; Arnofi, E.; Casadio, R.; Ciurli, S. Structure-based computational study of the catalytic and inhibition mechanisms of urease. *J. Biol. Inorg. Chem.* **2001**, *6*, 300–314. [CrossRef]
120. Suárez, D.; Díaz, N.; Merz, K.M. Ureases: Quantum chemical calculations on cluster models. *J. Am. Chem. Soc.* **2003**, *125*, 15324–15337. [CrossRef]
121. Estiu, G.; Merz, K.M. Catalyzed decomposition of urea. Molecular dynamics simulations of the binding of urea to urease. *Biochemistry* **2006**, *45*, 4429–4443. [CrossRef]
122. Estiu, G.; Merz, K.M. Competitive hydrolytic and elimination mechanisms in the urease catalyzed decomposition of urea. *J. Phys. Chem. B* **2007**, *111*, 10263–10274. [CrossRef]
123. Carlsson, H.; Nordlander, E. Computational Modeling of the Mechanism of Urease. *Bioinorg. Chem. Appl.* **2010**, *2010*, 8. [CrossRef]

124. Mazzei, L.; Cianci, M.; Benini, S.; Ciurli, S. The impact of pH on catalytically critical protein conformational changes; the case of the urease, a nickel enzyme. *Chem. Eur. J.* **2019**. [CrossRef]
125. Mazzei, L.; Cianci, M.; Benini, S.; Ciurli, S. The structure of the elusive urease-urea complex unveils the mechanism of a paradigmatic nickel-dependent enzyme. *Angew. Chem. Int. Ed.* **2019**, *131*, 7493–7497. [CrossRef]
126. Desguin, B.; Goffin, P.; Viaene, E.; Kleerebezem, M.; Diaconescu, V.M.; Maroney, M.J.; Declercq, J.P.; Soumillion, P.; Hols, P. Lactate racemase is a nickel-dependent enzyme activated by a widespread maturation system. *Nat. Commun.* **2014**, *5*, 3615. [CrossRef]
127. Xu, T.; Bauer, G.; Hu, X. A novel nickel pincer complex in the active site of lactate racemase. *ChemBioChem* **2016**, *17*, 31–32. [CrossRef]
128. Zamble, D. It costs more than a nickel. *Science* **2015**, *349*, 35–36. [CrossRef]
129. Yu, M.J.; Chen, S.L. From NAD$^+$ to Nickel Pincer Complex: A Significant Cofactor Evolution Presented by Lactate Racemase. *Chem. Eur. J.* **2017**, *23*, 7545–7557. [CrossRef]
130. Zhang, X.; Chung, L.W. Alternative Mechanistic Strategy for Enzyme Catalysis in a Ni-Dependent Lactate Racemase (LarA): Intermediate Destabilization by the Cofactor. *Chem. Eur. J.* **2017**, *23*, 3623–3630. [CrossRef]
131. Qiu, B.; Yang, X.Z. A bio-inspired design and computational prediction of scorpion-like SCS nickel pincer complexes for lactate racemization. *Chem. Commun.* **2017**, *53*, 11410–11413. [CrossRef]
132. Wang, B.; Shaik, S. The Nickel-Pincer Complex in Lactate Racemase Is an Electron Relay and Sink that acts through Proton-Coupled Electron Transfer. *Angew. Chem. Int. Ed.* **2017**, *56*, 10098–10102. [CrossRef]
133. Zhang, T.H.; Zhang, X.Y.; Chung, L.W. Computational Insights into the Reaction Mechanisms of Nickel-Catalyzed Hydrofunctionalizations and Nickel-Dependent Enzymes. *Asian J. Organ. Chem.* **2018**, *7*, 522–536. [CrossRef]
134. Pelmenschikov, V.; Siegbahn, P.E.M. Nickel superoxide dismutase reaction mechanism studied by hybrid density functional methods. *J. Am. Chem. Soc.* **2006**, *128*, 7466–7475. [CrossRef]
135. Prabhakar, R.; Morokuma, K.; Musaev, D.G. A DFT study of the mechanism of Ni superoxide dismutase (NiSOD): Role of the active site cysteine-6 residue in the oxidative half-reaction. *J. Comput. Chem.* **2006**, *27*, 1438–1445. [CrossRef]
136. Barondeau, D.P.; Kassmann, C.J.; Bruns, C.K.; Tainer, J.A.; Getzoff, E.D. Nickel superoxide dismutase structure and mechanism. *Biochemistry* **2004**, *43*, 8038–8047. [CrossRef]
137. Wang, Q.; Chen, D.; Liu, X.; Zhang, L.F. Theoretical mechanisms of the superoxide radical anion catalyzed by the nickel superoxide dismutase. *Comput. Theor. Chem.* **2011**, *966*, 357–363. [CrossRef]
138. Shearer, J.; Schmitt, J.C.; Clewett, H.S. Adiabaticity of the Proton-Coupled Electron-Transfer Step in the Reduction of Superoxide Effected by Nickel-Containing Superoxide Dismutase Metallopeptide-Based Mimics. *J. Phys. Chem. B*, **2015**, *119*, 5453–5461. [CrossRef]

© 2019 by the authors. Licensee MDPI, Basel, Switzerland. This article is an open access article distributed under the terms and conditions of the Creative Commons Attribution (CC BY) license (http://creativecommons.org/licenses/by/4.0/).

Article

Preliminary Characterization of a Ni^{2+}-Activated and Mycothiol-Dependent Glyoxalase I Enzyme from *Streptomyces coelicolor*

Uthaiwan Suttisansanee [1,2] and John F. Honek [1,*]

1. Department of Chemistry, University of Waterloo, 200 University Avenue West, Waterloo, ON N2L 3G1, Canada
2. Institute of Nutrition, Mahidol University, 25/25 Phutthamonthon 4 Rd., Salaya, Phutthamonthon, Nakhon Pathom 73170, Thailand
* Correspondence: jhonek@uwaterloo.ca; Tel.: +1-1519-888-4567 (ext. 35817)

Received: 6 July 2019; Accepted: 9 August 2019; Published: 14 August 2019

Abstract: The glyoxalase system consists of two enzymes, glyoxalase I (Glo1) and glyoxalase II (Glo2), and converts a hemithioacetal substrate formed between a cytotoxic alpha-ketoaldehyde, such as methylglyoxal (MG), and an intracellular thiol, such as glutathione, to a non-toxic alpha-hydroxy acid, such as D-lactate, and the regenerated thiol. Two classes of Glo1 have been identified. The first is a Zn^{2+}-activated class and is exemplified by the *Homo sapiens* Glo1. The second class is a Ni^{2+}-activated enzyme and is exemplified by the *Escherichia coli* Glo1. Glutathione is the intracellular thiol employed by Glo1 from both these sources. However, many organisms employ other intracellular thiols. These include trypanothione, bacillithiol, and mycothiol. The trypanothione-dependent Glo1 from *Leishmania major* has been shown to be Ni^{2+}-activated. Genetic studies on *Bacillus subtilis* and *Corynebacterium glutamicum* focused on MG resistance have indicated the likely existence of Glo1 enzymes employing bacillithiol or mycothiol respectively, although no protein characterizations have been reported. The current investigation provides a preliminary characterization of an isolated mycothiol-dependent Glo1 from *Streptomyces coelicolor*. The enzyme has been determined to display a Ni^{2+}-activation profile and indicates that Ni^{2+}-activated Glo1 are indeed widespread in nature regardless of the intracellular thiol employed by an organism.

Keywords: glyoxalase; nickel; streptomyces; mycothiol; metalloenzyme

1. Introduction

Intracellularly generated cytotoxic alpha-ketoaldehydes, such as methylglyoxal (MG; Figure 1), are highly reactive electrophiles that can inhibit protein synthesis, form adducts with protein, DNA and RNA, and promote Advanced Glycation End-products (AGEs) [1–8]. Detoxification of these molecules is a critical cellular process. The enzymes Glyoxalase I (Glo1) and Glyoxalase II (Glo2) work in tandem and convert the hemithioacetal, formed between the cytotoxic α-ketoaldehyde and an intracellular thiol, such as glutathione (GSH), into a non-toxic alpha-hydroxy acid, such as D-lactate (in the case of MG), and the regenerated thiol (Figure 1) [9–13]. Reduction of cellular toxicity is also dependent upon S-lactoylglutathione levels as this product of the Glo1 reaction can also control bacterial potassium efflux pumps and cytosolic acidification [14]. The metalloenzyme Glo1 catalyzes the isomerization reaction resulting in the production of the thioester product. Hydrolysis of the thioester is catalyzed by the second metalloenzyme, Glo2. An additional type of glyoxalase enzyme, Glo3, utilizes an active site cysteine to convert MG to D-lactate directly [15–18].

Glo1 is a metalloenzyme that can be divided into two different metal-activation classes [19]. The Zn^{2+}-activated class is exemplified by the enzyme from *Homo sapiens* (homodimeric), and the

Ni^{2+}-activated Glo1 class is exemplified by the enzyme from *Escherichia coli* (homodimeric) [20–25]. In the case of the *E. coli* enzyme, the Ni^{2+}-active form (Protein Database (PDB): 1F9Z) is situated in an octahedral ligand arrangement (residues from chain A: His5 and Glu56; residues from chain B: His74 and Glu122) with two water molecules as non-proteinaceous ligands, but the inactive Zn^{2+}-bound complex (PDB: 1FA5) has a trigonal bipyramidal ligand arrangement and has only one water molecule as a non-proteinaceous ligand (Figure 2) [26,27]. A similar octahedral ligand arrangement (residues from chain A: Gln34 and Glu100; residues from chain B: His127 and Glu173) is observed for the catalytically active Zn^{2+}-bound form of the *H. sapiens* Glo1 (PDBN: 1QIN) [28,29]. Thus, only octahedral metal environments appear to be active for this enzyme, regardless of the metal-activation class the Glo1 belongs to, and this arrangement appears critical to the enzyme's catalytic mechanism [30–32].

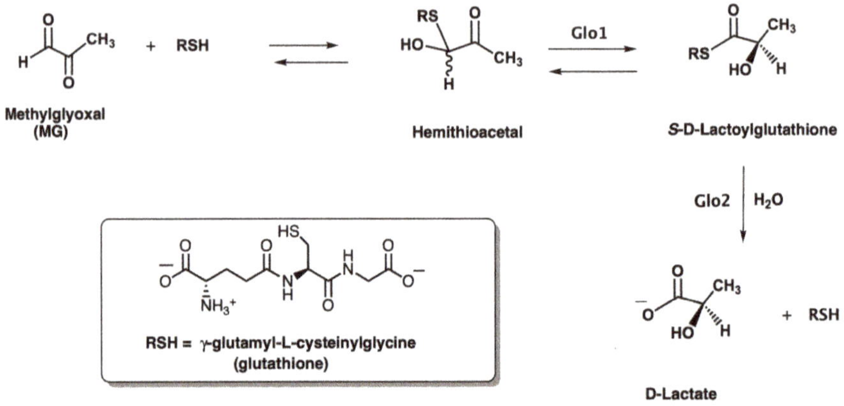

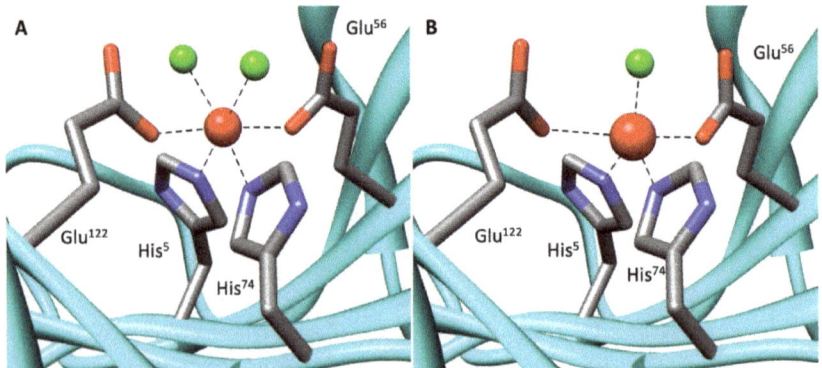

Figure 1. The overall reaction of the enzymes glyoxalase I (Glo1) and glyoxalase II (Glo2). The intracellular thiol utilized in many, but not all, organisms is the tripeptide glutathione.

Figure 2. The ribbon structure of (**A**) the active Ni^{2+}-bound *Escherichia coli* Glo1 (PDB ID: 1F9Z) forming an octahedral geometry with four metal-binding residues (His^5, Glu^{56}, His^{74} and Glu^{122}) and two nearby water molecules around the metal center and (**B**) the inactive Zn^{2+}-bound *E. coli* Glo1 (PDB ID: 1FA5) forming a trigonal bipyramidal metal coordination with the same four metal-binding residues but with only one water molecule [27]. The metal-binding residues are shown in sticks, and active site water molecules are shown in green. The divalent metals in the catalytic pocket are represented as red spheres. The 3D structures were generated by using the UCSF Chimera program (University of California, San Francisco, CA, USA) [33].

Structural studies have been reported on several Glo1, and these reports have provided additional insight into nature's control of active site geometry. As mentioned, the *H. sapiens* and *E. coli* enzymes are homodimeric, and each of their two active sites is formed by residues contributed by each of the two subunits. The molecular structure of the *Clostridium acetobutylicum* Glo1 (Ni^{2+}-activation class) is also homodimeric, but each of the two active sites is formed by protein residues solely from a *single* subunit [34]. Furthermore, larger single chain Glo1 are also known [35–41], and the recent report on the X-ray structure of the maize enzyme shows two metal-binding sites formed by the single protein chain with one site being catalytically active [42]. For the homodimeric Glo1, as well as the maize enzyme, detailed studies employing metal activation, NMR and X-ray experiments have provided unambiguous evidence that only a single active site is required for maximal activity [42–44]. In addition, deletional mutagenesis experiments have recently provided insight into the underlying structural factors involved in metal selectivity among the Glo1 enzymes [45].

To further add to the complex but fascinating biochemistry of Glo1, it has become clear that Glo1 enzymes have evolved to employ whatever major intracellular thiol [46–50] is available to them within a particular organism (Figure 3). In addition to glutathione (GSH), which is present in eukaryotes and most Gram-negative bacteria, protozoans, such as *Leishmania* and *Trypanosoma* employ N^1,N^8-bis (L-γ-glutamyl-L-hemicystinyl-glycyl) spermidine, trypanothione, for their cellular biochemistry, which includes their Glo1 enzymes [51–53]. The *Leishmania major* Glo1 has been determined to be a Ni^{2+}-activated class enzyme and lacks catalytic activity with Zn^{2+} [51,54]. The Glo1 from *Trypanosoma cruzi* has also been determined to be Ni^{2+}-activated enzymes [55]. The intracellular thiol, (2S)-2-[[2-(L-cysteinylamino)-2-deoxy-α-D-glucopyranosyl]oxy]succinic acid, known as bacillithiol (Figure 3), is present in bacilli, such as *Bacillus subtilis* and *B. anthracis*, and has also been identified in *Staphylococcus aureus*, as well as *Deinococcus radiodurans* [49,56–59]. Genetic studies on *Bacillus subtilis* have indicated the likely existence of a Glo1 enzyme employing bacillithiol, although no reports on the isolation and characterization of a bacillithiol-dependent Glo1 have been reported [60].

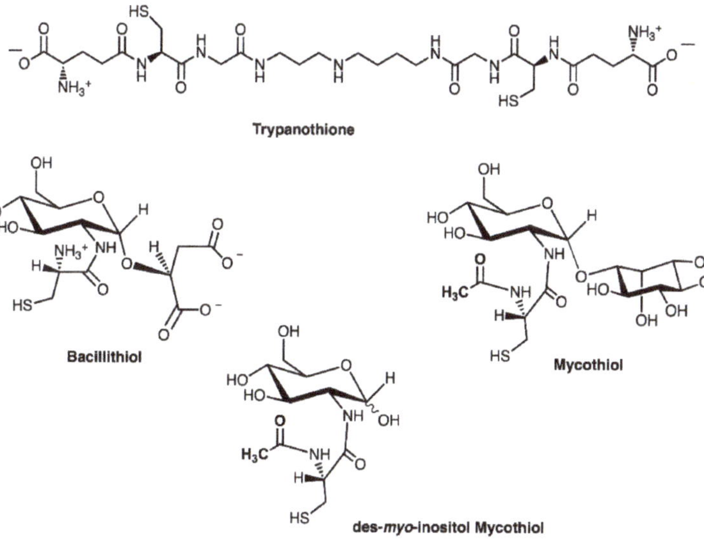

Figure 3. Several important intracellular thiols found in nature (trypanothione, bacillithiol, mycothiol). Des-*myo*-inositol mycothiol (tMSH) is an analog of mycothiol also employed in the current study.

Actinomycetes and mycobacteria biosynthesize the thiol, 1-D-myo-inositol-2-(N-acetyl-L-cysteinyl) amido-2-deoxy-α-D-glucopyranoside, mycothiol (Figure 3) [61–64]. Although substantial information is now available on the biochemical pathways employing mycothiol, and the functions of mycothiol and its metabolic pathways appear to parallel those of glutathione, to date no reports have appeared on the characterization of a mycothiol-dependent Glo1. However, mycothiol-null mutants in *Corynebacterium glutamicum* have been reported to endow this organism with MG sensitivity, indicating the possible existence of a mycothiol-dependent Glo1 [63].

In order to extend our knowledge of the Glo1 metalloenzymes, an investigation was undertaken to identify and provide a preliminary characterization of a mycothiol-dependent Glo1 (Figure 4), if it indeed existed. The current investigation reports on the identification of such an enzyme in *Streptomyces coelicolor* and provides a preliminary characterization of the isolated Glo1. The Glo1 from *Streptomyces coelicolor* has been determined to exhibit a homodimeric quaternary structure and to be a member of the Ni^{2+}-activation class of glyoxalase I enzymes.

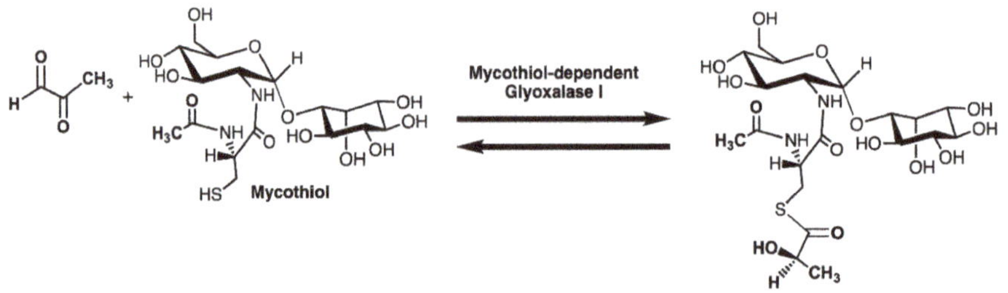

Figure 4. The overall reaction of a mycothiol-dependent glyoxalase I enzyme.

2. Results and Discussion

2.1. Sequence Analysis

As mycothiol is the key intracellular thiol in Streptomyces, sequence searching of a number of sequence databases, including the Streptomyces Annotation Server (http://streptrdb.streptomyces.org.uk), for a possible Glo1 enzyme employing various known Glo1 sequences was undertaken for the organism Streptomyces coelicolor A3(2) [65,66]. The search revealed two genes of interest with annotations of a putative dioxygenase/glyoxalase/ bleomycin resistance family gene product (SCO1970; EMBLCAC42744; NCBI reference sequence: NP_626233.1), which was termed putative dioxygenase (PDO), and a putative lyase (SCO2237; EMBLCAC37263; NCBI reference sequence: NP_626487.1), which was termed putative lyase (PLA). Multiple sequence alignments of PDO and PLA with other Glo1 from various sources, including Gram-negative bacteria (*E. coli* [24], *Yersinia pestis* [19], *Neisseria meningitides* [19,67], *Pseudomonas aeruginosa* GloA2 [68], *P. aeruginosa* GloA3 [68], and *P. putida* [22]), protozoa (*Leishmania major* [51]), human (*H. sapiens* [23]), and the previously identified gene [60] correlated with possible bacillithiol-dependent Glo1 activity in Bacillus subtilis are shown in Figure 5. In this group, the known metal-specificities of the Glo1 are as follows: Zn^{2+}-dependent (*P. putida, H. sapiens, P. aeruginosa* GloA3) and Ni^{2+}-dependent (*E. coli, Y. pestis, N. meningitidis, P. aeruginosa* GloA2, *L. major*). The metal specificity of the *B. subtilis* enzyme is unknown as the protein has not been isolated nor characterized. PDO possesses four potential metal-binding residues (His^{15}, Asp^{63}, His^{89}, Glu^{142}) corresponding to metal-binding residue positions in other Glo1 enzymes (asterisks in Figure 5). The corresponding residues in PLA (Val^7, Asp^{62}, Tyr^{75}, Gln^{127}) do not map as well, and the alignment suggests that PDO would be the more likely candidate to demonstrate Glo1 activity, although both gene products were investigated. Based on the alignment, PDO has a 21.7% identity and 20.2% identity to the *N. meningitidis* Glo1 and the *B. subtilis* gene product associated with the bacillithiol-dependent Glo1 activity, respectively (Figure S1). Based on previous sequence alignments and deletional mutagenesis experiments, it has been suggested that shorter Glo1 sequences tend to be Ni^{2+}-activated class enzymes [19]. In the case of PDO, if indeed a Glo1, a Ni^{2+}-activated class enzyme was hypothesized.

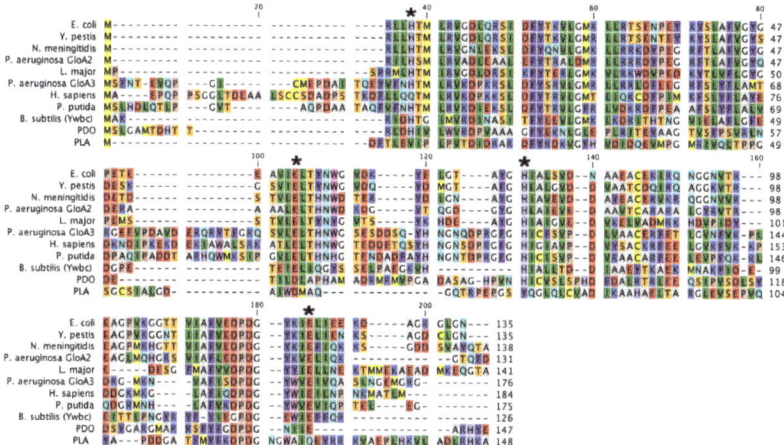

Figure 5. The multiple sequence alignment of the putative dioxygenase (PDO, CAC42744) and the putative lyase (PLA, CAC37263) from Streptomyces coelicolor A3(2) with glyoxalase I from other organisms (organism name followed by National Center for Biotechnology Information (NCBI) accession number), including *E. coli* (NP_310387), *Y. pestis* (ZP_01887743.1), *N. meningitides* (CAA74673), *P. aeruginosa* GloA2 (ATE47122.1), *L. major* (AAT98624.1), *P. aeruginosa* GloA3 (AAG08496.1), *H. sapiens* (AAB49495), *P. putida* (AAN69360), *B. subtilis* bacillithiol-dependent MG resistance proposed Glo1 (P39586.1). The metal-binding residues are highlighted with asterisks. Three different Glo1 have been identified in *P. aeruginosa* PA01: GloA1 (Ni^{2+}-activated), GloA2 (Ni^{2+}-activated and GloA3 (Zn^{2+}-activated) [68]. The alignment was created using CLC Main Workbench (version 8.1.2) with the accurate alignment algorithm (http://www.qiagenbioinformatics.com).

2.2. Overproduction, Isolation and Characterization of PDO and PLA

The putative dioxygenase gene (*pdo*) and putative lyase gene (*pla*) from chromosomal DNA from *S. coelicolor* A3(2) (NC_003888) were cloned into pET-28b(+) expression vectors utilizing NdeI and BamHI restriction endonuclease enzymes and polymerase chain reaction (PCR) to generate the purified proteins with an N-terminal His-tag followed by a thrombin protease cleavage site. The protein purification of PDO and PLA made use of the N-terminal His-tag, which was eventually removed by thrombin protease, resulting in three extra amino acids at the N-terminus (Gly-Ser-His) fused to the N-terminal Met residues. These modifications changed the predicted molecular weights of PDO and PLA to 16569.5 Da and 16749.9 Da, respectively, which were confirmed by electrospray ionization mass spectrometry (ESI-MS) analysis (Figures S2 and S3). No post-translational modification of either purified protein was observed. Analysis of a gel permeation chromatographic separation suggested a homodimeric quaternary structure for both PDO (1.5 mg mL^{-1}) and PLA (0.6 mg mL^{-1}) in 50 mM Tris (pH 8.0) and 150 mM KCl with a molecular weight of approximately 27.51 ± 2.46 kDa and 33.15 ± 2.89 kDa, respectively (Figure S4). The thermal stabilities of PDO and PLA as isolated were performed using differential scanning calorimetry (DSC) analysis, which suggested the estimated T_m of PDO (3 mg mL^{-1}) and PLA (1.2 mg mL^{-1}) in 50 mM MOPS (pH 7.0) and 10% *v/v* glycerol to be 57.8 and 58.5 °C, respectively (Figure S5). Their secondary structures were investigated by circular dichroism (CD) analysis which showed that both PDO (5.6 mg mL^{-1}) and PLA (3.6 mg mL^{-1}) in buffer containing 50 mM potassium phosphate buffer (KPB; pH 7.0) and 200 mM KCl possessed a negative maximum at 222 nm and a small shoulder at 208 nm, suggesting the presence of predominantly β-sheet structures (Figure S6), which are consistent with the previous reports on the X-ray crystallographic structures of other Glo1. The K2D3 secondary structural prediction program (http://cbdm-01.zdv.uni-mainz.de/~{}andrade/k2d3/) also estimated the secondary structural contents of PDO to be 1–2% α-helix, 37–41% β-sheet and 58–62% random coil and of PLA to be 2–6% α-helix, 34–40% β-sheet and 57–62% random coil. Other external factors, including pH (5–9), types of buffer (HEPES, KPB, Tris, MOPS and MOPSO), ionic strength (0–500 mM NaCl) and additive (0–30% v/v glycerol) did not significantly influence the secondary structures of either protein (data not shown).

2.3. Enzyme Assay for Thiol Cofactors

Optimization of the Glo1 enzyme assays employed the simpler and more readily available [69] truncated form of mycothiol (tMSH; des-myo-inositol mycothiol) (Figure 3) [70]. This analog has been employed in the study of mycothiol-utilizing enzymes [71]. Mycothiol was also employed at various stages in this study, especially with respect to final enzyme kinetic studies [70]. Experiments to study the details of the non-enzymatic formation of the hemithioacetal upon MG and tMSH reaction in solution were undertaken initially and included enzyme kinetic assays and ^{1}H NMR time course studies, benchmarked to previously studied MG and glutathione adduct formation studies (Figures S7–S10 and Tables S1 and S2) [24,72–79]. Analysis of the results of these experiments indicated that a 30 min equilibration time would be appropriate for the MG-tMSH hemithioacetal equilibrium to be attained, a time somewhat longer than that for the equilibrium to be reached for the MG-GSH hemithioacetal equilibrium to be reached (15 min).

The dissociation constant (K_d) of the hemithioacetal (MG-tSH) was also determined, due to its significance in the calculation of the exact concentration of the substrate used in the Glo1 kinetic assays. After allowing the hemithioacetal to reach its equilibrium (15 min for MG-GSH and 30 min for MG-tMSH), the dissociation constant could be measured using different concentrations of MG with a fixed concentration of thiols, an approach previously employed in the literature for MG-GSH [24,72–77]. The dissociation profile of MG-GSH suggested its K_d of approximately 3.19 ± 0.29 mM (Figure S10A), which was in excellent agreement to the literature value (K_d of 3.1 mM). The K_d of the hemithioacetal forming non-enzymatically between MG and tMSH was determined similarly and estimated to be 3.33 ± 0.41 mM (Figure S10B).

The increase in UV absorbance at 240 nm observed during the reaction of Glo1 is due to the formation of the product, S-D-lactoylglutathione, from the Glo1 reaction using MG-GSH as a substrate. The investigation on the optimum detection wavelength for the reaction using MG-tMSH substrate was performed similarly, and the results suggested that the detection at 240 nm could also be employed to detect the thioester product from the MG-tMSH substrate (Figure S11). Furthermore, the expected product, S-D-lactoyl-des-myo-inositol mycothiol, produced from the Glo1 catalyzed reaction using Ni^{2+}-reconstituted PDO (25 µg in 600 µL assay) with MG-tMSH (5 mM, K_d of 3.3 mM and equilibrium time of 30 min), was isolated using reverse phase C18 HPLC (Figure S12) and was identified by ESI-MS analysis (Figure S13), consistent with PDO serving as a mycothiol-dependent Glo1. It should be noted however that although the mass of the product is consistent with the chemical structure of the expected product, the exact determination of the stereochemistry of the lactate as D-lactate has not been shown, although it is highly likely given the commonality of the stereochemistry found in the product from various Glo1 enzymes.

Based on the information obtained on hemithioacetal formation as stated above, the substrate specificity of the Glo1 reaction was investigated in relation to metal-incorporation and the presence of PDO, PLA or commercial yeast Glo1 (1.5–3.0 µg for PDO and PLA and 0.05 µg for yeast Glo1 in 200 µL assay). The detection of the substrates, including MG, GSH, tMSH, hemithioacetal of MG-GSH and hemithioacetal of MG-tMSH did not show any significant increase in signal at 240 nm without the corresponding enzyme. Neither was product formation detected for any of these substrates in the presence of added metal ions (without the addition of the enzyme), suggesting that substrates with and without additional metal ions do not interfere with the detection of the enzyme reaction, if any—nor do they contribute any significant non-enzyme background activity. No Glo1 activity was observed for reactions containing only MG, only GSH, only tMSH, or only MG-GSH with Ni^{2+}-reconstituted PDO. Neither were any reactions observed to occur in the reactions of apo-PDO and denatured PDO (boiled for 10 min) with hemiacetal MG-tMSH. However, trace activity was obtained in the reaction of isolated

PDO (purified PDO without any additional metals) with MG-tMSH, while high activity was observed in the reaction of Ni^{2+}-reconstituted PDO with the same substrate. These observations suggested that the incorporated metal into PDO is significant for the Glo1 reaction to occur and that the "as isolated" PDO is not entirely in its apo-form, and this was confirmed by inductively coupled mass spectrometry (ICP-MS) analysis for the presence of metal ions in the PDO protein (Table S3). The isolated PDO enzyme might accept some metals from the organism's growth environment during the expression and purification processes undertaken in the laboratory. Additionally, the reaction catalyzed by the PDO enzyme is specific to the hemithioacetal substrate, MG-tMSH, while the substrate of the GSH-dependent Glo1 reaction, MG-GSH (5 mM), did not act as a substrate for PDO (data not shown). PLA, on the other hand, did not exhibit any glutathione or tMSH-dependent Glo1 activity under the above conditions. The enzymatic reaction catalyzed by yeast Glo1 is specific to its substrate, MG-GSH, in which it was observed to exhibit no activity in the presence of the MG-tMSH hemithioacetal.

2.4. Metal Characterization and Kinetic Studies

The metal analysis determined by ICP-MS (ALS Laboratory Group, Waterloo, ON, Canada), undertaken utilizing protocols [24,25,34,45,68] previously employed, suggested that the purified "as isolated" (no exogenously added metal ions) PDO (0.088 mg/mL) binds to copper, cobalt, nickel and zinc ions (Tables S3 and S4). The metal per enzyme ratios of nickel-and zinc-bound PDO were approximately 0.16–0.17, while those of cobalt and copper-bound enzyme were approx. 0.02. Thus, the ratios of Ni^{2+} and Zn^{2+} contents were approximately ten times higher than those of Co^{2+} and Cu^{2+}, which might suggest that PDO has a higher metal-binding affinity for Ni^{2+} and Zn^{2+} than Co^{2+} and Cu^{2+} ions, or could reflect the availability of these metal ions in E. coli during protein production and isolation steps. The concentrations of these incorporated metals in the isolated protein were low and appeared almost completely in a de-metallated form. The ICP-MS analysis on this form of the PDO indicated that the enzyme contained no iron, a common metal found in dioxygenase enzymes, but it did contain ions, such as zinc and nickel bound to the isolate protein. Both Zn^{2+} and Ni^{2+}, as previously mentioned, are metals that are usually associated with Glo1 enzymes. These results suggest that PDO may bind either of these metals in intact S. coelicolor, although, of course, these metals could have been picked up by the protein in steps associated with the protein isolation.

Analysis of the metal activation profile for PDO suggested that the enzyme was activated in the presence of Ni^{2+}, Cu^{2+}, and to a much lower extent by Cd^{2+} and Ca^{2+} (activity of PDO: $Ni^{2+} > Cu^{2+} > Cd^{2+} > Ca^{2+}$) (Figure 6A). The enzyme with incorporated Ni^{2+} exhibited the highest activity of conversion of the MG-tMSH hemithioacetal into its thioester, while that of Cu^{2+}-reconstitution gave approximately 60% compared to that of the Ni^{2+}-reconstituted enzyme. The activities with Cd^{2+} and Ca^{2+}-were low, while other metal ions (Zn^{2+}, Mg^{2+}, Co^{2+}, Mn^{2+}) were not activating. Metal titration studies of PDO with Ni^{2+} and Cu^{2+} indicated that 1 mole of metal per mole dimeric enzyme could optimize enzymatic activity (Figure 6B), suggesting a tight binding of the metal, as well as one functional active site (possibly two active sites per dimeric enzyme as PDO exhibits homodimeric quaternary structure as previously mentioned). These data are consistent with previous reports on other Glo1 (such as E. coli Glo1, P. aeruginosa GloA1, GloA2, and many others), which also find that the metal per dimeric enzyme ratio is approximately one [19,24,44,68]. The pH dependency of PDO activity was also determined (Figure 6C).

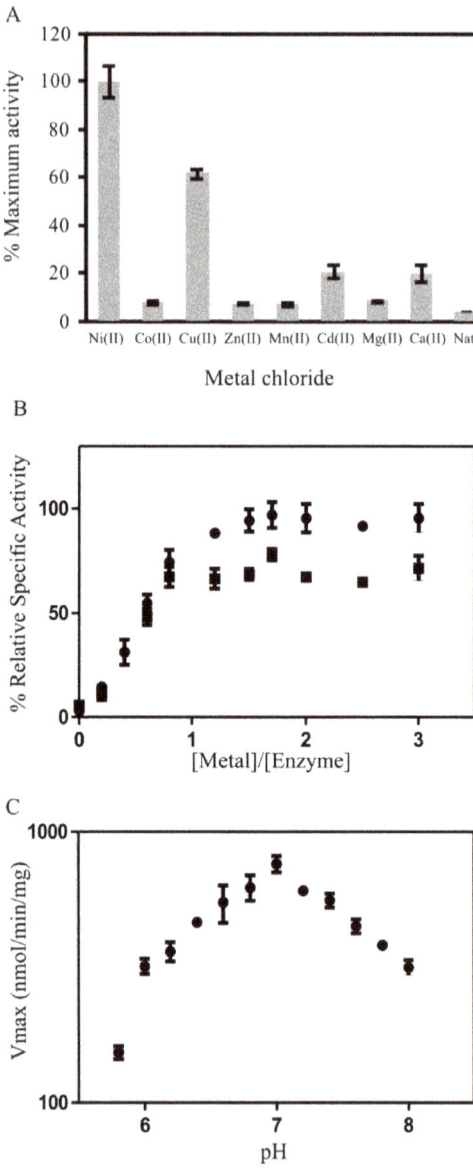

Figure 6. (**A**) PDO metal activation profile after metal ion (5 equivalents) preincubation of the apo-form of the enzyme with metal chloride (conditions: PDO (3.1 µg in 200 µL) incubated with metal chlorides overnight at 4 °C); (**B**) % Relative specific enzyme activity of PDO versus metal ion titration with (●) Ni^{2+} and (■) Cu^{2+}; (**C**) pH dependency of PDO (3.125 µg in 200 µL assay; 5 equivalents $NiCl_2$) with the substrate MG-tMSH (0.5 mM, K_d = 3.3 mM) that was incubated for 30 min in potassium phosphate buffer at various pH (5.8–8) at 25 °C.

The kinetics of Ni^{2+}-reconstituted PDO with MG-tMSH yielded a calculated V_{max} of 5.8 ± 1.0 µmol/min/mg, K_m of 1.25 ± 0.13 mM and k_{cat} of 3.2 s^{-1} (Table 1). The K_m is somewhat higher compared to other K_m values for the MG-GSH substrate processed by other Glo1 [24,34,68,80,81], which is probably due to the structure of the MG-tMSH hemithioacetal lacking the inositol moiety. This hypothesis was confirmed by the kinetics of Ni^{2+}-reconstituted PDO with the hemithioacetal of MG-MSH which yielded a calculated V_{max} of 11.5 ± 1.8 µmol/min/mg, K_m of 0.61 ± 0.06 mM and k_{cat} of 6.4 s^{-1} (Table 1).

Table 1. Kinetic data for the enzyme reaction catalyzed by Ni^{2+}-reconstituted PDO.

Thiol Cofactors	V_{max} (µmol/min/mg)	K_m (µM)	k_{cat} (s^{-1})	k_{cat}/K_m ($M^{-1}s^{-1}$)
MSH	11.5 ± 1.8	612 ± 56	6.4	10403
tMSH	5.8 ± 1.0	1247 ± 127	3.2	2545

Conditions: PDO (1.5–3.0 µg in 200 µL assay) using two different hemithioacetal substrates (0.08–1 mM, K_d = 3.3 mM, 30 min equilibrium time) that formed non-enzymatically between methylglyoxal and thiol cofactors, including isolated mycothiol (MSH) and synthesized truncated mycothiol (tMSH), in 50 mM KPB (pH 6.6) at 25 °C. The protein was prepared in its apo-form and incubated with five equivalents of $NiCl_2$ overnight at 4 °C prior to performing the assay.

Metal analysis performed on the "as isolated" PLA exhibited similar results, where copper, nickel and zinc were detected bound to various copies of the protein (Tables S3 and S5). The metal per dimeric enzyme ratios of nickel-and zinc-bound PLA were approx. 0.11–0.16, while the ratio of copper bound enzyme was 0.08. The concentrations of these metals were low, which indicated that PLA is isolated in predominantly apo-form. Glo1 activity of PLA, on the other hand, was undetectable under the studied conditions. The enzyme exhibited no activity with MG-tMSH nor MG-GSH (data not shown). Neither activity was observed in the presence of different divalent metals ($ZnCl_2$, $NiCl_2$, $CoCl_2$, $CuCl_2$, $MnCl_2$, $MgCl_2$, $CaCl_2$ and $CdCl_2$), suggesting that PLA does not function as a Glo1.

Glo1 is a member of the $\beta\alpha\beta\beta\beta$ superfamily of proteins, consisting of fosfomycin resistance protein (FosA), methylmalonyl-CoA epimerase (MMCE), extradiol dioxygenase (DIOX), mitomycin C resistance protein (MRP) and bleomycin resistance protein (BRP) [23,34,40,82–88]. Among these proteins, Glo1 and MMCE share high structural similarity, thiol cofactor (GSH), presence of divalent metal and tetradentated metal-binding protein ligands [34,85]. Due to these similarities, an investigation of possible MMCE activity in PDO and PLA with selected incorporated metals (Ni^{2+}, Co^{2+}, Cu^{2+}, and Zn^{2+}) was performed, similar to those previously reported for studies on the C. acetobutylicum Glo1 [34]. Ni^{2+}, Co^{2+} and Zn^{2+} are activating metal ions for particular Glo1 enzymes, while Co^{2+} is normally found in MMCE enzymes. Cu^{2+} was also chosen for our studies, due to its observed activation of PDO in the Glo1 reaction with Mg-tMSH. However, neither enzyme exhibited any MMCE activity understudied conditions (data not shown), and no divalent metal could activate these enzymes to accomplish the MMCE activity. Thus, PDO only functions as a Glo1 and not as an MMCE, while PLA is neither a Glo1 nor an MMCE. Previous investigations on the MMCE activity in GSH-dependent Glo1 from Clostridium acetobutylicum suggest that the enzyme is highly specific to its natural substrate (MG-GSH) even though its structure is more likely to resemble that of MMCE than other Glo1 [34]. These results suggest that Glo1 does not possess cross function with MMCE regardless of its different types of thiol cofactors that might be used across nature.

PDO is activated with Ni^{2+} and Cu^{2+} atoms. The unusual metal activation by Cu^{2+} in PDO may be anomalous and not related to any physiological role of this metal in the organism. However, it is interesting to speculate that it may be an important metal for this enzyme in the organism, since cytosolic copper has been found to be a major modulator of *Streptomyces coelicolor* germination, development and secondary metabolism [89]. It is also interesting to note that mycothiol has been found to be important in copper resistance in *C. glutamicum* [63], and perhaps there may be an advantage to mycothiol-dependent Glo1 to be able to utilize Cu^{2+} as an activating metal. However, this requires further research to confirm or dispute.

It had been found earlier that the isolated mammalian GloI exhibited high activity in the presence of Mg^{2+} and was thought to be a magnesium activated enzyme [21,90–92]. However, the naturally bound metal was later discovered to be zinc, which incorporated in the ratio of 1 mole metal per mole enzyme subunit [20]. The magnesium atoms were believed to be recovered from protein preparation processes and had no correlation with enzymatic activity. The analysis by electron paramagnetic resonance (EPR) of Co^{2+}-bound human Glo1, however, suggested octahedral metal coordination with four metal-binding protein residues and two water molecules for this ion, as well as other activating metals, such as magnesium [29]. Yeast Glo1 was also found to be partially activated by the presence of Mg^{2+} and Ca^{2+} [20,36,93]. It was possible that these metals could reactivate the enzyme, but play a major role in protein stability. As well, X-ray crystallographic structure of E. coli Glo1 with bound Cd^{2+} (PDB ID: 1FA7) suggested that the enzyme could be activated with this metal and its metal coordination forms an octahedral geometry [27]. However, the activity was low compared to the enzyme with Ni^{2+} and Co^{2+}, which might be related to the size of the metal that fits into the active site of the enzyme. Thus, the low activity observed in PDO with Ca^{2+} and Cd^{2+} might be explained in a similar fashion.

It is hoped that this preliminary characterization of a mycothiol-dependent Glo1 will provide important information for future studies on Glo1 enzymes from other microorganisms employing mycothiol, as well as bacillithiol, a closely related intracellular thiol.

3. Materials and Methods

3.1. DNA Cloning and Manipulation

All DNA manipulations and purifications were performed according to the protocols by Sambrook and Russell [94]. Putative dioxygenase gene (*pdo*) and putative lyase gene (*pla*) from S. coelicolor chromosome (NC_003888) were cloned into the pET-28b(+) expression vector utilizing *Nde*I and *Bam*HI restriction endonuclease enzymes and the polymerase chain reaction (PCR) to generate the protein with the N-terminal His-tag followed by thrombin protease cleavage site. The forward and reverse primers (Sigma Genosys, Oakville, Ontario, Canada) were designed as follows: (+) 5′-CCGAAGCTTCATATGAGCCTGGGAGCC-3′ and (−) 5′-GGCGAATTCGGATCC TACTCGTAGTG CCGG-3′ for *pdo* cloning, and (+) 5′-CCGAAGCTTCATATGGACTTCACGCTCG-3′ and (−) 5′-GGCGAATTCGGATCCTAGGCCTTGTGCCGG-3′ for *pla* cloning. The plasmid was heat shock transformed into the competent E. coli DH5α cells. Its DNA sequence was verified (Molecular Biology Core Facility, University of Waterloo, Waterloo, ON, USA) followed by heat shock transforming into E. coli BL21 (DE3) cells for protein expression purposes. Sequence alignments and percent identities were computed using CLC Main Workbench (version 8.1.2) (http://www.qiagenbioinformatics.com).

3.2. Protein Induction, Expression, and Purification

Bacterial cultures (1 L) containing kanamycin (30 µg/mL LB) were grown and shaken in an air incubator (220 rpm) at 37 °C until an OD_{600} of 0.6 was reached. Proteins were then induced with 0.5 mM IPTG for 4 h. Cell pellets were harvested by centrifugation at 6000× *g* for 10 min and flash frozen in liquid nitrogen before storing at −80 °C. The purifications of both PDO and PLA were performed as previously reported for the C. acetobutylicum Glo1 [34] using HisTrap HP Ni^{2+}-affinity (1 mL) and HiTrap Benzamidine FF affinity (1 mL) columns (GE Healthcare, Piscataway, NJ, USA). The protein concentration was determined by the Bradford Assay using bovine serum albumin (BSA) as a standard. Apo-enzyme preparation and metal analysis by ICP-MS were performed as described previously for other Glo1 and other enzymes [25,95–97]. The existence and the molecular weights of the denatured proteins were confirmed by the analysis of SDS-PAGE and ESI-MS data using a Micromass Q-TOF Global Ultima mass spectrometer (Mass Spectrometry Facility, University of Waterloo, Waterloo, ON, USA) using handling approaches as previously reported [68,78]. The molecular weight of the native protein was determined by gel permeation chromatography (Superose6 10/300 GL column) utilizing 50 mM Tris buffer (pH 8.0) and 150 mM KCl with a flow rate of 0.5 mL/min. A standard curve was prepared using Bio-Rad protein standards (Bio-Rad Laboratories, Hercules, CA, USA) containing γ-globulin (158 kDa), BSA (66 kDa), ovalbumin (44 kDa), carbonic anhydrase (29 kDa), myoglobulin (17 kDa) and vitamin B_{12} (1.35 kDa).

3.3. Protein Secondary Structure and Stability Experiments

CD experiments were performed on a Jasco J-715 spectropolarimeter from Jasco Inc. (Easton, MD, USA) and a Jasco J-700 Standard Analysis Program. The CD experiments with variation in protein concentration, pH, ionic strength and various additives were performed as previously reported. Prediction of protein secondary structure from CD spectra employed the K2D3 program as previously mentioned [98]. The transition midpoint (T_m) determined by DSC analysis was performed on a MicroCal VP-DSC microcalorimeter with cell volumes of 0.5 mL and self-contained pressurizing system of 0–30 p.s.i. for scanning solutions above boiling points to prevent any degassing during heating. The protein was run against 50 mM MOPS (pH 7.0) and 10% v/v glycerol over a temperature range of 10–80 °C with a scanning rate of 1 °C min^{-1}. The Origin, scientific plotting software package, supplied by MicroCal was used for baseline subtraction and T_m calculation by the integration of the heat capacity (C_p) versus temperature (t) curve.

3.4. Preparation of Hemithioacetal Substrate

tMSH was synthesized according to the protocol developed by Unson et al. [70]. Purified MSH was a kind gift from Dr. Gerald Newton (UCSD, San Diego, CA, USA) [70]. Two factors involved in the hemithioacetal substrate formation, including equilibrium time and dissociation constant (K_d) of MG-GSH, were previously reported, but none for MG-tMSH and MG-MSH. We assumed that the hemithioacetal formations of MG-tMSH and MG-MSH were in a similar fashion, thus only equilibrium time and K_d of MG-tMSH were determined as detailed in the Supplementary Materials using the protocols reported for MG-GSH. ^{1}H NMR experiments were performed on a Bruker (300 MHz) spectrometer (Bruker Ltd., Milton, ON, Canada). Predicted chemical shifts and integrations for substrates and hemithioacetals were estimated using ChemBioDraw Ultra 12.0 (http://www.cambridgesoft.com) software (version 12.0, PerkinElmer, Austin, TX, USA).

3.5. Enzyme Assays

An enzymatic assay was performed in 50 mM KPB (pH 6.6 or stated otherwise) using the hemithioacetal, a non-enzymatic product of MG and thiol cofactors (GSH, tMSH and MSH), as substrates utilizing previous Glo1 assay protocols [24,78]. The enzyme activity was measured as an increase in absorption at 240 nm (ε_{240} = 2860 M^{-1} cm^{-1}) for the formation of S-D-lactoylglutathione using a 96-well UV-visible plate on a plate reader. Since the product of the mycothiol dependent Glo1 reaction is unknown, its maximum absorbance and identification were investigated to confirm the existence of the reaction product thioester (Supplementary Materials). Enzyme kinetics were evaluated using the initial rate that was fitted by the Michaelis-Menten equation with least squares fit parameters using GraphPad Prism software version 5.00 (GraphPad Software, Inc., La Jolla, CA, USA). Typical conditions for Glo1 activity studies on PDO were: PDO (1.5–3.0 µg in 200 µL assay) using two different hemithioacetal substrates (0.08–1 mM, K_d = 3.3 mM, 30 min equilibrium time) that formed non-enzymatically between methylglyoxal and thiol cofactors, including isolated MSH and tMSH, in 50 mM KPB (pH 6.6) at 25 °C. The protein was prepared in its apo-form and incubated with five equivalents of NiCl$_2$ overnight at 4 °C prior to performing the assay.

Additionally, the investigation on the function of MMCE, a closely structural related protein to Glo1 in the same $\beta\alpha\beta\beta\beta$ superfamily, was performed using metal-substituted PDO and PLA (1 µg in 500 µL assay) in 20 mM Tris (pH 7.0) and 150 mM NaCl with the presence of 5 equivalents of Ni^{2+}, Co^{2+}, Cu^{2+} and Zn^{2+} as previously described for another Glo1 enzyme study [34].

4. Conclusions

Identification and isolation of a glyoxalase I (Glo1) enzyme from *Streptomyces coelicolor* A3(2) was accomplished in this study. A preliminary investigation of several properties of the protein, termed PDO (thermal stability, pH dependency on activity, secondary structure) was undertaken. PDO was found to be catalytically active as a glyoxalase I enzyme in the presence of Ni^{2+}, and converted the hemithioacetals formed from methylglyoxal and des-*myo*-inositol mycothiol (tMSH) to the identified thioester product. No enzymatic activity was observed using glutathione, suggesting PDO is specific to tMSH and after further experimentation, mycothiol itself. From the metal activation profile, PDO functions as a mycothiol-dependent Glo1 of the Ni^{2+}-activated class of glyoxalase I enzymes. This is the first protein characterization of an MSH-dependent Glo1.

Supplementary Materials: The following are available online at http://www.mdpi.com/2304-6740/7/8/99/s1, Figure S1: Percent identities calculated for various Glo1, Figure S2: SDS-PAGE and ESI mass spectrum of purified PDO, Figure S3: SDS-PAGE and ESI mass spectrum of purified PLA, Figure S4: Gel permeation chromatographic profile for PDO and PLA, Figure S5: DSC plots for PDO and PLA, Figure S6: CD plots for PDO and PLA, Figure S7: Enzymatic assay using various incubation times, Figure S8: ^{1}H NMR for various time incubations of MG and GSH, Table S1: Chemical Shifts and Integrations of 1H NMR signals (experimental and calculated) for GSH and MG-GSH, Figure S9: ^{1}H NMR for various time incubations of MG and tMSH, Table S2: Chemical Shifts and Integrations of 1H NMR signals (experimental and calculated) for tMSH and MG-tMSH, Figure S10: Plots for the determination of *Kd* for MG-GSH and MG-tMSH hemithioacetals, Figure S11: UV detection of MG-tMSH thioester product, Figure S12: HPLC chromatogram of products of MG-tMSH Glo1 activity by PDO, Figure S13: ESI mass spectrum of isolated thioester product, Table S3–S5: ICP-MS element analyses of "as isolated" PDO and PLA.

Author Contributions: Conceptualization: J.F.H.; Investigation and draft writing: U.S. and J.F.H.; writing—review and editing: J.F.H. and U.S.

Funding: This research was funded by NSERC (Canada) (JH) and the University of Waterloo (JH), and the Government of Thailand for a graduate scholarship (US).

Acknowledgments: The authors would like to thank Richard Smith for electrospray ionization MS analysis of the protein sample, Elisabeth Daub for *Streptomycetes* handling and David Ward, Michele Cossette and Vincent Azhikannickal for support of tMSH synthesis and enzyme purification. As well, we thank Zhengding Su, Christine Hand and Nicole Sukdeo for sharing their knowledge and suggestions. Gerard Wright (McMaster University) is gratefully acknowledged for supplying the *Streptomyces coelicolor* A3(2) strain. The authors gratefully acknowledge Gerald Newton (University of California Sand Diego) for the kind gift of purified mycothiol used in this investigation.

Conflicts of Interest: The authors declare no conflict of interest. The funders had no role in the design of the study; in the collection, analyses, or interpretation of data; in the writing of the manuscript, or in the decision to publish the results.

References

1. Thornalley, P.J. Dietary AGEs and ALEs and risk to human health by their interaction with the receptor for advanced glycation endproducts (RAGE)—An introduction. *Mol. Nutr. Food Res.* **2007**, *51*, 1107–1110. [CrossRef] [PubMed]
2. Krautwald, M.; Munch, G. Advanced glycation end products as biomarkers and gerontotoxins—A basis to explore methylglyoxal-lowering agents for Alzheimer's disease? *Exp. Gerontol.* **2010**, *45*, 744–751. [CrossRef] [PubMed]
3. Rabbani, N.; Thornalley, P.J. Dicarbonyl stress in cell and tissue dysfunction contributing to ageing and disease. *Biochem. Biophys. Res. Commun.* **2015**, *458*, 221–226. [CrossRef] [PubMed]
4. Schalkwijk, C.G. Vascular AGE-ing by methylglyoxal: The past, the present and the future. *Diabetologia* **2015**, *58*, 1715–1719. [CrossRef] [PubMed]
5. Honek, J.F. Glyoxalase biochemistry. *Biomol. Concepts* **2015**, *6*, 401–414. [CrossRef] [PubMed]
6. Thornalley, P.J. Protein and nucleotide damage by glyoxal and methylglyoxal in physiological systems—Role in ageing and disease. *Drug Metabol. Drug Interact.* **2008**, *23*, 125–150. [CrossRef] [PubMed]
7. Trellu, S.; Courties, A.; Jaisson, S.; Gorisse, L.; Gillery, P.; Kerdine-Romer, S.; Vaamonde-Garcia, C.; Houard, X.; Ekhirch, F.P.; Sautet, A.; et al. Impairment of glyoxalase-1, an advanced glycation end-product detoxifying enzyme, induced by inflammation in age-related osteoarthritis. *Arthritis Res. Ther.* **2019**, *21*, 18. [CrossRef] [PubMed]
8. De Bari, L.; Atlante, A.; Armeni, T.; Kalapos, M.P. Synthesis and metabolism of methylglyoxal, S-D-lactoylglutathione and D-lactate in cancer and Alzheimer's disease. Exploring the crossroad of eternal youth and premature aging. *Ageing Res. Rev.* **2019**, *53*, 100915. [CrossRef] [PubMed]

9. Sousa Silva, M.; Gomes, R.A.; Ferreira, A.E.; Ponces Freire, A.; Cordeiro, C. The glyoxalase pathway: The first hundred years... and beyond. *Biochem. J.* **2013**, *453*, 1–15. [CrossRef] [PubMed]
10. Sukdeo, N.; Honek, J.F. Microbial glyoxalase enzymes: Metalloenzymes controlling cellular levels of methylglyoxal. *Drug Metabol. Drug Interact.* **2008**, *23*, 29–50. [CrossRef]
11. Rabbani, N.; Xue, M.; Thornalley, P.J. Dicarbonyls and glyoxalase in disease mechanisms and clinical therapeutics. *Glycoconj. J.* **2016**, *33*, 513–525. [CrossRef] [PubMed]
12. Rabbani, N.; Thornalley, P.J. Glyoxalase Centennial conference: Introduction, history of research on the glyoxalase system and future prospects. *Biochem. Soc. Trans.* **2014**, *42*, 413–418. [CrossRef] [PubMed]
13. Thornalley, P.J. Glyoxalase I—Structure, function and a critical role in the enzymatic defence against glycation. *Biochem. Soc. Trans.* **2003**, *31*, 1343–1348. [CrossRef] [PubMed]
14. Ozyamak, E.; Black, S.S.; Walker, C.A.; Maclean, M.J.; Bartlett, W.; Miller, S.; Booth, I.R. The critical role of S-lactoylglutathione formation during methylglyoxal detoxification in *Escherichia coli*. *Mol. Microbiol.* **2010**, *78*, 1577–1590. [CrossRef] [PubMed]
15. Misra, K.; Banerjee, A.B.; Ray, S.; Ray, M. Glyoxalase III from *Escherichia coli*: A single novel enzyme for the conversion of methylglyoxal into D-lactate without reduced glutathione. *Biochem. J.* **1995**, *305 Pt 3*, 999–1003. [CrossRef]
16. Choi, D.; Kim, J.; Ha, S.; Kwon, K.; Kim, E.H.; Lee, H.Y.; Ryu, K.S.; Park, C. Stereospecific mechanism of DJ-1 glyoxalases inferred from their hemithioacetal-containing crystal structures. *FEBS J.* **2014**, *281*, 5447–5462. [CrossRef] [PubMed]
17. Zhao, Q.; Su, Y.; Wang, Z.; Chen, C.; Wu, T.; Huang, Y. Identification of glutathione (GSH)-independent glyoxalase III from *Schizosaccharomyces pombe*. *BMC Evol. Biol.* **2014**, *14*, 86. [CrossRef] [PubMed]
18. Subedi, K.P.; Choi, D.; Kim, I.; Min, B.; Park, C. Hsp31 of *Escherichia coli* K-12 is glyoxalase III. *Mol. Microbiol.* **2011**, *81*, 926–936. [CrossRef] [PubMed]
19. Sukdeo, N.; Clugston, S.L.; Daub, E.; Honek, J.F. Distinct classes of glyoxalase I: Metal specificity of the *Yersinia pestis*, *Pseudomonas aeruginosa* and *Neisseria meningitidis* enzymes. *Biochem. J.* **2004**, *384*, 111–117. [CrossRef]
20. Aronsson, A.C.; Marmstal, E.; Mannervik, B. Glyoxalase I, a zinc metalloenzyme of mammals and yeast. *Biochem. Biophys. Res. Commun.* **1978**, *81*, 1235–1240. [CrossRef]
21. Han, L.P.; Schimandle, C.M.; Davison, L.M.; Vander Jagt, D.L. Comparative kinetics of Mg^{2+}-, Mn^{2+}-, Co^{2+}-, and Ni^{2+}-activated glyoxalase I. Evaluation of the role of the metal ion. *Biochemistry* **1977**, *16*, 5478–5484. [CrossRef] [PubMed]
22. Saint-Jean, A.P.; Phillips, K.R.; Creighton, D.J.; Stone, M.J. Active monomeric and dimeric forms of *Pseudomonas putida* glyoxalase I: Evidence for 3D domain swapping. *Biochemistry* **1998**, *37*, 10345–10353. [CrossRef] [PubMed]
23. Cameron, A.D.; Olin, B.; Ridderstrom, M.; Mannervik, B.; Jones, T.A. Crystal structure of human glyoxalase I—Evidence for gene duplication and 3D domain swapping. *EMBO J.* **1997**, *16*, 3386–3395. [CrossRef] [PubMed]
24. Clugston, S.L.; Barnard, J.F.; Kinach, R.; Miedema, D.; Ruman, R.; Daub, E.; Honek, J.F. Overproduction and characterization of a dimeric non-zinc glyoxalase I from *Escherichia coli*: Evidence for optimal activation by nickel ions. *Biochemistry* **1998**, *37*, 8754–8763. [CrossRef] [PubMed]
25. Clugston, S.L.; Yajima, R.; Honek, J.F. Investigation of metal binding and activation of *Escherichia coli* glyoxalase I: Kinetic, thermodynamic and mutagenesis studies. *Biochem. J.* **2004**, *377*, 309–316. [CrossRef] [PubMed]
26. Davidson, G.; Clugston, S.L.; Honek, J.F.; Maroney, M.J. XAS investigation of the nickel active site structure in *Escherichia coli* glyoxalase I. *Inorg. Chem.* **2000**, *39*, 2962–2963. [CrossRef] [PubMed]
27. He, M.M.; Clugston, S.L.; Honek, J.F.; Matthews, B.W. Determination of the structure of *Escherichia coli* glyoxalase I suggests a structural basis for differential metal activation. *Biochemistry* **2000**, *39*, 8719–8727. [CrossRef] [PubMed]
28. Cameron, A.D.; Ridderstrom, M.; Olin, B.; Kavarana, M.J.; Creighton, D.J.; Mannervik, B. Reaction mechanism of glyoxalase I explored by an X-ray crystallographic analysis of the human enzyme in complex with a transition state analogue. *Biochemistry* **1999**, *38*, 13480–13490. [CrossRef]
29. Sellin, S.; Eriksson, L.E.; Aronsson, A.C.; Mannervik, B. Octahedral metal coordination in the active site of glyoxalase I as evidenced by the properties of Co(II)-glyoxalase I. *J. Biol. Chem.* **1983**, *258*, 2091–2093.

30. Himo, F.; Siegbahn, P.E. Catalytic mechanism of glyoxalase I: A theoretical study. *J. Am. Chem. Soc.* **2001**, *123*, 10280–10289. [CrossRef]
31. Richter, U.; Krauss, M. Active site structure and mechanism of human glyoxalase I—An ab initio theoretical study. *J. Am. Chem. Soc.* **2001**, *123*, 6973–6982. [CrossRef] [PubMed]
32. Davidson, G.; Clugston, S.L.; Honek, J.F.; Maroney, M.J. An XAS investigation of product and inhibitor complexes of Ni-containing GlxI from *Escherichia coli*: Mechanistic implications. *Biochemistry* **2001**, *40*, 4569–4582. [CrossRef] [PubMed]
33. Pettersen, E.F.; Goddard, T.D.; Huang, C.C.; Couch, G.S.; Greenblatt, D.M.; Meng, E.C.; Ferrin, T.E. UCSF Chimera—A visualization system for exploratory research and analysis. *J. Comput. Chem.* **2004**, *25*, 1605–1612. [CrossRef] [PubMed]
34. Suttisansanee, U.; Lau, K.; Lagishetty, S.; Rao, K.N.; Swaminathan, S.; Sauder, J.M.; Burley, S.K.; Honek, J.F. Structural variation in bacterial glyoxalase I enzymes: Investigation of the metalloenzyme glyoxalase I from *Clostridium acetobutylicum*. *J. Biol. Chem.* **2011**, *286*, 38367–38374. [CrossRef] [PubMed]
35. Deponte, M.; Sturm, N.; Mittler, S.; Harner, M.; Mack, H.; Becker, K. Allosteric coupling of two different functional active sites in monomeric *Plasmodium falciparum* glyoxalase I. *J. Biol. Chem.* **2007**, *282*, 28419–28430. [CrossRef] [PubMed]
36. Frickel, E.M.; Jemth, P.; Widersten, M.; Mannervik, B. Yeast glyoxalase I is a monomeric enzyme with two active sites. *J. Biol. Chem.* **2001**, *276*, 1845–1849. [CrossRef] [PubMed]
37. Inoue, Y.; Maeta, K.; Nomura, W. Glyoxalase system in yeasts: Structure, function, and physiology. *Semin. Cell Dev. Biol.* **2011**, *22*, 278–284. [CrossRef] [PubMed]
38. Iozef, R.; Rahlfs, S.; Chang, T.; Schirmer, H.; Becker, K. Glyoxalase I of the malarial parasite *Plasmodium falciparum*: Evidence for subunit fusion. *FEBS Lett.* **2003**, *554*, 284–288. [CrossRef]
39. Turra, G.L.; Agostini, R.B.; Fauguel, C.M.; Presello, D.A.; Andreo, C.S.; Gonzalez, J.M.; Campos-Bermudez, V.A. Structure of the novel monomeric glyoxalase I from *Zea mays*. *Acta Crystallogr. D Biol. Crystallogr.* **2015**, *71*, 2009–2020. [CrossRef]
40. Kaur, C.; Vishnoi, A.; Ariyadasa, T.U.; Bhattacharya, A.; Singla-Pareek, S.L.; Sopory, S.K. Episodes of horizontal gene-transfer and gene-fusion led to co-existence of different metal-ion specific glyoxalase I. *Sci. Rep.* **2013**, *3*, 3076. [CrossRef]
41. Mustafiz, A.; Ghosh, A.; Tripathi, A.K.; Kaur, C.; Ganguly, A.K.; Bhavesh, N.S.; Tripathi, J.K.; Pareek, A.; Sopory, S.K.; Singla-Pareek, S.L. A unique Ni^{2+}-dependent and methylglyoxal-inducible rice glyoxalase I possesses a single active site and functions in abiotic stress response. *Plant J.* **2014**, *78*, 951–963. [CrossRef]
42. Gonzalez, J.M.; Agostini, R.B.; Alvarez, C.E.; Klinke, S.; Andreo, C.S.; Campos-Bermudez, V.A. Deciphering the number and location of active sites in the monomeric glyoxalase I of *Zea mays*. *FEBS J.* **2019**. [CrossRef]
43. Su, Z.; Sukdeo, N.; Honek, J.F. 15N-1H HSQC NMR evidence for distinct specificity of two active sites in *Escherichia coli* glyoxalase I. *Biochemistry* **2008**, *47*, 13232–13241. [CrossRef]
44. Bythell-Douglas, R.; Suttisansanee, U.; Flematti, G.R.; Challenor, M.; Lee, M.; Panjikar, S.; Honek, J.F.; Bond, C.S. The crystal structure of a homodimeric *Pseudomonas* glyoxalase I enzyme reveals asymmetric metallation commensurate with half-of-sites activity. *Chemistry* **2015**, *21*, 541–544. [CrossRef]
45. Suttisansanee, U.; Ran, Y.; Mullings, K.Y.; Sukdeo, N.; Honek, J.F. Modulating glyoxalase I metal selectivity by deletional mutagenesis: Underlying structural factors contributing to nickel activation profiles. *Metallomics* **2015**, *7*, 605–612. [CrossRef]
46. Hand, C.E.; Honek, J.F. Biological chemistry of naturally occurring thiols of microbial and marine origin. *J. Nat. Prod.* **2005**, *68*, 293–308. [CrossRef]
47. Wang, M.; Zhao, Q.; Liu, W. The versatile low-molecular-weight thiols: Beyond cell protection. *Bioessays* **2015**, *37*, 1262–1267. [CrossRef]
48. Newton, G.L.; Arnold, K.; Price, M.S.; Sherrill, C.; Delcardayre, S.B.; Aharonowitz, Y.; Cohen, G.; Davies, J.; Fahey, R.C.; Davis, C. Distribution of thiols in microorganisms: Mycothiol is a major thiol in most actinomycetes. *J. Bacteriol.* **1996**, *178*, 1990–1995. [CrossRef]
49. Gaballa, A.; Newton, G.L.; Antelmann, H.; Parsonage, D.; Upton, H.; Rawat, M.; Claiborne, A.; Fahey, R.C.; Helmann, J.D. Biosynthesis and functions of bacillithiol, a major low-molecular-weight thiol in Bacilli. *Proc. Natl. Acad. Sci. USA* **2010**, *107*, 6482–6486. [CrossRef]
50. Jothivasan, V.K.; Hamilton, C.J. Mycothiol: Synthesis, biosynthesis and biological functions of the major low molecular weight thiol in actinomycetes. *Nat. Prod. Rep.* **2008**, *25*, 1091–1117. [CrossRef]

51. Vickers, T.J.; Greig, N.; Fairlamb, A.H. A trypanothione-dependent glyoxalase I with a prokaryotic ancestry in *Leishmania major*. *Proc. Natl. Acad. Sci. USA* **2004**, *101*, 13186–13191. [CrossRef]
52. Oza, S.L.; Shaw, M.P.; Wyllie, S.; Fairlamb, A.H. Trypanothione biosynthesis in *Leishmania major*. *Mol. Biochem. Parasitol.* **2005**, *139*, 107–116. [CrossRef]
53. Greig, N.; Wyllie, S.; Patterson, S.; Fairlamb, A.H. A comparative study of methylglyoxal metabolism in trypanosomatids. *FEBS J.* **2009**, *276*, 376–386. [CrossRef]
54. Ariza, A.; Vickers, T.J.; Greig, N.; Armour, K.A.; Dixon, M.J.; Eggleston, I.M.; Fairlamb, A.H.; Bond, C.S. Specificity of the trypanothione-dependent *Leishmania major* glyoxalase I: Structure and biochemical comparison with the human enzyme. *Mol. Microbiol.* **2006**, *59*, 1239–1248. [CrossRef]
55. Greig, N.; Wyllie, S.; Vickers, T.J.; Fairlamb, A.H. Trypanothione-dependent glyoxalase I in *Trypanosoma cruzi*. *Biochem. J.* **2006**, *400*, 217–223. [CrossRef]
56. Fahey, R.C. Glutathione analogs in prokaryotes. *Biochim. Biophys. Acta* **2013**, *1830*, 3182–3198. [CrossRef]
57. Sharma, S.V.; Arbach, M.; Roberts, A.A.; Macdonald, C.J.; Groom, M.; Hamilton, C.J. Biophysical features of bacillithiol, the glutathione surrogate of *Bacillus subtilis* and other firmicutes. *ChemBioChem* **2013**, *14*, 2160–2168. [CrossRef]
58. Perera, V.R.; Newton, G.L.; Pogliano, K. Bacillithiol: A key protective thiol in *Staphylococcus aureus*. *Expert Rev. Anti Infect. Ther.* **2015**, *13*, 1089–1107. [CrossRef]
59. Chandrangsu, P.; Loi, V.V.; Antelmann, H.; Helmann, J.D. The Role of Bacillithiol in Gram-Positive Firmicutes. *Antioxid. Redox Signal.* **2018**, *28*, 445–462. [CrossRef]
60. Chandrangsu, P.; Dusi, R.; Hamilton, C.J.; Helmann, J.D. Methylglyoxal resistance in *Bacillus subtilis*: Contributions of bacillithiol-dependent and independent pathways. *Mol. Microbiol.* **2014**, *91*, 706–715. [CrossRef]
61. Sao Emani, C.; Williams, M.J.; Wiid, I.J.; Baker, B. The functional interplay of low molecular weight thiols in *Mycobacterium tuberculosis*. *J. Biomed. Sci.* **2018**, *25*, 55. [CrossRef]
62. Sharma, S.V.; Van Laer, K.; Messens, J.; Hamilton, C.J. Thiol Redox and pKa Properties of Mycothiol, the Predominant Low-Molecular-Weight Thiol Cofactor in the Actinomycetes. *ChemBioChem* **2016**, *17*, 1689–1692. [CrossRef]
63. Liu, Y.B.; Long, M.X.; Yin, Y.J.; Si, M.R.; Zhang, L.; Lu, Z.Q.; Wang, Y.; Shen, X.H. Physiological roles of mycothiol in detoxification and tolerance to multiple poisonous chemicals in *Corynebacterium glutamicum*. *Arch. Microbiol.* **2013**, *195*, 419–429. [CrossRef]
64. Newton, G.L.; Buchmeier, N.; Fahey, R.C. Biosynthesis and functions of mycothiol, the unique protective thiol of Actinobacteria. *Microbiol. Mol. Biol. Rev.* **2008**, *72*, 471–494. [CrossRef]
65. Bentley, S.D.; Chater, K.F.; Cerdeno-Tarraga, A.M.; Challis, G.L.; Thomson, N.R.; James, K.D.; Harris, D.E.; Quail, M.A.; Kieser, H.; Harper, D.; et al. Complete genome sequence of the model actinomycete *Streptomyces coelicolor* A3(2). *Nature* **2002**, *417*, 141–147. [CrossRef]
66. Park, J.H.; Cha, C.J.; Roe, J.H. Identification of genes for mycothiol biosynthesis in *Streptomyces coelicolor* A3(2). *J. Microbiol.* **2006**, *44*, 121–125.
67. Kizil, G.; Wilks, K.; Wells, D.; Ala'Aldeen, D.A. Detection and characterisation of the genes encoding glyoxalase I and II from *Neisseria meningitidis*. *J. Med. Microbiol.* **2000**, *49*, 669–673. [CrossRef]
68. Sukdeo, N.; Honek, J.F. *Pseudomonas aeruginosa* contains multiple glyoxalase I-encoding genes from both metal activation classes. *Biochim. Biophys. Acta* **2007**, *1774*, 756–763. [CrossRef]
69. Hamilton, C.J.; Finlay, R.M.; Stewart, M.J.; Bonner, A. Mycothiol disulfide reductase: A continuous assay for slow time-dependent inhibitors. *Anal. Biochem.* **2009**, *388*, 91–96. [CrossRef]
70. Unson, M.D.; Newton, G.L.; Davis, C.; Fahey, R.C. An immunoassay for the detection and quantitative determination of mycothiol. *J. Immunol. Methods* **1998**, *214*, 29–39. [CrossRef]
71. Patel, M.P.; Blanchard, J.S. Synthesis of des-myo-inositol mycothiol and demonstration of a mycobacterial specific reductase activity. *J. Am. Chem. Soc.* **1998**, *120*, 11538–11539. [CrossRef]
72. Vander Jagt, D.L.; Han, L.P.; Lehman, C.H. Kinetic evaluation of substrate specificity in the glyoxalase-I-catalyzed disproportionation of α-ketoaldehydes. *Biochemistry* **1972**, *11*, 3735–3740. [CrossRef]
73. Vince, R.; Daluge, S.; Wadd, W.B. Studies on the inhibition of glyoxalase I by S-substituted glutathiones. *J. Med. Chem.* **1971**, *14*, 402–404. [CrossRef]
74. Cliffe, E.E.; Waley, S.G. The mechanism of the glyoxalase I reaction, and the effect of ophthalmic acid as an inhibitor. *Biochem. J.* **1961**, *79*, 475–482. [CrossRef]

75. Creighton, D.J.; Migliorini, M.; Pourmotabbed, T.; Guha, M.K. Optimization of efficiency in the glyoxalase pathway. *Biochemistry* **1988**, *27*, 7376–7384. [CrossRef]
76. Griffis, C.E.; Ong, L.H.; Buettner, L.; Creighton, D.J. Nonstereospecific substrate usage by glyoxalase I. *Biochemistry* **1983**, *22*, 2945–2951. [CrossRef]
77. Vander Jagt, D.L.; Daub, E.; Krohn, J.A.; Han, L.P. Effects of pH and thiols on the kinetics of yeast glyoxalase I. An evaluation of the random pathway mechanism. *Biochemistry* **1975**, *14*, 3669–3675. [CrossRef]
78. Mullings, K.Y.; Sukdeo, N.; Suttisansanee, U.; Ran, Y.; Honek, J.F. Ni^{2+}-activated glyoxalase I from *Escherichia coli*: Substrate specificity, kinetic isotope effects and evolution within the betaalphabetabetabeta superfamily. *J. Inorg. Biochem.* **2012**, *108*, 133–140. [CrossRef]
79. Rae, C.; O'Donoghue, S.I.; Bubb, W.A.; Kuchel, P.W. Stereospecificity of substrate usage by glyoxalase 1: Nuclear magnetic resonance studies of kinetics and hemithioacetal substrate conformation. *Biochemistry* **1994**, *33*, 3548–3559. [CrossRef]
80. Akoachere, M.; Iozef, R.; Rahlfs, S.; Deponte, M.; Mannervik, B.; Creighton, D.J.; Schirmer, H.; Becker, K. Characterization of the glyoxalases of the malarial parasite *Plasmodium falciparum* and comparison with their human counterparts. *Biol. Chem.* **2005**, *386*, 41–52. [CrossRef]
81. Mannervik, B.; Ridderstrom, M. Catalytic and molecular properties of glyoxalase I. *Biochem. Soc. Trans.* **1993**, *21*, 515–517. [CrossRef]
82. Bergdoll, M.; Eltis, L.D.; Cameron, A.D.; Dumas, P.; Bolin, J.T. All in the family: Structural and evolutionary relationships among three modular proteins with diverse functions and variable assembly. *Protein Sci.* **1998**, *7*, 1661–1670. [CrossRef]
83. Armstrong, R.N. Mechanistic diversity in a metalloenzyme superfamily. *Biochemistry* **2000**, *39*, 13625–13632. [CrossRef]
84. Honek, J.F. Nickel Glyuoxalase I. In *The Biological Chemistry of Nickel*; Kozlowski, H., Zamble, D., Rowinska-Zyrek, M., Eds.; Royal Societ of Chemistry: London, UK, 2017; Volume 10.
85. McCarthy, A.A.; Baker, H.M.; Shewry, S.C.; Patchett, M.L.; Baker, E.N. Crystal structure of methylmalonyl-coenzyme A epimerase from *P. shermanii*: A novel enzymatic function on an ancient metal binding scaffold. *Structure* **2001**, *9*, 637–646. [CrossRef]
86. Dumas, P.; Bergdoll, M.; Cagnon, C.; Masson, J.M. Crystal structure and site-directed mutagenesis of a bleomycin resistance protein and their significance for drug sequestering. *EMBO J.* **1994**, *13*, 2483–2492. [CrossRef]
87. Martin, T.W.; Dauter, Z.; Devedjiev, Y.; Sheffield, P.; Jelen, F.; He, M.; Sherman, D.H.; Otlewski, J.; Derewenda, Z.S.; Derewenda, U. Molecular basis of mitomycin C resistance in streptomyces: Structure and function of the MRD protein. *Structure* **2002**, *10*, 933–942. [CrossRef]
88. Thompson, M.K.; Keithly, M.E.; Harp, J.; Cook, P.D.; Jagessar, K.L.; Sulikowski, G.A.; Armstrong, R.N. Structural and chemical aspects of resistance to the antibiotic fosfomycin conferred by FosB from *Bacillus cereus*. *Biochemistry* **2013**, *52*, 7350–7362. [CrossRef]
89. Gonzalez-Quinonez, N.; Corte-Rodriguez, M.; Alvarez-Fernandez-Garcia, R.; Rioseras, B.; Lopez-Garcia, M.T.; Fernandez-Garcia, G.; Montes-Bayon, M.; Manteca, A.; Yague, P. Cytosolic copper is a major modulator of germination, development and secondary metabolism in *Streptomyces coelicolor*. *Sci. Rep.* **2019**, *9*, 4214. [CrossRef]
90. Aronsson, A.C.; Mannervik, B. Characterization of glyoxalase I purified from pig erythrocytes by affinity chromatography. *Biochem. J.* **1977**, *165*, 503–509. [CrossRef]
91. Mannervik, B.; Lindstrom, L.; Bartfai, T. Partial purification and characterization of glyoxalase I from porcine erythrocytes. *Eur. J. Biochem.* **1972**, *29*, 276–281. [CrossRef]
92. Uotila, L.; Koivusalo, M. Purification and properties of glyoxalase I from sheep liver. *Eur. J. Biochem.* **1975**, *52*, 493–503. [CrossRef]
93. Takatsume, Y.; Izawa, S.; Inoue, Y. Identification of thermostable glyoxalase I in the fission yeast *Schizosaccharomyces pombe*. *Arch. Microbiol.* **2004**, *181*, 371–377. [CrossRef]
94. Sambrook, J.; Russell, D. *Molecular Cloning: A Laboratory Manual*; Cold Spring Harbor Laboratory Press: Cold Spring Harbor, NY, USA, 2001.
95. O'Young, J.; Sukdeo, N.; Honek, J.F. *Escherichia coli* glyoxalase II is a binuclear zinc-dependent metalloenzyme. *Arch. Biochem. Biophys.* **2007**, *459*, 20–26. [CrossRef]

96. Hunt, J.B.; Neece, S.H.; Ginsburg, A. The use of 4-(2-pyridylazo)resorcinol in studies of zinc release from *Escherichia coli* aspartate transcarbamoylase. *Anal. Biochem.* **1985**, *146*, 150–157. [CrossRef]
97. McCall, K.A.; Fierke, C.A. Colorimetric and fluorimetric assays to quantitate micromolar concentrations of transition metals. *Anal. Biochem.* **2000**, *284*, 307–315. [CrossRef]
98. Louis-Jeune, C.; Andrade-Navarro, M.A.; Perez-Iratxeta, C. Prediction of protein secondary structure from circular dichroism using theoretically derived spectra. *Proteins* **2011**, *80*, 374–381. [CrossRef]

© 2019 by the authors. Licensee MDPI, Basel, Switzerland. This article is an open access article distributed under the terms and conditions of the Creative Commons Attribution (CC BY) license (http://creativecommons.org/licenses/by/4.0/).

Article

pH Dependent Reversible Formation of a Binuclear Ni₂ Metal-Center within a Peptide Scaffold

Brenna C. Keegan, Daniel Ocampo and Jason Shearer *

Department of Chemistry, Trinity University, 1 Trinity Place, San Antonio, TX 78212, USA
* Correspondence: jshearer@trinity.edu

Received: 24 May 2019; Accepted: 3 July 2019; Published: 16 July 2019

Abstract: A disulfide-bridged peptide containing two Ni^{2+} binding sites based on the nickel superoxide dismutase protein, $\{Ni_2(SOD^{mds})\}$ has been prepared. At physiological pH (7.4), it was found that the metal sites are mononuclear with a square planar NOS_2 coordination environment with the two sulfur-based ligands derived from cysteinate residues, the nitrogen ligand derived from the amide backbone, and a water ligand. Furthermore, S K-edge X-ray absorption spectroscopy indicated that the two cysteinate sulfur atoms ligated to nickel are each protonated. Elevation of the pH to 9.6 results in the deprotonation of the cysteinate sulfur atoms, and yields a binuclear, cysteinate bridged Ni_2^{2+} center with each nickel contained in a distorted square planar geometry. At both pH = 7.4 and 9.6, the nickel sites are moderately air sensitive, yielding intractable oxidation products. However, at pH = 9.6, $\{Ni_2(SOD^{mds})\}$ reacts with O_2 at an ~3.5-fold faster rate than at pH = 7.4. Electronic structure calculations indicate that the reduced reactivity at pH = 7.4 is a result of a reduction in S(3p) character and deactivation of the nucleophilic frontier molecular orbitals upon cysteinate sulfur protonation.

Keywords: biological nickel sites; nickel-thiolates; dinuclear nickel metallopeptides; thiolate oxidative damage

1. Introduction

Nickel is an essential biological co-factor found at the active-sites of a number of microbial metalloenzymes and proteins (Chart 1) [1–5]. Broadly divided into redox active and non-active nickel metalloproteins, it has been recognized that the majority of known redox active nickel containing metalloenzymes contain cysteinate sulfur ligation to nickel. Cysteinate ligation appears necessary to poise nickel-based one-electron redox couples so as to be accessible under physiological conditions. It has also been demonstrated that several redox inactive nickel transport and regulatory proteins also possess cysteinate ligands to Ni^{2+} [6–11].

An interesting feature of the nickel-thiolate moiety is its ability to support ligand protonation without subsequent protonolysis [12–14]. This is most often observed in (near) square planar Ni^{2+} centers where the nucleophilic HOMO possesses significant $S(3p\pi)$ character, which effectively act as S-based lone-pairs. To date, two nickel containing metalloenzymes, nickel iron hydrogenase [NiFe]H₂ase and nickel containing superoxide dismutase (NiSOD), have been demonstrated to possess at least one cysteinate ligand that becomes protonated, forming a Ni–S(H⁺)–Cys moiety under physiological conditions [15,16]. Concerning the role of the Ni–S(H⁺)–Cys moiety in biochemical reactions, it has been proposed that these moieties can behave as proton donors/acceptors and sources of formal hydrogen atoms [17–22]. However, their exact role(s) in biological reactions is currently unknown.

NiSOD is a nickel containing homohexameric metalloenzyme that disproportionates O_2^- into H_2O_2 and O_2 by cycling between reduced Ni^{2+} and oxidized Ni^{3+} oxidation states [23–25]. Each monomer contains a mononuclear nickel site that is coordinated by two cis-cysteinate sulfur atoms from Cys2 and Cys6, an amidate nitrogen atom from Cys2, and the N-terminal amine nitrogen atom from His1 (Chart 1). Upon oxidation to Ni^{III}, the square planar nickel site ligates the N^ε imidazole from His1 forming a square pyramidal coordination geometry. Taking advantage of the fact that all of the ligating residues to nickel are found within the first six residues from the protein N-terminus, we, and others, have prepared functional NiSOD metallopeptide based mimics utilizing the first 6–12 residues from the NiSOD N-terminal primary protein sequence. These metallopeptide based mimics reproduce the key structural and spectroscopic properties of the metalloenzyme [17,26–33].

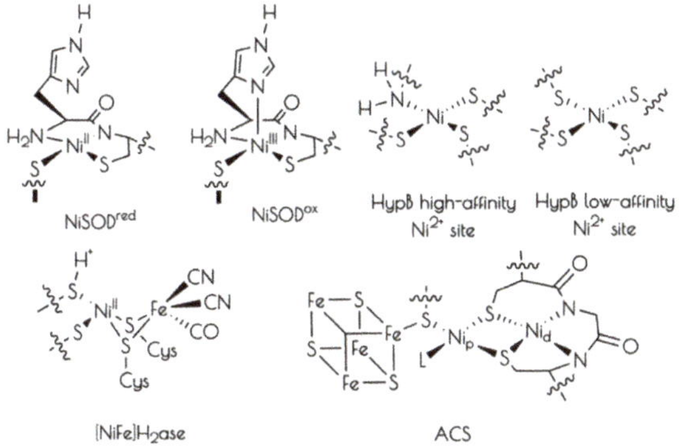

Chart 1. Representative structures of the active sites of cysteinate-ligated nickel containing metalloproteins.

As stated above, NiSOD itself has been shown to possess at least one Ni–S(H$^+$)–Cys moiety in its reduced form [15]. Although not without controversy [33], we have provided strong evidence based on sulfur K-edge X-ray absorption studies that like NiSOD itself, NiSOD metallopeptide based mimics possess a Ni–S(H$^+$)–Cys moiety at physiological pH as well [18,22]. Studies have also suggested that the pK_a of the Ni–S(H$^+$)–Cys proton within these active-sites is ~8.5, and can become reversibly deprotonated at high pH (>9.0) [22].

In addition to mimicking the NiSOD active site, derivatives of the NiSOD inspired metallopeptides are capable of not only mimicking NiSOD, but also mimicking the active-site of cobalt containing NHase, and coordinating Cu^{2+} [34,35]. This inspired us to further derivatize a NiSOD metallopeptide mimic, SOD^{m1} (SOD^{m1} = (SOD^{mds} = HCDLP-CGVYDA-PA), in order to generate different metal-site structures. The intent of such studies is not necessarily to generate NiSOD biomimetic metallopeptides, but instead to probe different metal coordination environments within a biologically derived scaffold. Herein, we present work on a nickel metallopeptide, {Ni$_2^{II}$(SOD^{mds})} (SOD^{mds} = (T^{a}CDLP-CGVYDA-PA)$_2$, where T^a is a 2-mercaptoacetate group). It will be demonstrated that at pH = 7.4 this metallopeptide possesses two mononuclear Ni^{2+} sites that support the formation of the Ni–S(H$^+$)–Cys moiety. Furthermore, we will show that the metallopeptide forms a dinuclear cysteinate bridged Ni_2^{2+} center upon elevation of the pH to 9.6. Lastly, it will be demonstrated that Ni–S–Cys protonation protects the metallopeptide from oxidative damage by O_2. The protection of the metallopeptide against oxidative damage will be rationalized in terms of an alteration of the electronic structure of the nickel-site upon protonation rendering the thiolate sulfurs relatively inert towards oxidation.

2. Materials and Methods

2.1. General Considerations

All manipulations were performed under an N_2/H_2 (97:3) atmosphere in a COY anaerobic chamber. Fmoc/O^{t}Bu protected amino acids and resins were obtained from Advanced Chemtech (Louisville, KY, USA). All other reagents obtained from commercial suppliers were of the highest purity available, and used as received. Analytical and semi-preparative reverse-phase HPLC were performed using Waters X-Bridge C-18 analytical (4.6 × 150 mm; 5 µm) and semi-preparative (30 × 150 mm; 5 µm) columns on a Waters Deltaprep 600 equipped with a photodiode array detector (detection wavelength set to 254 nm) (Waters Technology Corporation, Milford, MA, USA). Mass spectrometry was performed on either a Bruker Microflex MALDI-TOF mass spectrometer,(Bruker, Billerica, MA, USA) a ThermoFinnegan LCQ Deca XP ESI-MS,(Thermo Instrument Systems, Waltham, MA, USA) or a Waters Micromass 20 ESI-MS operating in positive ion mode (Waters Technology Corporation, Milford, MA, USA). NMR spectra were obtained on a 400 MHz Varian VNMRS NMR spectrometer (Agilent, Santa Clara, CA, USA). All chemical shifts (δ) are referenced to the residual protio solvent peak. Electronic absorption spectra were obtained on either a JASCO J-1500, CARY 50 or CARY 5000 UV-vis-NIR spectrometer (Agilent, Santa Clara, CA, USA). Circular dichroism spectra were obtained on a JASCO J-1500 spectropolarimeter (JASCO Inc, Easton, MD, USA). The simultaneous deconvolutions of the CD and electronic absorption spectra were performed using an in-house-written procedure for Igor Pro version 6 and 8 (Wavemetrics, Lake Oswego, OR, USA). Infrared spectra were collected using a Thermo Nicolet 6700 FTIR spectrometer with a diamond crystal ATR (Thermo Instrument Systems, Waltham, MA, USA). X-band EPR spectra were obtained on a Bruker EMXPlus EPR spectrometer equipped with a closed-cycle He cryostat (Bruker, Billerica, MA, USA).

2.2. Preparation of S-triphenylmethyl-thioglycolic Acid

S-triphenylmethyl-thioglycolic acid (T^A-trityl) was prepared by a modification of the procedure on Martingae et al. [36]. Briefly, 1.5 g (5.77 mmol) of triphenylmethanol was added to a 25 mL TFA solution of thioglycolic acid (400 µL, 5.77 mmol) and stirred for 5 h under argon at room temperature. The TFA was removed under vacuum, and the resulting orange solid was washed three times with toluene followed by three times with hexanes resulting in a white solid, which was analytically pure (1.46 g, 81% yield). ^{1}H NMR (CDCl$_3$, 400 MHz): δ 7.30 (m, 15H), 3.04 (s, 2H).

2.3. Preparation of {Ni$_2$(SOD^{mds})}

The peptide SOD^{mds} (T^{a}CDLP-CGVYD-PA) was prepared on an AAPTec Focus XC-2RV peptide synthesizer or by manual solid-phase peptide synthesis using HBTU/HOBt coupling strategies on a 0.12 mmole scale with alanine loaded Wang resin using a five-fold excess of activated protected amino acid. Following the coupling trityl protected Cys1 and removal of the Fmoc group, T^A-trityl was coupled to the N-terminus using standard HBTU/HOBt coupling strategies. Global peptide deprotection and peptide cleavage from the resin was performed under N$_2$ using a cleavage cocktail comprised of 84.5:5:5:5:2.5 (TFA:phenol:water:thioanisole:EDT) over the course of 12 h. Following removal of the cleavage solution by vacuum on a Schlenk line, the resulting glassy product was washed four times with cold freshly distilled diethyl ether. HPLC and MALDI-TOF studies of the crude peptide mixture demonstrated the complete formation of the disulfide bridged dimer. The resulting crude peptide was subsequently purified by preparative HPLC (9:1 water:acetonitrile–4:6 water:acetonitrile over the course of 30 min; rt = 14.6 min) resulting in the pure disulfide bridged dimer (m1S$_3$)$_2$ (14% yield). MALDI-TOF MS: [SOD^{mda}/Na]$^+$ exp: 2473.3 m/z; calcd: 2473.9 m/z.

Solutions of SOD^{mds} in 50 mM NEM buffer (pH 7.4 or 9.5) were prepared and 2.0 equiv of NiCl$_2$ (added from a pH 7.0 50 mM stock solution) per peptide were then added to solutions of the (m1S$_3$)$_2$. The number of free thiol groups per peptide was verified using an Ellman's assay compared to the peptide concentration as determined by the combined absorbance of the Y residue and disulfide moiety (combined ε = 1,525 M^{-1} cm^{-1} at 278 nm) [37]. ESI-MS data were obtained by injecting an air-free solutions of the pH 7.4 or 8.6 metallopeptide into the mass spectrometer using an air-tight syringe (ESI-MS: pH 9.6 [{Ni$_2$(SOD^{mds})}/Na]$^+$ exp: 2589.1; calcd. 2589.8; ESI-MS: pH 7.4 [{Ni$_2$(SOD^{mds})}/Na/H$_2$O]$^+$ exp: 2607.6; calcd. 2607.8).

2.4. Determination of Ni–S(H⁺)–Cys pK_a

Solutions of {Ni$_2$(SOD^{mds})} were formed in 50 mM NEM buffer at a pH of 6.5. To these, aliquots of NaOH or HCl (0.5 M) were then added to the solution and the pH measured using an Orion® micro-pH electrode (ThermoFisher Scientific, Waltham, MA, USA) and the electronic absorption measured following each addition. The resulting pH titration curve was constructed by monitoring the change in absorbance at λ = 320 nm, where the largest difference in absorbance between the two species is observed.

2.5. Kinetics of the Air Oxidation of {Ni$_2^{II}$(m1S$_3$)$_2$}

Air was bubbled for five minutes through the circular dichroism (CD) spectrum of anaerobically prepared solutions of 1.0 mM {Ni$_2$(SOD^{mds})} at pH 7.4 or 9.5 (50 mM NEM buffer). A CD spectrum of these solutions was subsequently taken every 10 min for 12 h with the sample continuously exposed to air. The oxidation kinetics was modeled using pseudo-first-order reaction kinetics using *KinTek Explorer v 5.2*. Second order rate constants are reported per nickel site—one at pH = 9.6 and two at pH = 7.4. From the deconvoluted spectra, kinetic traces are reported in the decay of the {Ni$_2$(SOD^{mds})} starting materials.

2.6. Nickel K-Edge X-ray Absorption Spectroscopy

Nickel K-edge X-ray absorption spectroscopic data were collected on the HXMA beamline (wiggler insertion device operating at 1.5 T) at the Canadian Light Source (Saskatoon, SA, Canada). Solutions of {Ni$_2$(SOD^{mds})} (1.0 mM in 1:1 50 mM NEM buffer:glycerol at a pH of 7.4 or 9.5) were injected between Kapton tape windows in aluminum sample holders and quickly frozen in liquid nitrogen. Data were collected at 20 K with sample temperatures maintained using an Oxford liquid He flow cryostat. Light was monochromatized using a Si(220) double crystal monochromator, which was detuned 50% for harmonic rejection, and focused using a Rh mirror. Spectra were obtained in fluorescence mode using a 32-element solid-state Ge detector on both lines with a 3-micron cobalt filter placed between the sample and detector, and spectra were calibrated against the first inflection point of Ni–foil (8333 eV), which was simultaneously recorded with the metallopeptide data. Data were collected in 10 eV steps from 8133–8313 eV (1 s integration time per point), 0.3 eV steps from 8313–8363 eV (3 s integration time per point), 2 eV steps from 8363–8633 eV (5 s integration time per point), and 5 eV steps from 8633 eV–16 k (5 s integration time per point). Total fluorescence counts were maintained under 30 kHz, and a deadtime correction yielded no appreciable change to the data. The reported spectra represent the averaged spectra from five individual data sets. Prior to data averaging, each spectrum and detector channel was individually inspected for data quality. Data were subsequently processed and analyzed as previously reported using in-house written procedures for *Igor Pro* and *FEFF 9.4* (University of Washington, Seattle, WA, USA) [18].

2.7. Sulfur K-Edge X-ray Absorption Spectroscopy

Solutions of {Ni$_2$(SOD^{mds})} were prepared at a pH of either 7.4 or 9.6 (~1 mM in 50 mM NEM buffer) and injected into Lucite sample holders with polypropylene windows. Data were obtained at room temperature (~20 °C) on beamline X-19a at the NSLS (Upton, NY, USA) in a He purged sample chamber using a passivated implanted planar silicon (PIPS) detector. The photon energies were calibrated against the first inflection point of S$_8$ recorded before and after each sample; it was found that there was no detectable monochromator drift throughout the data collection. Data were obtained in 5 eV steps in the pre-edge region, 0.1 eV steps in the edge region, 1 eV steps in the near edge region, and 5 eV steps in the far edge region. The reported data represents the average of five individual scans. Following data averaging and a baseline was applied to each spectrum by fitting the pre-edge region to a polynomial function. This baseline was then subtracted from the whole spectrum. The region above the edge jump was then fit to a two-knot cubic spline, and the data normalized to the edge height.

2.8. Electronic Structure Calculation

Electronic structure calculations were performed using ORCA v 4.1.0 (Max-Planck-Institut für Kohlenforschung, Mülheim a. d. Ruhr, Germany) [38]. Unless otherwise stated, all calculations employed Ahlrichs' def2-tzvp basis set [39] on all atoms and the atom pairwise dispersion correction with Becke-Johnson damping to account for dispersive interactions [40,41]. ORCA VeryTightSCF convergence criteria were used for the SCF cycles, with program defaults used for all other convergence criteria and settings. Geometry optimizations were performed at the BP86 level, and used the RI approximation and def2-tzvp/c auxiliary basis set [42]. Single point calculations were performed at the PBE0 level and used the RIJCOSX approximation and def2-tzvp/j auxiliary basis set [43].

Nickel and sulfur K-edge X-ray absorption spectra were simulated using TD-DFT calculations (PBE0/def2-tzvp(-f) and the ZORA relativistic approximation) examining the first 25 transitions originating from each sulfur atom or the nickel atom. A Gaussian function to each transition (FWHM = 0.75 eV for S K-edge calculations and 1.2 eV for Ni K-edge calculations) followed by summing the individual transitions was used to produce the calculated spectra. A +36.7 eV (sulfur K-edge) or +171.3 eV (nickel K-edge) energy correction was applied to each transition. Atomic orbital population analyses were performed using a Löwdin population analysis. Isosurface plots were generated using Chimera v 1.13.1 [44].

3. Results

3.1. Generation of {$Ni_2^{II}(SOD^{mds})$}

The apo-peptide SOD^{mds} (SOD^{mds} = (T^{a}CDLP-CGVYDA-PA)$_2$) was prepared by standard solid phase peptide synthesis using Fmoc/^{t}Bu protection strategies. Cleavage of the peptide from the resin and subsequent global deprotection of the side-chain residues was effected by so-called reagent K. Reagent K is a peptide cleavage mixture that is used for peptides containing readily oxidizable residues such as Cys and Tyr, and contains a number of scavengers that dramatically reduce oxidative side reactions [45]. Despite the use of reagent K, we found that the apo-peptide formed a disulfide bond upon cleavage from the resin; even workup under anaerobic conditions using rigorously purified diethyl ether yielded a crude peptide lacking monomeric disulfide free peptide as evidenced by mass spectrometry (MS) and an Ellman's assay (Figure 1). Figure 1 depicts an HPLC chromatogram of the crude product. LCMS examination of all components of the reaction mixture demonstrated that no monomeric peptide containing the N-terminal 2-mercaptoacetate group was produced.

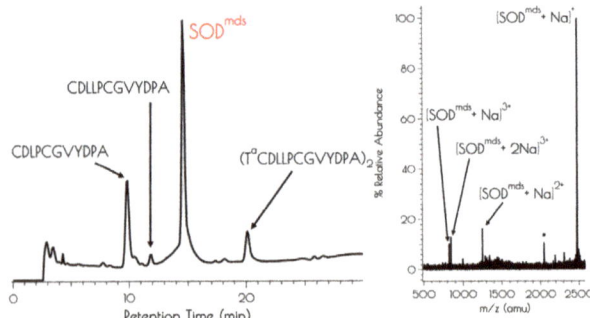

Figure 1. Left: Analytical HPLC chromatogram of the crude mixture resulting from the synthesis of SOD^{mds} with a detector cutoff of 0 intensity units. Identifiable products are highlighted. A mobile phase of a mixture of 0.1% TFA in water and 0.1% TFA in acetonitrile and a linear gradient of 9:1 water:acetonitrile–4:6 water:acetonitrile over the course of 30 min. Right: MALDI-TOF of the purified peptide SOD^{mds} (* indicates [SOD^{mds}]$^+$ with the YDPA residues cleaved from the C-terminus of one of the monomers).

An Ellman's assay [38] demonstrated that SOD^{mds} possessed four free thiol groups per dimeric apo-peptide. This is similar to what has been reported for all NiSOD metallopeptides generated to date. For these NiSOD inspired peptides lacking the N-terminal thiolate group, cleavage of the apo-peptide from the resin and subsequent aerobic work-up and purification yield purified monomeric peptides free of disulfide bonds between the two cysteinate sulfur atoms corresponding to Cys2 and Cys6 in the wild-type NiSOD sequence [26,27,30,34]. Considering the only major modification between SOD^{mds} and similar NiSOD inspired peptides generated to date is the presence of the 2-mercaptoacetate group, we suggest that the disulfide moiety within SOD^{mds} results from the oxidative S–S bond formation between two N-terminal thiol groups from two different peptides. Furthermore, all identifiable monomeric peptides by LCMS resulting from incomplete/mis-coupling events lacked the N-terminal 2-mercaptoacetate group, lending further support of the N-terminal disulfide bond in SOD^{mds}. This supposition will be shown to be further validated by computational modeling of the resulting nickel site (vide infra).

The addition of two equivalents of NiCl$_2$ to a pH 7.4 solution (50 mM N-ethylmorpholine, NEM) of SOD^{mds}, forming {Ni$_2$(SOD^{mds})}, yields a pinkish-tan solution. Raising the pH to 9.6 (50 mM NEM) causes a color change from the pinkish tan color to a more intensely colored reddish-brown solution. Although subtle changes in the far visible region of the electronic absorption spectra are noted at elevated solution pH, a much larger change is noted in the CD spectrum upon changing the solution pH from 7.4 to 9.6 (Figure 2). Monitoring the change in the electronic absorption spectrum vs. the change in pH shows a significant increase in the absorbance at 320 nm with two inflection points in the pH profile at a pH of 8.4 and 9.1. The change in the spectra in response to a change in pH is fully reversible. This suggests that there are two protonatable sites within {Ni$_2$(SOD^{mds})}. We note that the Y9 phenolic group will become deprotonated at the higher pH value; however, the change in absorbance at 320 nm upon deprotonation of phenol will be minimal and not dramatically influence the intensity of the electronic absorption spectrum at 320 nm. Thus, we do not attribute the second deprotonation/protonation event to Y9 deprotonation.

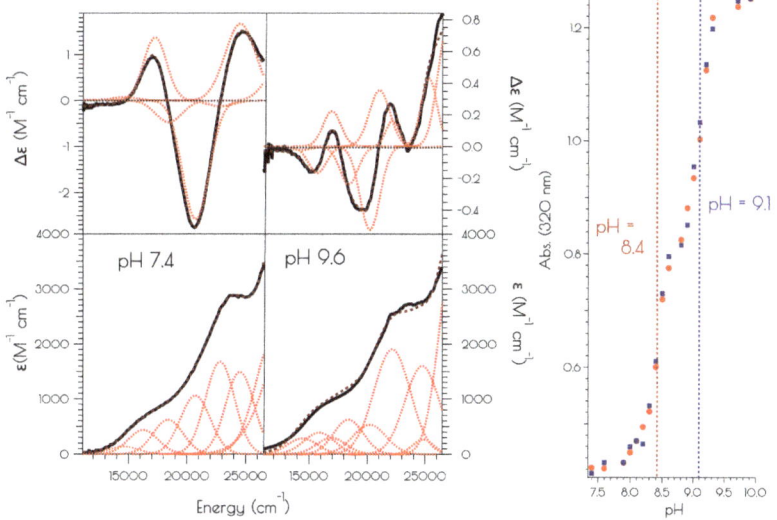

Figure 2. Left: electronic absorption spectra (bottom) and CD spectra (top) of {Ni$_2^{II}$(SOD^{mds})} at pH 7.4 and 9.6. The solid black spectra represent the experimental data, the red dashed curves represent the individual transitions deconvolved from the spectra and the dashed black spectra represent the convolution of the individual transitions. Right: pH profile showing the change in absorbance at 320 nm vs. change in pH upon going from pH 7.4 to 9.6 (red circles) and pH 9.6 to 7.4 (blue squares).

3.2. Nickel K-Edge X-ray Absorption Spectroscopy

The nickel K-edge X-ray absorption spectra of {Ni$_2^{II}$(SOD^{mds})} at pH 7.4 and 9.6 are depicted in Figure 3. The XANES at both pH 7.4 and 9.6 are consistent with Ni^{II} contained in a nomial square planar coordination environment; both display a weak Ni(1s → 3d) transition and a more intense low energy Ni(1s → 4p$_z$) transition [46]. However, the XANES spectra display notable differences at the two different solution pH values. At a pH of 7.4, the XANES region contains a poorly resolved, weak nominal Ni(1s → 3d) transition at 8333.4(3) eV and a higher energy more intense Ni(1s → 4p$_z$) transition at 8337.3(2) eV that is well separated from the edge. Raising the pH to 9.6 changes the overall shape of the edge indicating a change in coordination geometry about nickel. In addition, the Ni(1s → 3d) transition becomes more intense and red-shifts to 8331.0(2) eV while the Ni(1s → 4p$_z$) transition becomes less intense, blue-shifts to 8338.2(4) eV, and is now poorly resolved from the edge. These observed changes indicate that a more rigorous square-planar coordination environment about nickel is generated at pH 7.4, which undergoes a distortion towards nominal D_{2d} symmetry at pH 9.6 (vide infra).

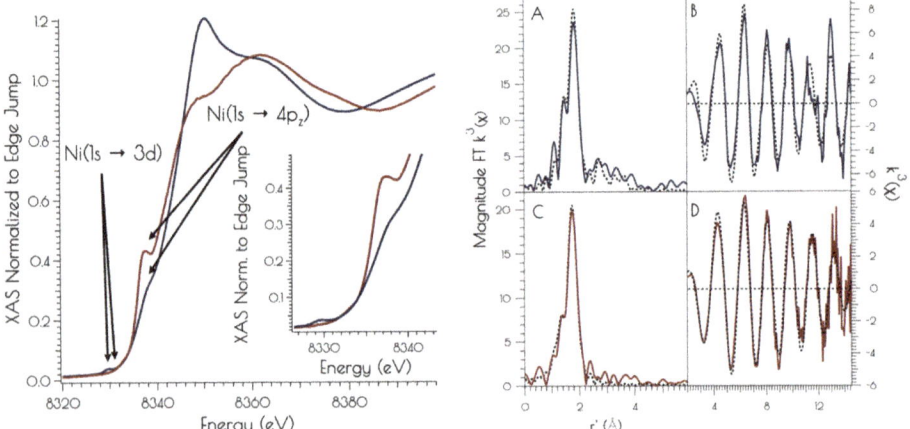

Figure 3. Left: XANES region of the Ni K-edge X-ray absorption spectrum for {Ni$_2^{II}$(SOD^{mds})} at pH 7.4 (red) and 9.6 (blue). The inset depicts a blow-up of the Ni(1s → 3d) and Ni(1s → 4p$_z$) transitions. Right: Magnitude Fourier Transformed k^3(χ) and unfiltered k^3(χ) for {Ni$_2^{II}$(SOD^{mds})} at pH 9.6 (**A** and **B**) and 7.4 (**C** and **D**). Refinements pH 7.4: (a) Ni–S: $n = 2$; $r = 2.1804(14)$ Å; $\sigma^2 = 0.0026(2)$ Å^2, (b) Ni–N: $n = 2$; $r = 1.907(16)$ Å; $\sigma^2 = 0.0013(6)$ Å^2; $\sigma^2 = 0.0019(6)$ Å^2; E$_o$ = 8347.1 eV; $\varepsilon^2 = 0.69$. Refinements pH 9.6: (a) Ni–S: $n = 3$; $r = 2.229(2)$ Å; $\sigma^2 = 0.0044(2)$ Å^2, (b) Ni–N: $n = 1$; $r = 1.889(9)$ Å; $\sigma^2 = 0.0013(8)$ Å^2, (c) Ni–Ni: $n = 1$; $r = 3.25(3)$ Å; $\sigma^2 = 0.0061(15)$ Å^2; E$_o$ = 8346.3 eV; $\varepsilon^2 = 1.47$.

At pH 7.4, the EXAFS region of the Ni K-edge X-ray absorption spectrum is best modeled as a four-coordinate Ni-center with two Ni–S scatterers at 2.18 Å and two Ni–N/O scatterer at 1.91 Å. This model is consistent with a nickel center ligated by two cysteinate S atoms, one amidate nitrogen atom and a water ligand, with the N/O scatterers modeled in one shell. At pH 9.6, we see the loss of one N/O scatterer with the subsequent coordination of an addition sulfur ligand; the best fit to the data contained three Ni–S scatterers at 2.23 Å and one Ni–N scatterer at 1.89 Å. In addition, a vector for an outersphere Ni–Ni scatterer could be located at 3.25 Å. These data are consistent with the structural models depicted in Scheme 1; at pH 7.4, the data is consistent with each peptide containing two mononuclear nickel sites that collapses into a binuclear cysteinate bridged Ni$_2$ center at high pH, the formation of which is well represented in small molecule Ni-thiolate chemistry [47,48].

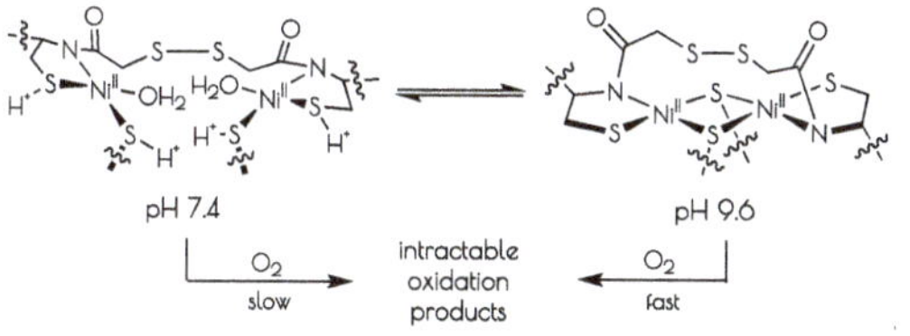

Scheme 1. Interconversion of the structures of the nickel-sites of {Ni$_2^{II}$(SOD^{mda})} at pH 7.4 and 9.6 and subsequent oxidative decomposition upon O$_2$ exposure.

3.3. Sulfur K-Edge X-ray Absorption Spectroscopy of {Ni$_2^{II}$(SOD^{mds})}

The sulfur K-edge X-ray absorption spectra of {Ni$_2^{II}$(SOD^{mds})} at pH 7.4 and 9.6 are depicted in Figure 4. At both pH values, there are higher energy edge features that are consistent with a disulfide bond, further demonstrating the presences of a bridging disulfide at both pH 7.4 and 9.6. At high pH, there is a pre-edge feature that possesses distinct asymmetry, and can be identified as terminal and bridging nickel-cysteinate S(1s → LUMO) transitions [15]. At pH 7.4, this feature is lost while the edge broadens and gains intensity. This suggests that the S(1s → LUMO) transitions have been blue-shifted into the edge, and is fully consistent with the EXAFS analysis of the metallopeptide at high and low pH; lowering the pH that induces the breaking of the bridging Ni–S^{Cys}–Ni bonds, generating two mononuclear nickel centers. Furthermore, the sulfur K-edge spectrum at pH 7.4 is consistent with the formation of protonated coordinated Ni–S(H$^+$)–Cys moieties at physiological pH; the reversible blue-shift of the S(1s → LUMO) transition upon lowering the pH is a hallmark of the formation of Ni–S(H$^+$)–Cys moieties [15]. This is further validated by time-dependent DFT calculations simulating the S K-edge X-ray absorption spectra (vide infra).

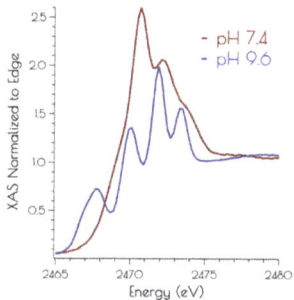

Figure 4. Sulfur K-edge XANES of {Ni$_2^{II}$(SOD^{mds})} at high 7.4 (red) and 9.6 (blue).

3.4. DFT Generated Structure for {Ni$_2^{II}$(SOD^{mds})} and TD-DFT Calculations of Ni and S K-Edge XAS Transitions

Based on the Ni and S K-edge XAS data, several computational models of the high and low pH forms of {Ni$_2^{II}$(SOD^{mds})} were generated (Figure 5). One is a mononuclear nickel site that contains a formal Ni^{II} center in an S$_2$NO coordination environment with ligands derived from two thiolate sulfur atoms, an amidate nitrogen atom, and a water ligand (see Supplementary Materials). Models of this mononuclear site with variable thiolate sulfur atom protonation were also investigated. As computational models for the pH = 9.6 form of {Ni$_2^{II}$(SOD^{mds})}, two dinuclear thiolate bridged Ni$_2$ models were also generated. One possessed two bridging thiolates, one terminal thiolate and one amidate nitrogen per nickel center. The other dinuclear model possessed an identical Ni$_2$S$_2$ core, but with a disulfide bond derived from the 2-mercaptoacetate groups bridging the two ligand sets together. Geometry optimizations were performed at the BP86/def2-tzvp level of theory with a dispersion correction; this functional generally yields computationally derived structures for transition metal complexes that are in excellent agreement with experimental data [49,50].

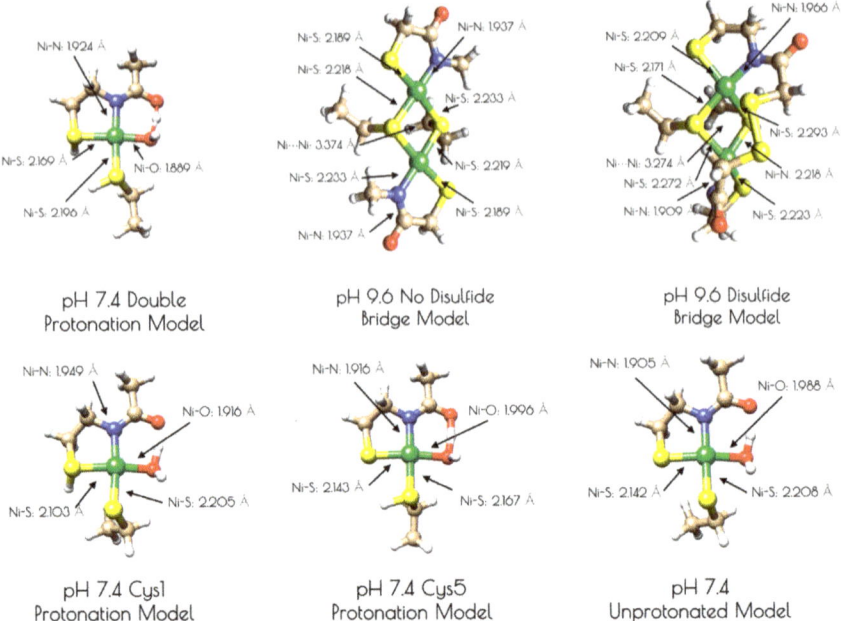

Figure 5. Computationally derived nickel site models of {Ni$_2^{II}$(SOD^{mds})}. Metric parameters are provided next to the Ni-ligand bonds.

An examination of the metric parameters derived for the mononuclear Ni-sites (pH 7.4 monomeric model) demonstrates that a model with each thiolate sulfur becoming protonated and a water ligand bound to nickel is most consistent with the experimental parameters derived from EXAFS. The two Ni–S bond lengths are calculated to be 2.169 Å (*trans* to the water ligand) and 2.196 (*trans* to the amidate nitrogen); the Ni–N bond length is calculated to be 1.925 Å and the Ni–O bond length is calculated to be 1.889 Å. These are fully consistent with the EXAFS derived average Ni–S and Ni–N/O bond lengths. The mono- and unprotonated models yield Ni–ligand bond lengths that are less consistent with the EXAFS data than the doubly protonated model. However, when one considers the inherent error in EXAFS-derived bond lengths, the mono and unprotonated models could still be considered valid structural models for {Ni$_2^{II}$(SOD^{mds})} based on bond-lengths alone. Nevertheless, the experimental vs. theoretical Ni and S K-edge studies strongly suggest that the doubly-protonated model is most valid (vide infra).

It was determined that inclusion of the N-terminal disulfide bridge is required to reproduce the metric parameters observed in the high pH form of {Ni$_2^{II}$(SOD^{mds})}. When the disulfide moiety is omitted from the computational model, the geometry about the nickel-sites is nearly planar: the S–Ni–S–Ni dihedral angles are all ~3°. This planar geometry yields a Ni–Ni distance that is 0.1 Å longer than what is observed by EXAFS (3.374 Å). Inclusion of the disulfide bond forces the geometry around the nickel sites to become distorted away from planarity. Furthermore, the two nickel sites are no longer equivalent to one another with one center distorted more toward a tetrahedral-like coordination environment than the other (the above measured dihedral angles are now 33° and 27°). This distortion brings the two Ni–Ni centers closer to one another by 0.1 Å, which reproduces the EXAFS derived Ni–Ni distance nicely (3.274 Å). Furthermore, the average Ni–S and Ni–N bond lengths are more consistent with the EXAFS-derived bond lengths. This lends further evidence for the presence of an N-terminal disulfide bond.

Correlating the experimental Ni K-edge XANES at high and low pH with the time-dependent DFT (TD-DFT; PBE0/zora-def2-tzvp(-f)/ZORA) calculated Ni K-edge spectra of the computational models above supports the proposed structures (Figure 6). For the monomeric structures, we find that the nearly rigorous square planar coordination environment about Ni^{II} yields Ni(1s → 3d) and Ni(1s → $4p_z$) transitions that correlate well with the experimental data. As might be expected, the pH 9.6 model lacking the disulfide bridge resembles that of the pH 7.4 data, but is inconsistent with the pH 9.6 experimental data. Once the disulfide bridge is included in the computational model, the Ni(1s → 3d) transition red-shifts and increases in intensity while the Ni(1s → $4p_z$) transition blue shifts and decreases in intensity. The reason for this is a consequence of 3d/4p mixing. In the D_{2d} distorted non-centrosymmetric coordination environment, the loss of the pseduo-inversion center allows for the Ni(1s → 3d) transition to acquire 4p character and the Ni(1s → $4p_z$) transition to acquire 3d character. The result is an increase in the intensity of the Ni(1s → 3d) transition by gaining the dipole-allowed Ni(1s → 4p) character while the Ni(1s → $4p_z$) transition losses intensity owing to an increase dipole-forbidden Ni(1s → 3d) character and a decrease in Ni(1s → 4p) character (Figure 6).

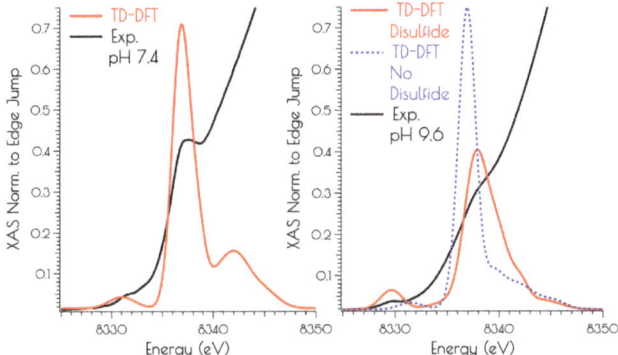

Figure 6. Comparison of the TD-DFT calculated Ni K-edge XANES with the experimental spectra for {$Ni_2(SOD^{mds})$} at pH 7.4 (left) and 9.6 (right).

Calculation of the S K-edge X-ray absorption spectra using time-dependent DFT (TD-DFT; PBE0/zora-def2-tzvp(-f)/ZORA) also supports the structural assignments proposed above. The experimental S K-edge X-ray absorption spectrum of {$Ni_2^{II}(SOD^{mds})$} obtained at pH 9.6 and the TD-DFT calculated S K-edge X-ray absorption spectrum using the disulfide bridged model match well, lending further credence to the validity of the proposed structural model (Figure 7). To determine the likely protonation state of the monomeric pH 7.4 form of {$Ni_2^{II}(SOD^{mds})$}, the doubly-protonated, monoprotonated, and unprotonated monomeric models outlined above were examined. When one or both of the cysteinate sulfur atom(s) is unprotonated, a low energy pre-edge feature corresponding to the nominal S(1s → LUMO) transitions is produced in the calculated spectrum. As would be predicted, this pre-edge feature is weaker for the mono-protonated model relative to the unprotonated model owing to the fact that only one S(1s → LUMO) transition comprises this feature as opposed to two; the S(1s → LUMO) transition for a protonated thiolate sulfur atom will blue shift into the edge. It is therefore only the doubly-protonated model that reproduces the edge feature of the pH 7.4 experimental spectrum of {$Ni_2^{II}(SOD^{mds})$}. The TD-DFT calculated S K-edge X-ray absorption spectrum for the doubly-protonated model has both S(1s → LUMO) transitions blue shifted by ~3.5 eV relative to the energy of the unprotonated sulfur atom S(1s → LUMO) transition, which reproduce the experimental spectrum well. Thus, we conclude that the pH 7.4 form of {$Ni_2^{II}(SOD^{mds})$} possesses two protonated cysteinate sulfur atoms coordinated to nickel.

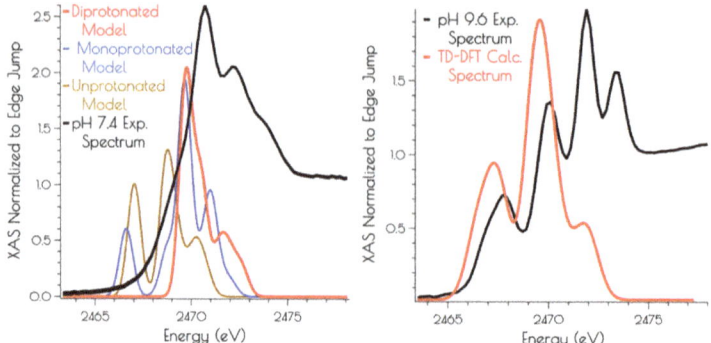

Figure 7. Left: Experimental (black spectrum) pH 7.4 and TD-DFT calculated S K-edge X-ray absorption spectra models of {Ni$_2^{II}$(SOD^{mds})} (unprotonated model: gold spectrum; monoprotonated model: blue spectrum; doubly-protonated model: red spectrum). Right: Experimental (black spectrum) pH 9.6 and calculated spectrum (disulfide bridged model: red spectrum) of {Ni$_2^{II}$(SOD^{mds})}.

3.5. Oxidation of {Ni$_2^{II}$(SOD^{mds})} at pH 7.4 and 9.6

{Ni$_2^{II}$(SOD^{mds})} reacts with O$_2$ at both pH 7.4 and 9.6. The kinetics of O$_2$ oxidation was followed by CD spectroscopy over the course of 12 h (Figure 8). It was found that the oxidation kinetics obey a pseudo-first order rate law under constant O$_2$ concentration, and proceeds at a faster rate at high vs. low pH. Extraction of the second order rate constant for the oxidation reactions demonstrates that the reaction at pH 9.6 proceeds with a rate that is over 3.5-fold faster than at pH 7.4 ($k = 1.8(3) \times 10^{-2}$ M^{-1} s^{-1} vs. $6.5(2) \times 10^{-2}$ M^{-1} s^{-1}). Based on an analysis of the products formed during the oxidation reaction, we suspect that following the initial oxidation step there are multiple oxidation pathways leading to different final oxidation products.

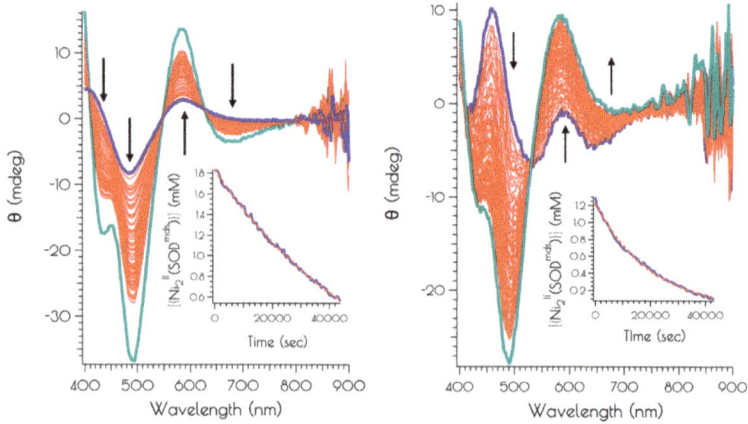

Figure 8. CD spectra following the air oxidation of {Ni$_2^{II}$(SOD^{mds})} at pH 7.4 (left) and 9.6 (right), with the blue spectra representing the trace at $t = 0$ s, the red spectra representing the traces recorded every 600 s (10 min) over the course of 12 h, and the teal spectra represent the CD spectra of the solutions following 24 h of O$_2$ exposure. The insets depict the kinetics traces highlighting the decay of {Ni$_2^{II}$(SOD^{mds})} (blue trace) and best fit of the kinetic trace to a first order rate law.

There are a number of items to note concerning the oxidation products formed at pH 7.4 and 9.6. First, under both pH conditions, at least one of the final soluble oxidation products contains nickel, and this product is identical by CD under both pH conditions. However, MS data of the solution and solid materials produced by O_2 initiated oxidation indicates a complex mixture of unidentifiable products. Furthermore, the EPR spectra of the resulting products are silent down to 10 K, indicating that the Ni-site is in the formal Ni^{2+} oxidation state. The resulting IR spectra of the produced solutions and solids showed no bands corresponding to S=O stretching frequencies, indicating that oxygen atom insertion reactions into the Ni-S moiety does not represent a major oxidation pathway. Instead, it is possible O_2 is initiating irreversible sulfur based ligand oxidation as has been observed in the work of Darensbourg, for example [51]. Validating this possibility is the observation that the metallopeptide cannot be cleanly oxidized; attempts to chemically oxidize {$Ni_2^{II}(SOD^{mds})$} at pH 7.4 and 9.6 by a 3% hydrogen peroxide solution, ethanolic I_2 or MnO_4^- lead to the rapid bleaching of the solution and subsequent formation of unidentifiable tan insoluble aggregates, all of which yielded EPR and IR spectra consistent with the above formed from O_2 oxidation of the solutions.

A complex mixture of soluble and insoluble nickel containing products is also noted by Ni K-edge X-ray absorption spectroscopy. At both solution pH values, the Ni K-edge XANES no longer contains the prominent Ni(1s → $4p_z$) transition, and is more reminiscent of a six coordinate Ni^{2+} species (Figures S1 and S2). Because of the low signal to noise at high k, the EXAFS regions could only be simulated to $k = 11$ Å^{-1} The EXAFS region of the decomposition product obtained from air oxidation at pH 7.4 was best modeled as containing 1.4 Ni–S interactions ($r = 2.22$ Å) and 4.8 N/O interactions ($r = 1.97$ Å). The EXAFS region of the air oxidation product generated at pH 9.6 was best modeled with 0.6 Ni–S interactions ($r = 2.24$ Å), 4.3 N/O interaction ($r = 1.94$ Å), and 2.1 Ni–Ni interactions ($r = 3.25$ Å). In both cases, the resulting fitting statistics are poor with ε^2 values greater than 3. As this represents a mixture of soluble and insoluble compounds in a number of coordination environments, formulating likely structures about the nickel center(s) is not warranted based on the available data.

Electronic structure calculations suggest the reason for the increased stability of {$Ni_2(SOD^{mds})$} at pH 7.4 vs. 9.6 results from the deactivation of the high-lying sulfur dominated Ni(3dπ)–S(3pπ) anti-bonding orbital upon protonation (Figure 9, Tables 1 and 2). For the monomeric species, the HOMO is identified as a nickel dominated Ni(3dπ)–N(2pπ) anti-bonding orbital. This is destabilized by 0.29 eV relative to the essentially non-bonding Ni(3d$_{z2}$) orbital (HOMO-1). The HOMO-2 is a water O(2p) dominated O(2pπ)–Ni(3dπ) antibonding orbital followed by the Ni(3d$_{xz}$) dominated HOMO-3. Thus, none of the frontier MOs (FMOs) possess significant S(3p) character, rendering the sulfur atoms reasonably unreactive towards oxidative damage by O_2. In contrast, deprotonation of the sulfur atoms of the mononuclear {$Ni_2^{II}(SOD^{mds})$} nickel-site dramatically alters the electronic structure of the complex. Electronic structure calculations reveal that the HOMO and HOMO-1 are significantly activated relative to the essentially non-bond Ni(3d$_{z2}$) HOMO-2 by 0.73 and 0.42 eV, respectively. Furthermore, these two orbitals are S(3p) dominated S(3pπ)–Ni(3dπ) anti-bonding orbitals. Therefore, if the deprotonated monomeric form of {$Ni_2(SOD^{mds})$} could be generated, we would predict it would be highly sensitive to O_2 damage owing to the activated S(3p) dominated HOMO. Given the slow rate of O_2 oxidation of {$Ni_2(SOD^{mds})$} at pH = 7.4, it is likely that deprotonation of the Ni–S(H$^+$)–Cys moieties are required for oxidative damage; as protonation is an equilibrium process, there will always be a small concentration of O_2 reactive unprotonated Ni-S-Cys bonds in solution. The disulfide bridged computational model was found to possess highly covalent Ni–ligand bonds. The HOMO, although containing less S(3p) than the HOMO of the unprotonated mononuclear computational model, still contains a significant amount of S(3p) character, and would therefore be predicted to be more susceptible to O_2 damage than the doubly-protonated mononuclear nickel-site. This is observed experimentally.

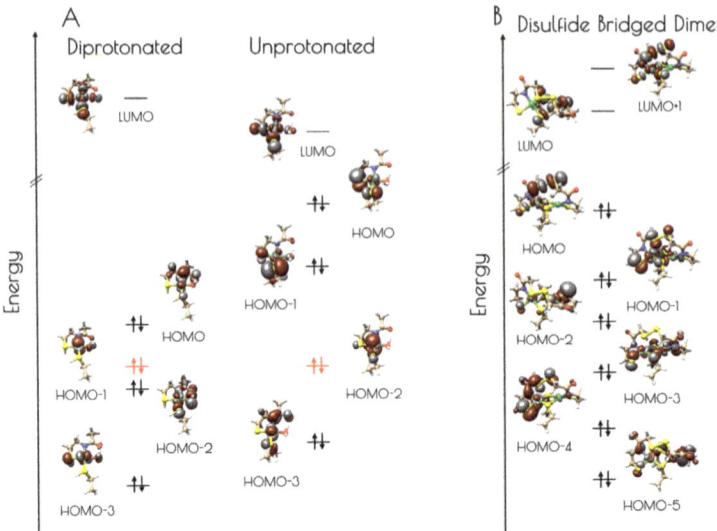

Figure 9. (**A**) Isosurface plots (0.03 a.u.) of the LUMO through HOMO-3 of the doubly-protonated (left) and unprotonated (right) computational models of the pH 7.4 form of the nickel-site of {Ni$_2$(SOD^{mds})}. The energies were normalized to the non-bonding Ni(3d$_{z2}$) orbital, highlighted in red. (**B**) Isosurface plots (0.03 a.u.) of the LUMO+1 through HOMO-5 of the disulfide bridged dinuclear {Ni$_2$(SOD^{mds})} computational model.

Table 1. Ni(3d), S(3p), N(2p), and water O(2p) Löwdin MO population analysis (%AO to MO) and energies (eV) relative to the Ni(3d$_{z2}$) orbital for the LUMO through HOMO-3 of the computational models for doubly-protonated and unprotonated monomeric {Ni$_2$(SOD^{mds})} computational models. Orbital compositions for the doubly-protonated model are given above the unprotonated model for each AO.

	LUMO	HOMO	HOMO-1	HOMO-2	HOMO-3
Ni doubly-protonated	61.4	48.9	76.8	29.9	80.5
unprotonated	51.7	29.9	16.2	76.3	61.7
S^1 doubly-protonated	6.0	1.6	2.4	7.4	0.0
unprotonated	11.5	9.9	55.7	5.6	0.3
S^2 doubly-protonated	3.9	0.2	1.1	0.1	6.4
unprotonated	11.6	46.1	10.1	6.5	4.9
N doubly-protonated	3.9	28.5	2.7	3.3	0.0
unprotonated	2.1	0.2	10.0	11.0	5.3
O doubly-protonated	7.4	0.2	6.7	41.0	3.7
unprotonated	3.6	0.6	3.0	0.9	4.8
E doubly-protonated	5.36	0.29	0.00	−0.07	−0.49
unprotonated	5.12	0.73	0.42	0.00	−0.22

^{1} *trans* to amidate nitrogen; ^{2} *trans* to water oxygen.

Table 2. Ni(3d), S(3p), and N(2p) Löwdin MO population analysis (%AO to MO) and energies (E, eV) relative to the HOMO for the LUMO+1 through HOMO-5 of the computational model for the disulfide bridged dinuclear {Ni$_2$(SOD^{mds})} nickel site computational model. Orbital compositions for the doubly-protonated model are given above the unprotonated model for each AO.

	LUMO+1	LUMO	HOMO	HOMO-1	HOMO-2	HOMO-3	HOMO-4	HOMO-5
Ni^1	0.3	21.1	28.2	4.8	11.8	20.3	7.5	18.5

Table 2. Cont.

	LUMO+1	LUMO	HOMO	HOMO-1	HOMO-2	HOMO-3	HOMO-4	HOMO-5
Ni^2	50.6	0.4	4.3	23.3	11.7	1.7	24.6	38.4
S^3	0.4	16.4	10.4	1.5	1.2	10.2	0.1	0.4
S^4	0.4	17.1	5.2	0.2	0.6	88	0.8	7.3
S^5	6.1	2.0	4.2	2.0	9.5	0.6	1.2	1.1
S^6	8.7	2.4	9.5	1.4	8.9	0.4	6.0	1.4
S^7	0.1	5.1	21.9	4.6	15.6	30.0	6.1	5.1
S^8	10.1	0.0	3.2	49.2	15.2	0.4	2.3	4.3
N^9	0.2	1.6	0.4	0.2	1.7	6.1	5.7	5.5
N^{10}	3.0	0.0	0.2	0.9	5.0	0.8	14.9	3.8
E	3.76	3.62	0.00	−0.30	−0.54	−0.69	−0.72	−1.07

[1] more distorted nickel site; [2] less distorted nickel site; [3] disulfide sulfur over Ni^1; [4] disulfide sulfur over Ni^2; [5] bridging thiolate sulfur; [6] bridging thiolate sulfur; [7] terminal thiolate sulfur ligated to Ni^1; [8] terminal thiolate ligated to Ni^2; [9] amidate nitrogen ligated to Ni^1; [10] amidate nitrogen ligated to Ni^2.

4. Discussion

In this study we have demonstrated the reversible formation of a dinuclear Ni$_2^{II}$ site within a peptide in response to pH. Dinuclear Ni$_2^{II}$ sites are not observed in other monomeric SOD metallopeptide-based mimics while the disulfide linked dimeric metallopeptide facilitates formation of a dinuclear Ni$_2$ center. We speculate that dimer formation in {Ni$_2$(SOD^{mds})} is facilitated by the close proximity of the two metal centers, which was enforced through a disulfide bond linkage near the individual nickel centers.

The driving force for conversion of the dinuclear Ni$_2^{II}$ site into two monomeric Ni^{II} sites within {Ni$_2$(SOD^{mds})} likely involves protonation of a terminal cysteinate sulfur atom. The nucleophilic HOMO of the dinuclear Ni$_2^{II}$ site possesses S(3p) character corresponding to the terminal thiolate sulfur coordinated to the more distorted nickel center, making it the likely protonation site. Protonation of that sulfur atom shortens the Ni–S(H$^+$)–Cys bond length relative to the unprotonated model with a concomitant increase in the bridging Ni–S bond length *trans* to the protonation site (Figure 10). Association of water to the nickel center followed by breaking of the weakened Ni–S bond would lead to the eventual formation of two mononuclear Ni^{II} centers.

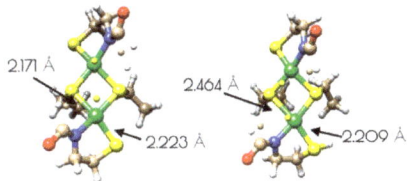

Figure 10. Structures of binuclear nickel-site models of {Ni$_2^{II}$(SOD^{mds})} and {Ni$_2^{II}$(SOD^{mds}–S(H$^+$)C1)}. The disulfide bridge and methylene groups have been represented as small spheres and wires for clarity.

This study also gives insight into an additional mechanism of protection of Ni-thiolate bonds from oxidative damage against reactive oxygen species (ROSs). Nickel thiolates are susceptible to oxidative damage by O$_2$ and H$_2$O$_2$ [52,53], yet the NiSOD active-site is robust against oxidative damage effected by such species. Possible explanations for protection of the NiSOD active-site against Ni–S oxidative damage by ROSs have been proposed, including electronic fine tuning of the Ni–S moiety via the mixed amine/amide coordination environment [54,55] and a fast rate of the O$_2^-$ disproportionation reaction relative to the rate of oxidation of the coordinate cysteinate sulfur atoms [45,48,49]. In this study we have shown that oxidation of the Ni–S–Cys bond by O$_2$ is slow for {Ni$_2$(SOD^{mds})} at pH = 7.4, with a half-life of nearly 7 h under the reaction conditions investigated (ambient O$_2$ concentration, 1.0 mM {Ni$_2$(SOD^{mds})}). This is the result of both the deactivation and a significant reduction in S(3p) character to the nucleophilic FMOs upon protonation; the FMOs of the doubly-protonated monomeric {$Ni_2(SOD^{mds})$} computational model possesses little S(3p) character, while the HOMO and HOMO-1 of the unprotonated {$Ni_2(SOD^{mds})$} computational model are both energetically activated and possess a large degree of S(3p) character. Thus, protonation will inherently protect the thiolate sulfur atoms against ROS induced oxidative damage. The dinuclear nickel site of {$Ni_2(SOD^{mds})$} produced at pH = 9.6 was found to undergo oxidation at an increased rate relative to the mononuclear pH = 7.4 form. This is to be expected as the degree of terminal cysteinate S(3p) character to the more covalent anti-bonding FMOs has increased relative to the mononuclear pH = 7.4 doubly-protonated Ni form, rendering it more susceptible to oxidative damage.

This study has also pointed to an additional role for the protonation of coordinated cysteinate sulfur atoms at metalloenzyme active-sites—poising the centers for reactivity via electronic fine tuning. As demonstrated above, the FMOs of the doubly-protonated mononuclear {$Ni_2(SOD^{mds})$} are biased to the nickel-site, while those of the unprotonated mononuclear {$Ni_2(SOD^{mds})$} computational model are biased to the thiolate sulfur atoms. Thus, one would expect that upon thiolate sulfur atom protonation reactivity would be shifted from the thiolate sulfur atoms to the nickel-site. For example, one could consider the active-site of [NiFe]H_2ase, which has been shown to possess a terminal protonated cysteinate residue ligated to a D_{2d} distorted nickel center. It has been proposed that a key intermediate in the reactivity of [NiFe]H_2ase is a formal Ni^{III}–H species. Protonation of the cysteinate sulfur atom would bias the reactivity towards nickel, making the nickel-center the site susceptible to subsequent protonation events. Thus, one may envision a dual role for the Ni–S(H$^+$)–Cys moiety in [NiFe]H_2ase—gating of reactivity and proton donation to the hydride ligand.

5. Conclusions

A disulfide bridged metallopeptide has been prepared inspired by the metalloenzyme NiSOD. This metallopeptide contains two nickel centers in close proximity owing to a disulfide bridge between two peptide monomers. In response to pH, the mononuclear nickel-sites found at pH 7.4, which contain Ni^{II} in an S_2NO coordination motif reversibly form a dinuclear cysteinate Ni_2^{II} center at elevated pH (pH 9.6). The driving force for the interconversion of the dinuclear nickel center to two mononuclear nickel sites is proposed to be cysteinate S-atom protonation, which results in two coordinated protonated Cys S atoms at lower pH. It was shown that these Ni–S(H$^+$)–Cys moieties reduce the O_2 initiated oxidative damage of the nickel-site, likely through the modulation of the electronic structure of the Ni-center rendering the S-atoms less nucleophilic upon protonation. This may have relevance in biological Ni systems, offering the Ni–S–Cys moiety protection against oxidative damage upon Ni–S(H$^+$)–Cys formation. Furthermore, the modulation of the electronic structure of the Ni-site upon Ni–S(H$^+$)–Cys formation suggests that reversible cysteinate sulfur atom protonation may be involved in the gating of biological reactivity at such metal-centers.

Supplementary Materials: The following are available online at http://www.mdpi.com/2304-6740/7/7/90/s1, Figures S1 and S2: nickel K-edge XANES of the O_2 decomposition products of {$Ni_2^{II}(SOD^{mds})$}, Tables S1–S7: xyz files for computational models, Tables S8 and S9: alternative EXAFS models for the high and low pH forms of {$Ni_2^{II}(SOD^{mds})$}.

Author Contributions: Conceptualization, J.S.; methodology, B.K. and J.S.; formal analysis, B.K and J.S..; investigation, J.S., D.O. and B.K.; writing, B.K. and J.S.; supervision, J.S.; project administration, J.S.; funding acquisition, J.S.

Funding: This research was funded by the National Science Foundation, grant number CHE-1565766, the National Institutes of Health, grant number R15-GM120641-01, and Trinity University. Molecular graphics and analyses performed with UCSF Chimera, developed by the Resource for Biocomputing, Visualization, and Informatics at the University of California, San Francisco, with support from NIH P41-GM103311.

Acknowledgments: Nickel K-edge X-ray absorption spectroscopy was performed at the Canadian Light Source, which is supported by the Canada Foundation for Innovation, Natural Sciences and Engineering Research Council of Canada, the University of Saskatchewan, the Government of Saskatchewan, Western Economic Diversification Canada, the National Research Council Canada, and the Canadian Institutes of Health Research.

Conflicts of Interest: The authors declare no conflict of interest.

References

1. Can, M.; Armstrong, F.A.; Ragsdale, S.W. Structure, function, and mechanism of the nickel metalloenzymes, CO dehydrogenase, and acetyl-CoA synthase. *Chem. Rev.* **2014**, *114*, 4149–4174. [CrossRef] [PubMed]
2. Maroney, M.J.; Ciurli, S. Nonredox nickel enzymes. *Chem. Rev.* **2014**, *114*, 4206–4228. [CrossRef] [PubMed]
3. Ragsdale, S.W. *Biochemistry of Methyl-Coenzyme M Reductase: The Nickel Metalloenzyme That Catalyzes the Final Step in Synthesis and the First Step in Anaerobic Oxidation of the Greenhouse Gas Methane*; Springer: Dordrecht, The Netherlands, 2014; pp. 125–145.
4. Ragsdale, S.W. Nickel biochemistry. *Curr. Opin. Chem. Biol.* **1998**, *2*, 208–215. [CrossRef]
5. Ragsdale, S.W. Nickel-based enzyme systems. *J. Biol. Chem.* **2009**, *284*, 18571–18575. [CrossRef] [PubMed]
6. Higgins, K.A.; Carr, C.E.; Maroney, M.J. Specific metal recognition in nickel trafficking. *Biochemistry* **2012**, *51*, 7816–7832. [CrossRef] [PubMed]
7. Chung, K.C.C.; Cao, L.; Dias, A.V.; Pickering, I.J.; George, G.N.; Zamble, D.B. A High-Affinity Metal-Binding Peptide from *Escherichia coli* HypB. *J. Am. Chem. Soc.* **2008**, *130*, 14056–14057. [CrossRef]
8. Dias, A.V.; Mulvihill, C.M.; Leach, M.R.; Pickering, I.J.; George, G.N.; Zamble, D.B. Structural and Biological Analysis of the Metal Sites of *Escherichia coli* Hydrogenase Accessory Protein HypB. *Biochemistry* **2008**, *47*, 11981–11991. [CrossRef]
9. Douglas, C.D.; Ngu, T.T.; Kaluarachchi, H.; Zamble, D.B. Metal Transfer within the *Escherichia coli* HypB-HypA Complex of Hydrogenase Accessory Proteins. *Biochemistry* **2013**, *52*, 6030–6039. [CrossRef]
10. Lacasse, M.J.; Douglas, C.D.; Zamble, D.B. Mechanism of Selective Nickel Transfer from HypB to HypA, *Escherichia coli* [NiFe]-Hydrogenase Accessory Proteins. *Biochemistry* **2016**, *55*, 6821–6831. [CrossRef]
11. Schreiter, E.R.; Sintchak, M.D.; Guo, Y.; Chivers, P.T.; Sauer, R.T.; Drennan, C.L. Crystal structure of the nickel-responsive transcription factor NikR. *Nat. Struct. Mol. Biol.* **2003**, *10*, 794–799. [CrossRef]
12. Clegg, W.; Henderson, R.A. Kinetic Evidence for Intramolecular Proton Transfer Between Nickel and Coordinated Thiolate. *Inorg. Chem.* **2002**, *41*, 1128–1135. [CrossRef] [PubMed]
13. Autissier, V.; Zarza, P.M.; Petrou, A.; Henderson, R.A.; Harrington, R.W.; Clegg, W.C. Proton Transfer to Nickel-Thiolate Complexes. 2. Rate-Limiting Intramolecular Proton Transfer in the Reactions of [Ni(SC$_6$H$_4$R-4)(PhP{CHCH$_2$PPh$_2$}$_2$)]$^+$ (R = NO$_2$, Cl, H, Me, or MeO). *Inorg. Chem.* **2004**, *43*, 3106–3115. [CrossRef] [PubMed]
14. Alwaaly, A.; Henderson, R.A. Sterics level the rates of proton transfer to [Ni(XPh){PhP(CH$_2$CH$_2$PPh$_2$)$_2$}]$^+$ (X = O, S or Se). *Chem. Commun.* **2014**, *50*, 9669–9671. [CrossRef] [PubMed]
15. Szilagyi, R.K.; Bryngelson, P.A.; Maroney, M.J.; Hedman, B.; Hodgson, K.O.; Solomon, E.I. S K-Edge X-ray Absorption Spectroscopic Investigation of the Ni-Containing Superoxide Dismutase Active Site: New Structural Insight into the Mechanism. *J. Am. Chem. Soc.* **2004**, *126*, 3018–3019. [CrossRef] [PubMed]
16. Ogata, H.; Nishikawa, K.; Lubitz, W. Hydrogens detected by subatomic resolution protein crystallography in a [NiFe] hydrogenase. *Nature* **2015**, *520*, 571. [CrossRef] [PubMed]
17. Shearer, J. Insight into the structure and mechanism of nickel-containing superoxide dismutase derived from peptide-based mimics. *Acc. Chem. Res.* **2014**, *47*, 2332–2341. [CrossRef]
18. Shearer, J.; Peck, K.L.; Schmitt, J.C.; Neupane, K.P. Cysteinate protonation and water hydrogen bonding at the active-site of a nickel superoxide dismutase metallopeptide-based mimic: Implications for the mechanism of superoxide reduction. *J. Am. Chem. Soc.* **2014**, *136*, 16009–16022. [CrossRef]
19. Shearer, J.; Schmitt, J.C.; Clewett, H.S. Adiabaticity of the Proton-Coupled Electron-Transfer Step in the Reduction of Superoxide Effected by Nickel-Containing Superoxide Dismutase Metallopeptide-Based Mimics. *J. Phys. Chem. B* **2015**, *119*, 5453–5461. [CrossRef]
20. Pelmenschikov, V.; Siegbahn, P.E.M. Nickel Superoxide Dismutase Reaction Mechanism Studied by Hybrid Density Functional Methods. *J. Am. Chem. Soc.* **2006**, *128*, 7466–7475. [CrossRef]
21. Krämer, T.; Kampa, M.; Lubitz, W.; van Gastel, M.; Neese, F. Theoretical Spectroscopy of the Ni^{II} Intermediate States in the Catalytic Cycle and the Activation of [NiFe] Hydrogenases. *ChemBioChem* **2013**, *14*, 1898–1905. [CrossRef]

22. Shearer, J. Use of a Metallopeptide-Based Mimic Provides Evidence for a Proton-Coupled Electron-Transfer Mechanism for Superoxide Reduction By Nickel-Containing Superoxide Dismutase. *Angew. Chem. Int. Ed.* **2013**, *52*, 2569–2572. [CrossRef] [PubMed]
23. Barondeau, D.P.; Kassmann, C.J.; Bruns, C.K.; Tainer, J.A.; Getzoff, E.D. Nickel Superoxide Dismutase Structure and Mechanism. *Biochemistry* **2004**, *43*, 8038–8047. [CrossRef] [PubMed]
24. Wuerges, J.; Lee, J.-W.; Yim, Y.-I.; Yim, H.-S.; Kang, S.-O.; Carugo, K.D. Crystal structure of nickel-containing superoxide dismutase reveals another type of active site. *Proc. Natl. Acad. Sci. USA* **2004**, *101*, 8569–8574. [CrossRef] [PubMed]
25. Ryan, K.C.; Guce, A.I.; Johnson, O.E.; Brunold, T.C.; Cabelli, D.E.; Garman, S.C.; Maroney, M.J. Nickel Superoxide Dismutase: Structural and Functional Roles of His1 and Its H-Bonding Network. *Biochemistry* **2015**, *54*, 1016–1027. [CrossRef] [PubMed]
26. Shearer, J.; Long, L.M. A Nickel Superoxide Dismutase Maquette That Reproduces the Spectroscopic and Functional Properties of the Metalloenzyme. *Inorg. Chem.* **2006**, *45*, 2358–2360. [CrossRef]
27. Neupane, K.P.; Shearer, J. The Influence of Amine/Amide versus Bisamide Coordination in Nickel Superoxide Dismutase. *Inorg. Chem.* **2006**, *45*, 10552–10566. [CrossRef] [PubMed]
28. Neupane, K.P.; Gearty, K.; Francis, A.; Shearer, J. Probing Variable Axial Ligation in Nickel Superoxide Dismutase Utilizing Metallopeptide-Based Models: Insight into the Superoxide Disproportionation Mechanism. *J. Am. Chem. Soc.* **2007**, *129*, 14605–14618. [CrossRef]
29. Shearer, J.; Neupane, K.P.; Callan, P.E. Metallopeptide Based Mimics with Substituted Histidines Approximate a Key Hydrogen Bonding Network in the Metalloenzyme Nickel Superoxide Dismutase. *Inorg. Chem.* **2009**, *48*, 10560–10571. [CrossRef]
30. Tietze, D.; Breitzke, H.; Imhof, D.; Koeth, E.; Weston, J.; Buntkowsky, G. New insight into the mode of action of nickel superoxide dismutase by investigating metallopeptide substrate models. *Chem. - Eur. J.* **2009**, *15*, 517–523. [CrossRef]
31. Tietze, D.; Voigt, S.; Mollenhauer, D.; Tischler, M.; Imhof, D.; Gutmann, T.; Gonzalez, L.; Ohlenschlaeger, O.; Breitzke, H.; Goerlach, M.; et al. Revealing the Position of the Substrate in Nickel Superoxide Dismutase: A Model Study. *Angew. Chem. Int. Ed.* **2011**, *50*, 2946–2950. [CrossRef]
32. Tietze, D.; Sartorius, J.; Koley Seth, B.; Herr, K.; Heimer, P.; Imhof, D.; Mollenhauer, D.; Buntkowsky, G. New insights into the mechanism of nickel superoxide degradation from studies of model peptides. *Sci. Rep.* **2017**, *7*, 1–15. [CrossRef] [PubMed]
33. Tietze, D.; Koley Seth, B.; Brauser, M.; Tietze, A.A.; Buntkowsky, G. Ni^{II} Complex Formation and Protonation States at the Active Site of a Nickel Superoxide Dismutase-Derived Metallopeptide: Implications for the Mechanism of Superoxide Degradation. *Chem. - Eur. J.* **2018**, *24*, 15879–15888. [CrossRef] [PubMed]
34. Shearer, J.; Callan, P.E.; Amie, J. Use of Metallopeptide Based Mimics Demonstrates That the Metalloprotein Nitrile Hydratase Requires Two Oxidized Cysteinates for Catalytic Activity. *Inorg. Chem.* **2010**, *49*, 9064–9077. [CrossRef] [PubMed]
35. Dutta, A.; Flores, M.; Roy, S.; Schmitt, J.C.; Hamilton, G.A.; Hartnett, H.E.; Shearer, J.M.; Jones, A.K. Sequential oxidations of thiolates and the cobalt metallocenter in a synthetic metallopeptide: Implications for the biosynthesis of nitrile hydratase. *Inorg. Chem.* **2013**, *52*, 5236–5245. [CrossRef] [PubMed]
36. Martinage, O.; Le Clainche, L.; Czarny, B.; Dugave, C. Synthesis and biological evaluation of a new triazole-oxotechnetium complex. *Org. Biomol. Chem.* **2012**, *10*, 6484–6490. [CrossRef] [PubMed]
37. Ellman, G.L. A colorimetric method for determining low concentrations of mercaptans. *Arch Biochem Biophys* **1958**, *74*, 443–450. [CrossRef]
38. Neese, F. The ORCA program system. *Wiley Interdiscip. Rev.: Comput. Mol. Sci.* **2012**, *2*, 73–78. [CrossRef]
39. Weigend, F.A. Reinhart, Balanced basis sets of split valence, triple zeta valance and quadruple zeta valence quality for H to Rn: Design and assessment of accuracy. *Phys. Chem. Chem. Phys.* **2005**, *7*, 3297–3305. [CrossRef]
40. Grimme, S.; Ehrlich, S.; Goerigk, L. Effect of the damping function in dispersion corrected density functional theory. *J. Comput. Chem.* **2011**, *32*, 1456–1465. [CrossRef]
41. Grimme, S.; Antony, J.; Ehrlich, S.; Krieg, H. A consistent and accurate ab initio parameterization of density functional dispersion correction (DFT-D) for the 94 elements H-Pu. *J. Chem. Phys.* **2010**, *132*, 154104. [CrossRef]

42. Hellweg, A.; Hättig, C.; Hofener, S.; Klopper, W. Optimized accurate auxiliary basis sets for RI-MP2 and RI-CC2 calculations for the atoms Rb to Rn. *Theor. Chem. Acc.* **2007**, *117*, 587. [CrossRef]
43. Weigend, F. Accurate Coulomb-fitting basis sets for H to Rn. *Chem. Phys. Phys. Chem.* **2006**, *8*, 1057–1065. [CrossRef] [PubMed]
44. Pettersen, E.F.; Goddard, T.D.; Huang, C.C.; Couch, G.S.; Greenblatt, D.M.; Meng, E.C.; Ferrin, T.E. UCSF Chimera—A visualization system for exploratory research and analysis. *J. Comput. Chem.* **2004**, *25*, 1605–1612. [CrossRef] [PubMed]
45. Thompson, P.E.; Keah, H.H.; Gomme, P.T.; Stanton, P.G.; Hearn, M.T. Synthesis of peptide amides using Fmoc-based solid-phase procedures on 4-methylbenzhydrylamine resins. *Int. J. Pept. Protein Res.* **1995**, *46*, 174–180. [CrossRef] [PubMed]
46. Colpas, G.J.; Maroney, M.J.; Bagyinka, C.; Kumar, M.; Willis, W.S.; Suib, S.L.; Mascharak, P.K.; Baidya, N. X-ray spectroscopic studies of nickel complexes, with application to the structure of nickel sites in hydrogenases. *Inorg. Chem.* **1991**, *30*, 920–928. [CrossRef]
47. Denny, J.A.; Darensbourg, M.Y. Metallodithiolates as Ligands in Coordination, Bioinorganic, and Organometallic Chemistry. *Chem. Rev.* **2015**, *115*, 5248–5273. [CrossRef]
48. Jenkins, R.M.; Singleton, M.L.; Leamer, L.A.; Reibenspies, J.H.; Darensbourg, M.Y. Orientation and Stereodynamic Paths of Planar Monodentate Ligands in Square Planar Nickel N_2S Complexes. *Inorg. Chem.* **2010**, *49*, 5503–5514. [CrossRef]
49. Neese, F. Prediction of molecular properties and molecular spectroscopy with density functional theory: From fundamental theory to exchange-coupling. *Coord. Chem. Rev.* **2009**, *253*, 526–563. [CrossRef]
50. Kirchner, B.; Wennmohs, F.; Ye, S.; Neese, F. Theoretical bioinorganic chemistry: The electronic structure makes a difference. *Curr. Opin. Chem. Biol.* **2007**, *11*, 134–141. [CrossRef]
51. Jenkins, R.M.; Singleton, M.L.; Almaraz, E.; Relbenspies, J.H.; Darensbourg, M.Y. Imidazole-Containing (N_3S)-Ni-II Complexes Relating to Nickel Containing Biomolecules. *Inorg. Chem.* **2009**, *48*, 7280–7293. [CrossRef]
52. Grapperhaus, C.A.; Darensbourg, M.Y. Oxygen Capture by Sulfur in Nickel Thiolates. *Acc. Chem. Res.* **1998**, *31*, 451–459. [CrossRef]
53. Green, K.N.; Brothers, S.M.; Jenkins, R.M.; Carson, C.E.; Grapperhaus, C.A.; Darensbourg, M.Y. An Experimental and Computational Study of Sulfur-Modified Nucleophilicity in a Dianionic NiN_2S_2 Complex. *Inorg. Chem.* **2007**, *46*, 7536–7544. [CrossRef] [PubMed]
54. Shearer, J.; Dehestani, A.; Abanda, F. Probing Variable Amine/Amide Ligation in $Ni^{II}N_2S_2$ Complexes Using Sulfur K-Edge and Nickel L-Edge X-ray Absorption Spectroscopies: Implications for the Active Site of Nickel Superoxide Dismutase. *Inorg. Chem.* **2008**, *47*, 2649–2660. [CrossRef] [PubMed]
55. Mullins, C.S.; Grapperhaus, C.A.; Kozlowski, P.M. Density functional theory investigations of NiN_2S_2 reactivity as a function of nitrogen donor type and N–H···S hydrogen bonding inspired by nickel-containing superoxide dismutase. *J. Biol. Inorg. Chem.* **2006**, *11*, 617–625. [CrossRef] [PubMed]

© 2019 by the authors. Licensee MDPI, Basel, Switzerland. This article is an open access article distributed under the terms and conditions of the Creative Commons Attribution (CC BY) license (http://creativecommons.org/licenses/by/4.0/).

Review

Nickel Metalloregulators and Chaperones

Khadine Higgins

Department of Chemistry, Salve Regina University, Newport, RI 02840, USA; khadine.higgins@salve.edu;
Tel.: +1-401-341-3215

Received: 11 June 2019; Accepted: 14 August 2019; Published: 19 August 2019

Abstract: Nickel is essential for the survival of many pathogenic bacteria. *E. coli* and *H. pylori* require nickel for [NiFe]-hydrogenases. *H. pylori* also requires nickel for urease. At high concentrations nickel can be toxic to the cell, therefore, nickel concentrations are tightly regulated. Metalloregulators help to maintain nickel concentration in the cell by regulating the expression of the genes associated with nickel import and export. Nickel import into the cell, delivery of nickel to target proteins, and export of nickel from the cell is a very intricate and well-choreographed process. The delivery of nickel to [NiFe]-hydrogenase and urease is complex and involves several chaperones and accessory proteins. A combination of biochemical, crystallographic, and spectroscopic techniques has been utilized to study the structures of these proteins, as well as protein–protein interactions resulting in an expansion of our knowledge regarding how these proteins sense and bind nickel. In this review, recent advances in the field will be discussed, focusing on the metal site structures of nickel bound to metalloregulators and chaperones.

Keywords: nickel; metalloregulator; chaperone; [NiFe]-hydrogenase; urease

1. Introduction

Nickel, the twenty-eighth element in the periodic table, is an essential metal for the functioning of many proteins in archaea, bacteria, plants, and some eukaryotes [1]. It is a cofactor for at least eight enzymes including urease and [NiFe]-hydrogenase [1]. Many pathogenic bacteria require nickel for their survival and pathogenicity, including *Helicobacter pylori* (*H. pylori*), [2,3]. Despite being an essential metal, high concentrations of nickel can be toxic as nickel can bind to other metalloproteins and displace the cognate metals resulting in an inactive metalloprotein [4]. As such nickel concentrations in the cell need to be regulated. One way in which nickel concentrations are controlled in the cell is by utilizing metalloregulators. Nickel-responsive metalloregulators function by binding to nickel in a specific coordination environment resulting in the transcriptional repression, activation, or depression of genes associated with nickel export or import. To maintain nickel concentration in the cell, *E. coli* utilizes two metalloregulators, NikR and RcnR [5]. NikR controls the expression of the genes associated with the ATP binding cassette (ABC)-type nickel importer, NikABCDE [6], and RcnR, regulates the expression of the genes associated with the export proteins RcnAB [7,8].

E. coli is a facultative anaerobe and under anaerobic conditions expresses four different [NiFe]-hydrogenases, hydrogenase 1, 2, 3, and 4, each associated with a specific metabolic pathway [9,10]. Hydrogenases are a group of metalloenzymes that catalyze the reversible oxidation of molecular hydrogen to protons and electrons [11]. [NiFe]-hydrogenases are composed of at least two subunits, a large and a small subunit [12,13]. The large subunit contains a complex NiFe(CN)$_2$CO center at the active site and the small subunit contains up to three iron sulfur clusters [12,13]. The assembly of [NiFe]-hydrogenase 3 is dependent on the *hyp* (hydrogenase pleiotropy) genes *hypABCDEF* and *slyD* genes (Figure 1) [9,14–17]. For the assembly of hydrogenases 1 and 2, HypA and HypC are replaced by HybF and HybG [18]. HypE and HypF form a complex and synthesize the cyanide ligands from carbamoylphosphate (Figure 1A) [12,13,19]. HypE transfers the cyanide ligand to the HypC–HypD complex, which acts as a scaffold for the assembly and delivery of the iron cofactor, Fe(CN)$_2$CO, to the precursor large subunit of [NiFe]-hydrogenase (Figure 1B) [12,13]. Finally, HypA, HypB, and SlyD insert nickel into the large, precursor subunit of [NiFe]-hydrogenase (Figure 1C) [12,13,17]. At least six review articles have been published within the last ten years that provide a more detailed description of the biosynthesis of hydrogenases [11–13,17,20,21].

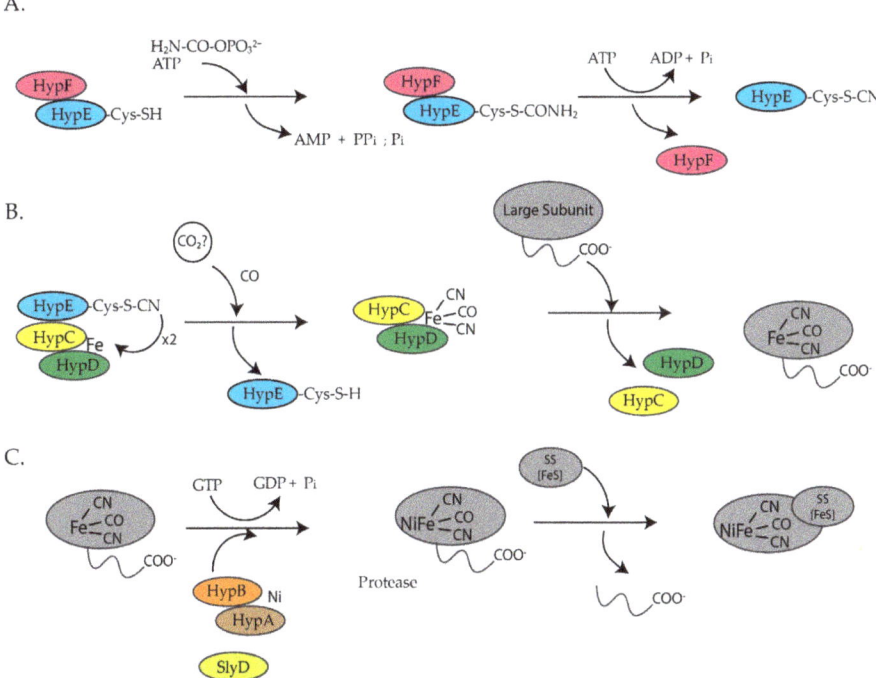

Figure 1. Model for the assembly [NiFe]-hydrogenase [12,13,17,20]. (**A**) HypE and HypF synthesize the cyanide ligands from carbamoylphosphate. (**B**) The assembly of the iron cofactor, Fe(CN)$_2$CO, and delivery to the precursor large subunit of [NiFe]-hydrogenase. (**C**) Nickel delivery and insertion into the large, precursor subunit of [NiFe]-hydrogenase and the peptide on the C-terminus of the large subunit is processed. The large subunit associates with the small subunit (SS) forming the mature [NiFe]-hydrogenase.

H. pylori is a human pathogen that causes gastritis, peptic ulcers, and some types of gastric cancer [22–25]. Mouse studies determined that *H. pylori* requires [NiFe]-hydrogenase for efficient colonization of the gut [26]. *H. pylori* also requires urease to survive the acidic environment of the stomach [1]. Urease catalyzes the hydrolysis of urea to ammonia and carbamate, which reacts with water to yield a second molecule of ammonia and bicarbonate [1,27]. The ammonia produced is used to neutralize the cytoplasmic and periplasmic pH of the bacteria under acidic conditions [24]. The enzyme urease has been discussed extensively in the past ten years [1,20,27–30]. *H. Pylori* urease has two distinct types of subunits that combine to form a trimer of dimers [28]. Four of these trimer of dimers combine to form a super molecular complex of the form ((UreAB)$_3$)$_4$ [28]. The maturation of *H. pylori* urease is dependent on the accessory proteins UreIEFGH [28,29]. Unlike urease from *H. pylori*, urease from *Klebsiella aerogenes* (*K. aerogenes*) is composed of three subunits that form a trimer of trimers, (UreABC)$_3$, structure [28]. In *K. aerogenes* the accessory proteins are UreDEFG, where UreD is homologous to UreH in *H. pylori* [27,28]. UreI is a proton-gated urea channel that regulates urease activity as it permits urea entry into *H. pylori* [24]. UreI is located in the inner membrane of the bacteria;

at pH 7.0 the channels are closed and at pH 5.0 the channels are open [24]. Figure 2 shows the current model for the activation of *K. aerogenes* urease [27,28]. The exact role of UreD/UreH is unknown [28], UreE is the chaperone that delivers nickel to urease [3,29], and UreF has been suggested to be an activator of the GTPase activity of UreG. Formation of the active enzyme requires CO_2 to carbamylate the lysine in the active site [27]. The active site of urease contains two nickel ions that are bridged by the oxygen atoms of a carbamylated lysine residue and a hydroxide ion [1]. Both nickel ions are bound by two different histidines [1]. One nickel ion is bound by aspartate and the other is bound by a water molecule [1]. Additionally, in *H. pylori* the [NiFe]-hydrogenase accessory proteins, HypA and HypB, are important for urease maturation [31–33]. Deletion of *hypA* or *hypB* resulted in a urease activity that was forty- and two hundred-fold lower than the wild-type strain, respectively [32]. Urease activity was restored with nickel supplementation in the growth medium [32].

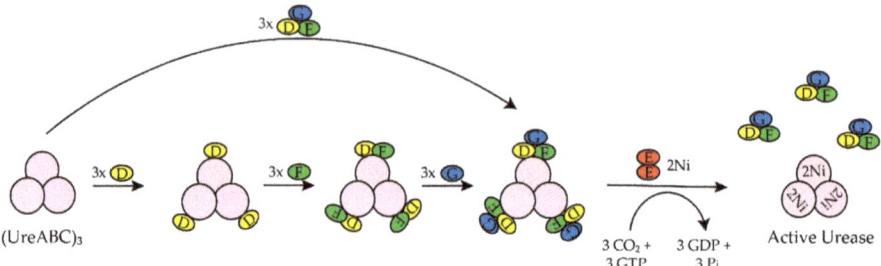

Figure 2. A simplified scheme of *K. aerogenes* urease activation [20,27,28]. The trimer of trimers (UreABC)$_3$ sequentially binds UreD, UreF, UreG or the UreDFG complex. The isolated UreDFG complex contains two UreD, UreF, and UreG promoter but only one monomer is shown for simplicity. Lys 217 is carbamylated by CO_2, GTP is hydrolyzed by UreG, and nickel is delivered by UreE to form the active enzyme and the UreDFG proteins are released.

A recent study by Jones et al. determined that urease is activated under acid shock conditions [34]. *H. pylori* cells were exposed to different pHs (6.8, 5.6, and 2.0) and the cytosolic pH of the bacteria was measured in the presence and absence of nickel sulfate [34]. Minor changes in the internal pH were observed when the bacteria was exposed to mild acidic conditions (pH 6.8 and 5.6) but when the bacteria was exposed to an external pH of 2.0 (pH of the stomach) the internal pH dropped to 5.1 initially and rapidly recovered to a pH of ~6.0 [34]. The drop in pH was less when nickel was supplemented in the *H. pylori* cultures [34]. It is thought that this apparent increase in buffering capacity is due to an increase in urease activation with increased nickel uptake [34,35]. Additionally, they determined that under acid shock conditions the following genes were regulated by *H. pylori* NikR: *ureA, nikR, amiE, amiF, arsR, ureG, nixA, frpB2* [34]. These results demonstrate that *H. pylori* NikR is responsive to pH changes in the cytosol.

In this review, several nickel metalloregulators and chaperones from various organisms will be discussed. We will focus on how nickel binding to metalloregulators results in conformational changes in the protein that influence DNA binding. We will also examine the metal site structures, binding affinities, and the associated conformational changes that accompany nickel binding to these proteins. The roles that the nickel chaperone proteins play in the maturation of [NiFe]-hydrogenase and urease as well as how these proteins interact with other chaperone proteins and the target proteins will be discussed.

2. Nickel-Responsive Metalloregulators

Metalloregulators can be divided into two groups depending on whether cognate metal ion binding results in the downregulation of metal uptake systems or the upregulation of metal efflux/sequestration systems [36]. The nickel-responsive metalloregulators NikR and Nur belong to the first group as nickel binding to the protein results in conformational changes that favors DNA binding leading to the repression of the genes encoding proteins associated with import. RcnR, InrS, DmeR, and NmtR belong to the second group as nickel binding to the protein causes conformational changes in the protein that disfavors DNA binding allowing for the expression of the genes encoding proteins associated with export. Nickel-responsive metalloregulators have been characterized in five of the seven major families of transcriptional metalloregulators in prokaryotes [36]. These include NikR (NikR family), RcnR (RcnR/CsoR family), InrS (RcnR/CsoR family), DmeR (RcnR/CsoR family), NmtR (ArsR family), KmtR (ArsR family), NimR (MerR family), Nur (Fur family). These metalloregulators have been reviewed at least three times in the last ten years [5,20,37]. The respective K_ds and the pH at which the K_d was determined, the coordination number (CN #), and the ligands used to coordinate the metal ion are included in Table S1.

2.1. NikR

NikR is a Ni(II)-responsive metalloregulator that inhibits the expression of the genes associated with Ni(II) import in the presence of excess Ni(II) [6]. In *E. coli*, NikR controls the expression of the *nik* operon that encodes the *E. coli* Ni(II) importer NikABCDE [6]. Ni(II)-NikR binds to a palindromic operator sequence within the *nikABCDE* promoter, GTATGA-N_{16}-TCATAC, with nanomolar affinity [38–40] Like *E. coli*, *Brucella abortus* also has a putative nickel importer NikABCDE and NikR controls the expression of *nikA* [41]. *H. pylori* NikR controls the expression of genes encoding the Ni(II) import protein NixA as well as the urease structural genes (*ureA* [42]-*ureB*), Ni(II) uptake factors (*fecA3*, *frpB4*, and *exbB/exbD*), Ni(II) storage genes (*hpn* and *hpn*-like) and genes associated with iron uptake (*fur* and *pfr*) [34,43]. A recent study by Vannini et al. determined that NikR controlled the expression of the genes encoding two outer membrane proteins that are predicted to be metal ion transporters (*hopV* and *hopW*), and an outer membrane-absorbed protein (*hcpC*) [44]. Additionally, it was determined that NikR regulated the expression of genes predicted to be associated with toxin-antitoxin systems (*dvnA*, *mccB*), a gene predicted to be associated with metabolism (*phbA*), and a putative component of an ABC transporter (*fecD*) as well as non-coding RNAs (nrr1, nrr2 and isoB) [44].

Several crystal structures of NikR proteins from *E. coli* [45–48], *Pyrococcus horikoshii* (*P. horikoshii*) [49], and *H. pylori* [50–54] show that the protein is a homotetramer. The protein has two ribbon-helix-helix DNA binding domains (DBDs) attached by a flexible linker at either end of the tetrameric C-terminal metal-binding domain (MBD) [45]. NikR binds one Ni(II) per monomer in the MBD and is the sole member of the ribbon-helix-helix (β-α-α) family of prokaryotic DNA-binding proteins that is regulated by a metal [55]. The crystal structures of NikR homologues show NikR in different conformations (open, trans, and cis) depending on the orientation of the DBDs with respect to the MBD (Figure 3). An open conformation where the DBDs are linearly placed in each side of the MBD was observed for the apo structures from *E. coli* (Figure 3A) and *P. horikoshii* [45,49]. The nickel-bound *P. horikoshii* and *H. pylori* (Figure 3B) structures feature a closed trans-conformation, where the DBDs are on opposite sides of the MBD [49,54]. The *P. horikoshii* NikR protein was crystallized from a solution containing the apo protein and nickel chloride, while the *H. pylori* NikR protein was obtained by soaking apo *H. pylori* NikR crystals with nickel sulfate [49,54]. The *E. coli* nickel- and DNA-bound NikR structure, obtained from Ni(II)-NikR using the hanging drop vapor diffusion method in a drop consisting of 2 µL of NikR–DNA complex, reveals a closed cis conformation with the DNA-binding domains on the same side of the C-terminal metal binding domain (Figure 3C) [46]. It is still unclear if these conformational changes exist in solution. Nuclear magnetic resonance (NMR) experiments coupled with molecular dynamic simulations suggest that NikR is capable of interconverting between the open, trans, and cis conformations and that these interconversions are facilitated by Ni(II) [56]. However, Small-angle X-ray Scattering (SAXS) experiments done on *H. pylori* NikR do not support these large conformational changes [50].

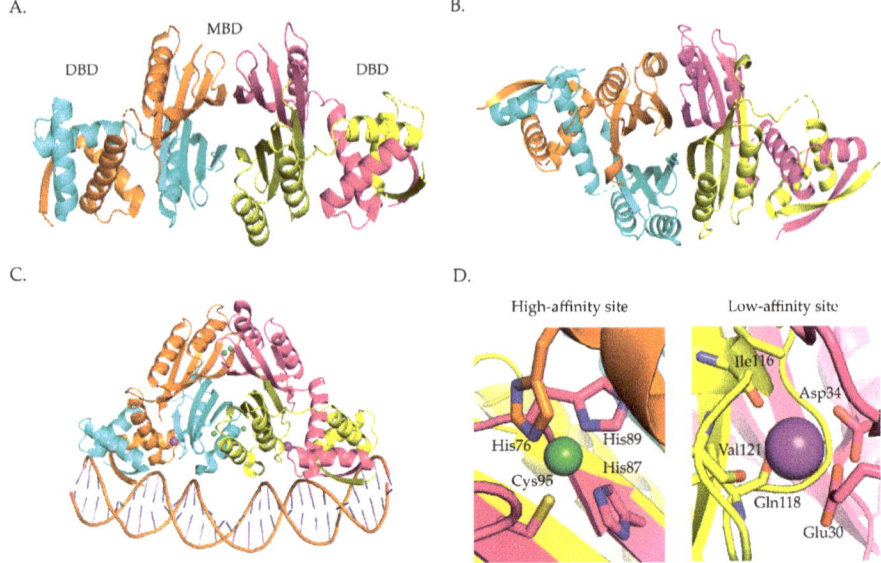

Figure 3. NikR proteins from *E. coli*, and *H. pylori* showing different conformations. The NikR monomers are colored in orange, blue, purple, and yellow, the nickel ions are shown as green spheres and the potassium ions are shown as purple spheres. (**A**) Apo *E. coli* NikR (PDB ID 1Q5V) [45] in an open conformation where the DBDS are placed linearly on each side of MBD. (**B**) *H. pylori* Ni(II)-NikR (PDB ID 2CAD) [54] depicting a closed trans-conformation. (**C**) *E. coli* NikR bound to DNA, nickel ions and potassium ions (PDB 2HZV) [46]. (**D**) Closeup of the high- and low-affinity sites where the nickel ion and potassium ion adopt square planar and octahedral geometries, respectively. (Figure adapted with permission from Higgins, K.A.; Giedroc, D.P. Metal Specificity of Metallosensors. In *Encyclopedia of Inorganic and Bioinorganic Chemistry*; John Wiley and Sons, Ltd.: Chichester, UK, 2013; pp. 209–224. Copyright (2013) John Wiley and Sons).

The *E. coli* NikR DNA bound crystal structure shows that there are two distinct metal-binding sites, four "high affinity" metal binding sites located in the MBD and two "low affinity" metal-binding sites located near the interface of the C-terminus and the DBD (Figure 3C) [46]. Nickel ions bind to the high affinity sites in a four-coordinate planar geometry [46,57] (Figure 3D). The nickel ion is bound by the side chains of His87, His89, and Cys95 from one NikR monomer, and by His76 from an adjacent monomer (Figure 3D) [46]. The structure of *E. coli* Ni(II) NikR bound to DNA shows that potassium ions are coordinated in a bidentate manner by Glu30 and Asp34, and by the backbone carbonyl oxygens of Ile116, Gln118, and Val121 (Figure 3D) [46]. XAS studies reveal that Ni(II) ions bind to the "low affinity site" site with octahedral geometry [57]. Like, *E. coli* NikR, *H. pylori* NikR also binds four nickel ions in the high-affinity sites with four-coordinate planar geometry [51]. Isothermal titration calorimetry (ITC) studies determined that the low-affinity sites in *H. pylori* NikR can bind up to ten nickel ions [58].

The binding affinities of Ni(II) ion to the "high-affinity site" of NikR have been determined in both *E. coli* at pH 7.5 and 7.6, and *H. pylori* at pHs 6.5, 7.0, 7.5 and 8.0 by competition studies and ITC, respectively [38,58,59]. A K_d of 10^{-12} M was determined for *E.coli* NikR, while *H. pylori* NikR had dissociation constants of 10^{-9} M [38,58,59]. The differences in affinities determined for NikR may be attributed to the differences in the experimental conditions and the techniques as *E. coli* NikR ITC experiments that used conditions identical to those used for *H. pylori* NikR determined a dissociation constant of 10^{-9} M [37]. A K_d of 29×10^{-9} M was determined for the "low-affinity site" by titrating

NikR loaded with 4 Ni(II)/monomer with excess Ni(II) ions [39]. A similar K_d was determined using ITC but features 4 Ni(II) binding to the "low-affinity" site [37].

There is no thermodynamic metal preference for Ni(II) ions binding to *E. coli* NikR as it binds to a variety of transition metal ions in vitro in the "high-affinity" site, and the binding affinities follow the Irving Williams series (Co(II) < Ni(II) < Cu(II) > Zn(II) [47,57,59]. Nonetheless, in vivo, the protein only responds to the binding of Ni(II) ions. Ni(II) loaded NikR is less susceptible to chemical and thermal denaturation than Mn(II)-, Co(II)-, Cd(II)-, Cu(II)- or Zn(II)-NikR [59]. X-ray absorption spectroscopy (XAS) studies done on NikR showed that the non-cognate metals, Co(II), Cu(I), Cu(II), and Zn(II) adopt six-coordinate octahedral, three-coordinate trigonal, four-coordinate planar, and four-coordinate tetrahedral geometries, respectively [57]. Similar to Ni(II), the Cu(II) site is also four-coordinate planar but has a slightly different ligand set [57]. Based on the XAS studies it can be concluded that in *E. coli* NikR metal ion selectivity is achieved by the coordination number and geometry of the metal protein complex as well as ligand selection [57].

2.2. Nur

Streptomyces coelicolor (*S. coelicolor*) Nur is a member of the Fur family of metalloregulators that controls nickel homeostasis and oxidative stress response [60]. Nur responds to the binding of Ni(II) and represses the *sodF* gene that encodes an iron-containing superoxide dismutase (FeSOD) and the *nikABCDE* gene cluster encoding components of a nickel transporter [60,61]. Additionally, Nur induces the transcription of *sodN* that encodes a nickel-containing superoxide dismutase (NiSOD) [60,61]. Nur was crystallized from solutions containing the apo protein with zinc chloride or nickel chloride. The crystal structure shows that Nur is a homodimeric protein with two metal binding sites per monomer [62] (Figure 4A). The first site, the M site, binds nickel ions using four histidine ligands in a four-coordinate square planar geometry (Figure 4B) [62]. The second site, the nickel site, coordinates Ni(II) with octahedral geometry using three histidine ligands, two oxygen ligands from malonate, and one oxygen ligand from ethylene glycol (Figure 4C) [62]. It is important to note that this site may be an artifact of crystallization conditions and may not reflect the actual coordination environment of Ni(II) binding to this site of Nur [62]. ITC studies determined that 4 Ni(II) ions bind per dimer with two binding events that have dissociation constants of 10 nM and 280 nM, respectively [37].

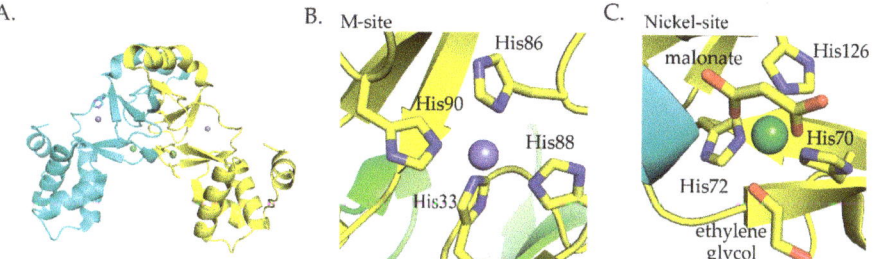

Figure 4. (**A**) The crystal structure of *S. coelicolor* Nur (PDB ID 3EYY) [62]. The monomers are colored cyan and yellow. Zinc ions and nickel ions are shown as slate spheres and green spheres, respectively. (**B**) Close-up of the M-site. (**C**) Close-up of the Ni-site. (Figure adapted with permission from Higgins, K.A.; Carr, C.E.; Maroney, M.J. Specific Metal Recognition in Nickel Trafficking. *Biochemistry* **2012**, *51*, 7816–7832. Copyright (2012) American Chemical Society.).

2.3. RcnR/Csor Family

Three members of the RcnR/CsoR family of metalloregulators are nickel responsive. The crystal structures of the Cu(I) bound *Mycobacterium tuberculosis* (*M. tuberculosis*) CsoR, a Cu(I)-responsive metalloregulator, and *Synechocystis* sp. PCC68 InrS (Figure 5) with no metal bound show that proteins in this family are an all α-helical dimer of dimers [63,64]. In *M. tuberculosis*, Cu(I) is bound in a trigonal geometry by Cys36 from one subunit, and His61 and Cys65 from another subunit [63]. Sequence alignment (Figure 6) shows that these residues are the same in InrS, but in RcnR, and DmeR Cys65 of CsoR corresponds to a His. The CsoR crystal structure showed that there was a hydrogen bonding network between His61 and Glu81 from one subunit and Tyr35 from another subunit [63]. This hydrogen bonding network was shown to be important for allosterically coupling Cu(I) binding to DNA binding [63,65].

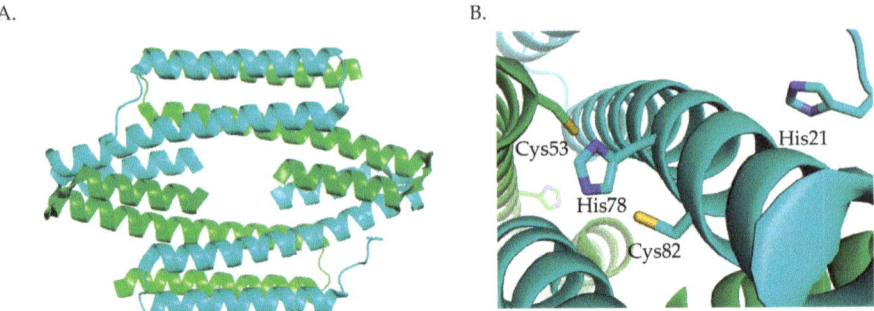

Figure 5. (**A**) The crystal structure of *Synechocystis* PCC6803 InrS (PDB ID 5FMN) [64]. The monomers in each dimer are colored in blue and green. (**B**) Highlights the metal binding pocket of InrS and important residues for Ni(II) binding to the protein. (Figure adapted with permission from Springer Nature Customer Service Centre GmbH: Springer Nature, Nature Chemical Biology. Foster, A.W.; Pernil, R.; Patterson, C.J.; Scott, A.J.P.; Pålsson, L.O.; Pal, R.; Cummins, I.; Chivers, P.T.; Pohl, E.; Robinson, N.J. A tight tunable range for Ni(II) sensing and buffering in cells. *Nat. Chem. Biol.* **2017**, *13*, 409–414. Copyright (2017)).

2.3.1. RcnR

E. coli RcnR is a Ni(II)- and Co(II)-responsive metalloregulator that regulates the expression of the Ni(II) and Co(II) exporter and periplasmic protein RcnAB [7,8,66]. Apo-RcnR binds to a TACT-G$_6$-N-AGTA sequence of which two are located in the *rcnA-rcnR* intergenic region [67,68]. Computational models suggest that RcnR is composed of three α-helices in each monomer; these monomers come together to form a dimer, and the tetrameric oligomer is formed by the dimerization of two dimers [37]. RcnR binds one Ni(II) or Co(II) ion per monomer with nanomolar affinity [69]. Similar to NikR, ITC studies determined that the 4 Ni(II) ions bind to the protein tetramer in two binding events [37]. The protein binds the cognate metals, Ni(II) and Co(II), with six-coordinate octahedral geometry and noncognate metals, Cu(I) and Zn(II), with three- and four-coordinate geometry [69,70]. Likewise, NikR metal ion selectivity is achieved by the coordination number and geometry of the metal protein complex.

```
                                      W
E. coli, RcnR              1  MSHTIRDKQKLKARASKIQGQVVALKKMLDEP    32
M. tuberculosis, CsoR      4  -ELTAKKRAALN-RLKTVRGHLDGIVRMLESD    33
Synechocystis, InrS       19  HVHSQESLQKLVNRLSRIEGHIRGVKTMVQEN    50
R. leguminosarum, DmeR     1  MSHTTLQKKKLVARISRLKGQMEAVERALESE    32


    X                          Y     Z
   HECAAVLQQIAAIRGAVNGLMREVIKGHLTEHIVHQGDELKREEDLDVVLKVLDSYIK-----   90
   AYCVDVMKQISAVQSSLERANRVMLHNHLETCFSTAVLDGHGQAAIEELIDAVKFTPALTGPH  96
   RPCPEVLIQVAAVRGALDRVARLILDDHMNECITRAAAEGNIEQELAELKEALDRFL------ 107
   RPCGEILQLLASVRGALTGLTGEVLDDHLREHVLNAADDAARAEAVEDISEVLRTYMR-----  90
```

Figure 6. Sequence alignment of *E. coli* RcnR with *M. tuberculosis* CsoR, *Synechocystis* PCC 6803 InrS, and *R. leguminosarum* DmeR. Fingerprint residues (W,X,Y,Z) found in all four proteins are shown in orange while those that are unique to CsoR and RcnR are in red and green, respectively. The residues that form a hydrogen bonding network with His61, Tyr35 and Glu81, in *M. tuberculosis* CsoR are shown in blue. The sequence alignment was generated with Clustal Omega [71].

Mutagenesis studies coupled with XAS studies suggest that the N-terminal amine, His3, Glu34, Cys35, Glu63, and His64 bind to Co(II) forming a CoSNH$_2$(N$_{Im}$)$_2$(O)$_2$ site, this arrangement of the cobalt site is supported by computational models [70,72,73]. Similarly, the Ni(II) site is composed of the N-terminal amine, Cys35, Glu63, His64, and two other N/O ligands, one of which may be Glu34, forming a NiSNH$_2$ON$_{Im}$(N/O)$_2$ site [70,72,73]. Computational models show that most of the metal binding residues, the N-terminal amine, His3, Glu63, His64, are derived from one monomer, while Glu34 and Cys35 are derived from another monomer [73]. The Ni(II) and Co(II) sites also differ in M–S bond distances, the number of histidine ligands, and the number of Glu ligands. XAS studies determined that the M–S bond distance is longer for Ni(II) compared to Co(II), 2.62(3) Å versus 2.31(2) Å, respectively [70].

Mass spectrometry, NMR, metal binding studies, and modeling studies suggest that cognate metal binding remodels the shape of the tetramer in a way that does not allow for the relevant positively charged residues to interact with the DNA [36,74,75]. Hydrogen-deuterium exchange studies coupled with mass spectrometry (HDX-MS) was used to probe the RcnR structure in the presence of DNA, Ni(II), Co(II), or Zn(II) [76]. These experiments determined that Ni(II) and Co(II) binding to RcnR orders the N-terminus, decreases the flexibility of helix 1, and induces conformational changes in the protein that restricts DNA interactions with Arg14 and Lys17 [76].

2.3.2. InrS

The second Ni(II) metalloregulator, InrS, belonging to the RcnR/CsoR family was identified in *Synechocystis* sp. PCC6803 [77]. InrS regulates the transcription of the genes encoding the Ni(II) efflux protein NrsD [77]. *Synechocystis* lacks the *nikR* gene, however, InrS also plays a role in Ni(II) import as it binds upstream of the *nik* operon and enhances the expression of the genes encoding nickel-import machinery [64].

InrS binds one stoichiometric equivalent of Ni(II) per monomer [77]. Surprisingly, InrS possesses all the residues known to bind Cu(I) in *M. tuberculosis* CsoR, Cys53, Cys82, and His78 (equivalent to Cys36, Cys65 and His61 in *M. tuberculosis* CsoR) (Figures 5 and 6) but InrS is responsive to Ni(II) [63,77]. It was determined by mutagenesis studies that these residues are important for Ni(II) binding to InrS [64,78]. The crystal structure (Figure 5) shows that these residues are located in the α2 helix [64]. His21 (analogous to His3 in RcnR) (Figure 6) is located on the flexible N-terminal extension and is capable of approaching the Ni(II)-binding site [64]. XAS spectroscopy determined that the nickel site is four-coordinate planar and that His21 is a Ni(II) ligand in InrS [79]. The protein binds Ni(II) using His21, Cys53, His78, and Cys82 forming a Ni(N$_{Im}$)$_2$(S)$_2$ complex [79]. The protein can also bind 1 Cu(I) per monomer or 1–1.5 Co(II) per monomer [77]. The metal site structure of InrS are different from those observed for RcnR and are similar to those of CsoR as they feature a square planar geometry for Ni(II) and tetrahedral geometry for Co(II) [63,70,77,80].

Competition experiments with Ni(II) chelators EDTA, NTA, and EGTA determined that InrS binds Ni(II) tighter than RncR with an affinity of 10^{-14} M [77]. However, dissociation constants obtained from ITC determined that the binding affinities of InrS to Ni(II) are similar to those determined for RcnR with values of 70×10^{-9} M and 4.5×10^{-6} M [37]. Like NikR and RcnR, InrS binds 4 Ni(II) in two binding events. InrS also binds Cu(I) and Zn(II) tightly with affinities of 10^{-18} and 10^{-13} M, respectively [78]. Like CsoR, InrS also has a glutamate residue, Glu98, that aligns with Glu81 of CsoR, which was shown to be important for the hydrogen bonding network that links Cu(I) binding to DNA binding [63,65]. Work done by Foster et al. showed that an E98A mutant InrS protein binds Ni(II) like the wild-type protein and binds DNA tighter than the wild-type protein in the presence of Ni(II) [78]. This result suggests that, unlike CsoR, Glu98 is not essential for allostery in InrS.

2.3.3. DmeR

DmeR is a Ni(II)- and Co(II)-responsive metalloregulator that has been identified in *Rhizobium leguminosarum* (*R. leguminosarum*), *Agrobacterium tumefaciens* (*A. tumefaciens*), *Sinorhizobium meliloti* [81–83]. The *dmeRF* operon encodes both DmeR and DmeF, a cation diffusion facilitator (CDF) [81]. DmeR binds to an AT-rich inverted repeats sequence. DmeR has a RcnR-like finger print, HCHH (Figure 6), and is therefore, expected to bind Ni(II) and Co(II) with six-coordinate octahedral geometry as seen in RcnR. It is interesting to note that cobalt stress induces the expression of iron-responsive genes in *A. tumefaciens* [82]. Additionally, RirA, an iron regulator, is essential for the bacteria to cope with nickel and cobalt toxicity [82].

2.4. NmtR

M. tuberculosis NmtR is a Ni(II)- and Co(II)-responsive transcriptional metalloregulator that belongs to the ArsR/SmtB family [84]. NmtR controls the expression of the *nmt* operon that encodes a P-type ATPase metal efflux pump, NmtA [85]. NmtR binds 2 Ni(II) per dimer with K_{ds} of 8.7×10^{-11} M and 1.4×10^{-10} M [86]. The NMR structure of apo-KmtR (Figure 7) shows the winged-helix fold (α1-α2-α3-αR-β1-β2-α5) that is typical of the ArsR/SmtB family of transcriptional metalloregulators [87]. Additionally, NmtR has long N-terminal and C-terminal extensions (Figure 7) [87]. Like RcnR, NmtR binds its cognate metals with higher coordination numbers than noncognate metals. NmtR binds Ni(II) with six-coordinate geometry and Co(II) with five- or six-coordinate geometry [85]. Additionally, Zn(II) binds to NmtR with four-coordinate geometry [85]. NmtR is the third example in this review of a metalloregulator that achieves metal ion selectivity through the coordination number and geometry of the metal site structure. A combination of molecular dynamics simulations, quantum chemical calculations, metal binding studies and mutagenesis studies confirms that the Ni(II) binds to NmtR with six-coordinate geometry binding to the N-terminal amine, and the side chains of His3, Asp91, His93, His104, and His107 (Figure 7B) [86,87].

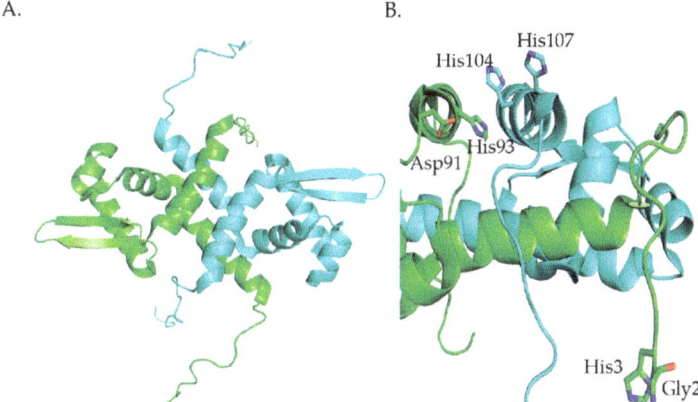

Figure 7. (**A**) NMR structure of apo *M. tuberculosis* NmtR (PDB ID 2LKP) [87]. NmtR monomers are colored in green and blue (**B**) Close-up of one of the nickel sites showing the nickel binding site. (Figure adapted with permission from Lee, C.W.; Chakravorty, D.K.; Chang, F-M.J.; Reyes-Caballero, H.; Ye, Y.; Merz, K.M.; Giedroc, D.P. Solution structure of *Mycobacterium tuberculosis* NmtR in the apo state: Insights into Ni(II)-mediated allostery. *Biochemistry* **2012**, *51*, 2619–2629. Copyright (2012) American Chemical Society).

2.5. Other Transcriptional Metalloregulators

Nickel responsive transcriptional metalloregulators have been identified in other bacteria but the metal site structures have not been characterized. These include: SrnRQ, KmtR, NcrB, Mua, and NimR. In *Streptomyces griseus*, the expression of *sodF*, which encodes an iron- and zinc-containing superoxide dismutase is regulated by two nickel responsive regulators SrnR and SrnQ that work together [88]. SrnR is a member of the ArsR/SmtB family while SrnQ does not show any similarities to any known proteins [88]. SrnR and SrnQ form a complex both in the presence and absence of nickel with maximum interaction shown when they are in a 1:1 stoichiometric ratio [88]. The two proteins come together to form an octamer that is composed of four subunits from each protein [88]. A second Ni(II)- and Co(II)-responsive transcriptional regulator, KmtR, was identified in *M. tuberculosis* [89]. Apo KmtR represses the expression of a putative CDF metal exporter [89]. *Leptospirillum ferriphilum* UBK03 NcrB is a histidine rich protein that regulates the expression of genes associated with nickel export *ncrAC* [90]. Mua binds nickel and modulates urease activity in *H. pylori* [91]. The protein binds two nickel per dimer and represses urease activity at high concentrations [91]. *Haemophilus influenzae* NimR is a regulator belonging to the MerR family that binds one Ni(II) ion per dimer [92]. NimR regulates the expression of a Ni(II) uptake transporter, NikLMQO [92].

3. Ni(II) Chaperones Associated with [NiFe]-Hydrogenase and Urease

The nickel chaperones HypA, HypB, SlyD, and UreE have been characterized individually and in complexes from various organisms including *E. coli* and *H. pylori*. Several studies have shown that these proteins interact with each other to deliver nickel to [NiFe]-hydrogenase. These studies are discussed in this section. Table S2 lists the respective K_ds and the pH at which the K_d was determined, the coordination number (CN #), and the ligands used to coordinate the metal ion in HypA, HypB, SlyD, and UreE.

3.1. HypA

HypA is a nickel chaperone that is involved in delivering nickel to [NiFe]-hydrogenase and in the case of *H. pylori* it is involved in nickel delivery to urease [32]. Homologous HypA proteins have been characterized from *E. coli*, *H. pylori*, and *Thermococcus kodakarensis* (*T. kodakarensis*). *E. coli* HypA and its homologous protein, HybF are essential for the maturation of [NiFe]-hydrogenases [18,93]. HypA proteins have been reported as being a dimer in solution based on elution volumes [94,95], however, size exclusion chromatography in combination with multiple light scattering (SEC-MALS) determined that the protein is a monomer in solution [96]. HypA proteins feature two metal binding sites, a N-terminal nickel site and a zinc site [94,95,97–100]. The HypA structures reveal that the nickel binding site and the zinc binding site are separated by a long flexible linker (Figures 8 and 9) [98–100]. The N-terminal nickel site features a highly conserved MHE motif, which includes the first three amino acids at the N-terminus of the protein, that is used to bind two Ni(II) ions per dimer with micromolar affinity [93–96,98,101]. A K_d of 75 ± 46 nM was reported by Douglas et al. for a Strep-tagged HypA protein binding to nickel in competition with a well-established metal binding indicator, Mag-fura 2 [102]. Mutagenesis studies showed that His2 located in the MHE motif is critical for nickel binding, [NiFe]-hydrogenase activity, and, in the case of *H. pylori*, it is important for urease activity [95]. HypA utilizes two conserved CXXC motifs to bind one zinc ion per monomer [93,94,97–101] with nanomolar affinity [94].

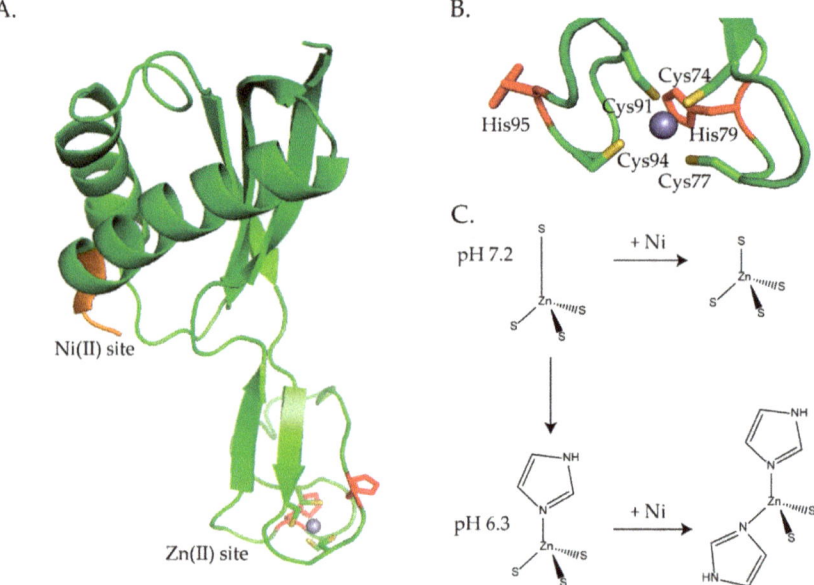

Figure 8. NMR structure of *H. pylori* HypA at pH 7.2 (PDB ID 6G81) [100]. HypA was expressed in growth medium containing zinc sulphate and purified as ZnHypA [100]. (**A**) The nickel binding MHE residues are in orange and the zinc site is shown at the base of the protein where the zinc ion is coordinated by 4S-donors forming a tetrahedral zinc site. The histidine residues flanking the CXXCX$_n$CXXC motif are shown in red. (**B**) Close-up of the zinc site. (**C**) Scheme depicting the changes in the zinc site when the pH changes and nickel binds to HypA (This scheme is adapted with permission from Herbst, R.W.; Perovic, I.; Martin-Diaconescu, V.; O'Brien, K.; Chivers, P.T.; Pochapsky, S.S.; Pochapsky, T.C.; Maroney, M.J. Communication between the zinc and nickel sites in dimeric HypA: metal recognition and pH sensing. *J. Am. Chem. Soc.* **2010**, *132*, 10338–10351. Copyright (2010) American Chemical Society).

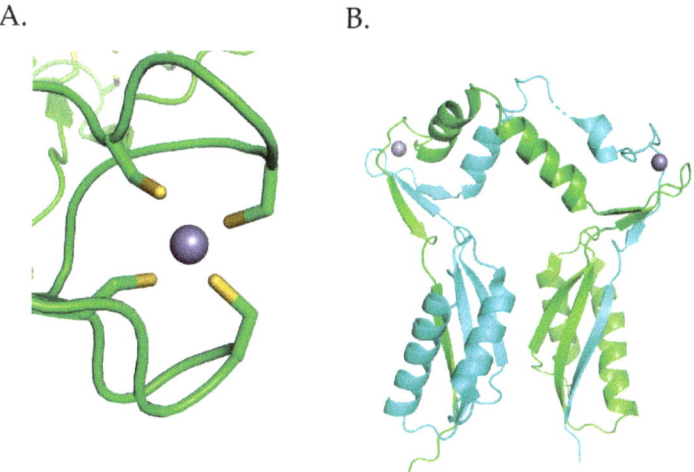

Figure 9. Crystal structure of HypA from *T. kodakarensis* [99]. The protein monomers are shown in cyan and green and the zinc ion is shown as a slate sphere. (**A**) The zinc site from the monomeric protein (PDB ID 3A43) [99]. (**B**) Crystal structure of the domain swapped dimer (PDB ID 3A44) [99]. (Figure adapted with permission from Higgins, K.A.; Carr, C.E.; Maroney, M.J. Specific Metal Recognition in Nickel Trafficking. *Biochemistry* 2012, 51, 7816–7832. Copyright (2012) American Chemical Society).

H. pylori HypA binds Ni(II) with a K_d of ~1 µM at pH 7.2 and this binding is weaker at pH 6.3 [96,97,103]. There are some discrepancies regarding the coordination number and geometry of the nickel biding site determined by XAS and X-ray crystallography. XAS studies determined that the Ni(II) site is six-coordinate and Ni(II) is ligated by 6 N/O-donors of which 1–2 are histidines [97,104]. The first NMR structure of the monomeric protein from *H. pylori* features an N-terminus that is modified by two additional residues, Gly and Ser, that were left over after thrombin cleavage of a histidine tag [98]. UV-vis studies done on this protein determined that nickel binds to *H. pylori* HypA in a planar coordination with 4N/O-donors. Initial NMR studies revealed that the Ni(II) binding site is located at the N-terminus and that nickel is coordinated by the backbone nitrogen of His2, Glu3, and Asp40 as well as the imidazole from His2 [98]. Additional NMR studies on the unmodified HypA protein revealed nickel binding to HypA resulted in dipolar broadening in the N-terminal region. This result is consistent with a paramagnetic ($S = 1$) five- or six-coordinate nickel site [97]. The latter six-coordinate Ni(II) site is supported by density functional theory (DFT) calculations, which suggest that Ni(II) is bound by the N-terminal amine, the amide nitrogen of His2 and Glu3, and the sidechains of Glu3, Asp40, and His2, forming a $NiNH_2(N)_2O_2N_{Im}$ site [100]. XAS studies done by Hu et al. on the unmodified protein confirmed that the nickel site is six-coordinate and that the nickel ion is coordinated by 2N/O-donors, 1 imidazole, the N-terminal amine, and 2 backbone amide [96]. Additionally, the binding of the N-terminal amine to nickel was confirmed with a L2* HypA mutant protein. This mutation features the insertion of Leu into position two so that the MHE motif becomes MLHE [96]. The L2*HypA protein changes the coordination environment of the Ni(II) ion, the Ni(II) site is five-coordinate Ni(II) site with a Br^- ligand from the buffer [96]. The presence of the Br^- ligand bound to Ni(II) suggests that the MLHE mutant HypA protein results in an open coordination position on Ni(II), which is occupied by an anion from the buffer.

The zinc site consists of a loop with a $CXXCX_nCXXC$ motif that coordinates Zn(II) ions in a four-coordinate tetrahedral geometry (Figure 8A,B and Figure 9A) [93,94,97–101]. The *H. pylori* NMR structures show that the Zn(II) ion is coordinated by 4 S-donors from Cys74, Cys77, Cys91, and Cys94 (Figure 8A,B) [98,100]. The crystal structure of HypA from *T. kodakarensis* was solved in both the monomeric and dimeric forms (Figure 9) [99]. The zinc site in the monomeric form confirms that zinc is coordinated by the CXXCX$_n$CXXC motif (Figure 9A), however, the structure of the dimeric protein shows that there is a domain swap in the homodimeric protein that affects the cysteine coordination of the zinc site (Figure 9B) [99]. The physiological role of this domain swap is unknown [99]. In the dimeric form the zinc is coordinated by only 1 CXXC motif from each monomer [99]. Additionally, the crystallographic studies revealed that the HypA zinc site is dynamic [99].

XAS studies also revealed that the zinc site is dynamic and senses both pH changes and nickel binding to the protein (Figure 8C) [97]. In agreement with the NMR structures, the XAS studies determined that at pH 7.2, the zinc site is four-coordinate, tetrahedral, with 4 S-donors, (Zn(S)$_4$) (Figure 8B,C) [97,98,100]. Similarly, at pH 6.3 (~the internal pH of *H. pylori* under acid shock), the zinc site is also four-coordinate but features 3 S-donors and 1 imidazole ligand from a histidine (Zn(S)$_3$(N$_{Im}$)) (Figure 8C) [97]. ITC studies determined that this structural change altered the nickel binding stoichiometry and binding affinity. At pH 6.3, ZnHypA binds 1 Ni(II) ion per dimer with a K$_d$ of 17 µM versus 1 Ni(II) ion per monomer at pH 7.2 with a K$_d$ of 1 µM [97]. When ZnHypA binds Ni(II) at pH 7.2 the structural changes in the zinc site are subtle, the zinc site is still four-coordinate, but 1 of the Zn–S bonds is shortened (Figure 8C) [97]. At pH 6.3, the changes in the zinc site are more pronounced when Ni(II) binds to ZnHypA; 2 of the S-donor ligands at the zinc site are replaced by 2 imidazole ligands from histidines forming a (Zn(S)$_2$(N$_{Im}$)$_2$) structure (Figure 8C) [97]. In *H. pylori* the two CXXC motifs are flanked by two histidine residues, His79 and His95 (Figure 8A,B). Mutating any of the Cys residues in CXXC or His79 or His95 resulted in a static zinc site that no longer senses pH changes or Ni(II) binding to HypA [97]. Cys to Ala or Asp mutations resulted in a zinc site that was four-coordinate with 2 S-donors and 2 imidazoles ((Zn(S)$_2$(N$_{Im}$)$_2$) [97]. This zinc site is indistinguishable from the ZnHypA site with Ni(II) bound at pH 6.3 (Figure 8C) [97]. Alanine substitutions for His79 and His95 also resulted in a zinc site that featured 4 S-donor ligands, (Zn(S)$_4$), as seen for the wild-type HypA zinc site in the presence of Ni(II) at pH 7.2, which suggests that His79 and His95 are the residues that bind the wild-type NiZnHypA protein at pH 6.3 (Figure 8C) [97].

In vivo studies done on *H. pylori* mutant strains determined that the dynamic zinc site in HypA is important for the acid viability of the bacterium, urease activity, and [NiFe]-hydrogenase activity [33,105]. However, the conformational changes in the zinc site observed by XAS studies when Ni(II) binds to HypA are not supported by NMR studies [100]. *H. pylori* mutant strains were created that involved mutating individual residues in the CXXC motifs, Cys74, Cys77, Cys91, and Cys94 to Asp and Ala, as well as His79 and His95 to Ala [33]. *H. pylori* strains of these 10 *hypA* mutants, a *hypA* mutant (*hypA::kan-sacB*), a *hypA*-restorant (hypA-R) used as a control for any defects that may have resulted from genetic manipulation, and a ureB mutant (ΔureB, used as a control) all survived when resuspended in phosphate buffered saline (PBS) at pH 6 both in the absence and presence of urea [33]. However, in PBS buffer at pH 2.3 a significant decrease in acid resistance was observed for C74D, C91A, and C94A *hypA* mutant strains compared to the wild type strain [33]. All the *H. pylori* mutant strains involving the Cys residues resulted in a decrease in urease activity (15% activity compared with the *hypA*-restorant strain) [33,97]. The C77D, C91D, and C94D mutant strains had moderate effects on urease activity (12–13% compared with the hypA-restorant strain), while C74A, C74D, C77A, C91A, and C94A had severe effects on urease activity (similar to the ΔhypA strain and >4% compared to the hypA-restorant strains) [33]. The His mutations had no effect on urease activity [33]. Johnson et al. suggest that the cysteine residues play a critical role in acid viability and mutating these residues affects the ability of HypA to supply the urease maturation pathway with nickel [33]. Studies done by Blum et al. determined that [NiFe]-hydrogenase activity does not contribute to acid resistance in *H. pylori*. *H. pylori* mutant strains similar to the ones used above in the acid viability and urease study were used to measure [NiFe]-hydrogenase activity [105]. The C74A, C74D, C91A, and C94A mutations all resulted in a severe decrease (9–15% of wild type activity) in [NiFe]-hydrogenase activity, whereas, only a moderate decrease (36–60% of wild type activity) in [NiFe]-hydrogenase activity was observed for C77A, C77D, C91D, and C94D mutant strains [105]. The H79A and H95A mutant strains had little effect on [NiFe]-hydrogenase activity, 90% and 110% of wild type activity, respectively [105]. These results coupled with the XAS studies [97,104] suggest a regulatory role for the zinc site [33,97,104,105] and it has been suggested that changes in the zinc site may affect interactions between HypA and other nickel proteins [33,97] and the delivery of nickel to [NiFe]-hydrogenase and urease [33,105]. This dynamic zinc site appears to be unique to *H. pylori* HypA as other HypA homologues, including *T. kodakarensis* HypA, do not have conserved histidine residues flanking the CXXCX$_n$CXXC motif [17,101]. Additionally, the crystal structures of *T. kodakarensis* HypA show that there is no change in the structure of the zinc binding domain when Ni(II) binds to the protein [101].

Nickel binding to the MHE motif is also important for both [NiFe]-hydrogenase and urease activity [33,95,105]. A H2A *hypA* mutant strain of *H. pylori* was created by replacing the His codon with the codon for Ala. [95]. This mutant strain of *H. pylori* lacked [NiFe]-hydrogenase activity and only had 2% of the wild-type urease activity [95]. NMR studies showed that the H2A mutant protein had a structure that was similar to the wild-type protein but nickel binding experiments determined that this mutant protein did not bind nickel [95]. Additionally, the L2* *hypA* mutant strain of *H. pylori* is sensitive to acid and has decreased urease activity [96]. Furthermore, in *H. pylori hypA*, the L2* mutation resulted in a decrease (14% of wild type activity) in [NiFe]-hydrogenase activity [105].

HypA has been shown to form complexes with HypB in *E. coli*, *H. pylori*, and *T. kodakarensis* [94,95,101,106]. Three crystal structures have been solved of the HypAB complex from *T. kodakarensis* (vide infra). One structure is in the presence of adenosine 5'-O-(3-thiotriphosphate) (ATPγS) and nickel ions, the other with beta,gamma-methyleneadenosine-5'triphosphate (AMPPCP) (a nonhydrolyzed ATP analog) and Ni(II) ions, and the third with just AMPPCP [101]. The overall architecture of the proteins is similar with small differences in the local conformations. Additionally, there are crystal structures of HypA complexed with the [NiFe]-hydrogenase large subunit in an immature state, HyhL, from *T. kodakarensis* [107]. The HyhL–HypA complex was prepared by mixing HyhL and HypA in a 1:1.4 molar ratio [107]. The proteins were incubated for 6 days at 20 °C, followed by further incubation for 1 day at 4 °C [107]. Analysis of the HyhL–HypA structure revealed that the nickel metal binding sites of the two proteins are in close proximity and that the N-terminal tail of HyhL interacts with the nickel binding site on HypA [107].

3.2. HypB

HypB is a member of the G3E family of P-loop GTPases [108]. All HypB proteins have a conserved GTPase domain (G-domain), which contains a low affinity metal binding site than can bind either nickel or zinc [109–114]. The GTPase activity of HypB is required for nickel insertion and maturation of [NiFe]-hydrogenase and plays a role in urease maturation [95,115,116]. In *E. coli*, GTPase hydrolysis controls nickel but not zinc binding to HypB and protein–protein interactions between HypA and HypB [117]. *E. coli* HypB binds two metal ions per monomer in two very distinct metal sites, both of which are critical for [NiFe]-hydrogenase maturation [109,110]. The high affinity site, which is unique to *E. coli* HypB, binds Ni(II) with a K_d in the subpicomolar range, is located at the N-terminus of the protein, and features a CXXCGC motif [109]. The second site, a low affinity site, located in the GTPase domain, has a higher affinity for Zn(II) (K_d 1 µM vs 12 µM for Ni(II)) [109]. The GTPase activity of *E. coli* HypB is low, with a k_{cat} 0.17 min^{-1} and a k_m of 4 µM [116].

XAS studies determined that Ni(II) binds to the high affinity site and low affinity site in a four-coordinate planar geometry and six-coordinate octahedral geometry, respectively [110]. Nickel binds to the high affinity site utilizing the 3 S-donors from the 3 N-terminal cysteines, Cys2, Cys5, and Cys7, as well as 1 N/O donor from the N-terminal amine to form a Ni(S)$_3$NH$_2$ structure [109,110,118]. Nickel binds to the low affinity site using Cys166, His167, and 4 other N/O-donating residues [109,110]. XAS studies revealed that metal binding to the high affinity site results in structural changes at the low affinity site [110]. Likewise, metal binding to the low affinity site results in structural changes at the high affinity site [110]. X-ray absorption near-edge spectroscopy (XANES) analysis shows that when Zn(II) binds to the low affinity site, a less intense 1s → 4p$_z$ transition is observed, and the peak at 8342 eV is slightly more intense [110]. This indicates that there is a slight distortion in the planar geometry and a reduction in the S-donor ligand content of the high affinity site [110]. Extended X-ray absorption fine structure (EXAFS) analysis of Ni(II) in the high affinity site in the absence of Zn(II) bound to the low affinity site shows the Ni(II) coordinated by 3 S-donors at a distance of 2.17 Å and 1 N/O-donor at 1.87 Å, forming a Ni(S)$_3$(N/O) structure [110]. When zinc binds to the low affinity site these bond distances are altered, the Ni–S bond distance is slightly shorter at 2.15 Å and the Ni–N/O bond distance is longer at 2.02 Å [110]. A Ni(II) site with one less S-donor, Ni(S)$_2$(N/O)$_2$, that is in agreement with the XANES analysis of a decrease in the S-donor ligand content involving 2 S-donors at 2.17 Å and 2 N/O-donors at 2.02 Å, was also obtained for Ni(II)Zn(II) HypB but this fit had a similar F-factor (a goodness of fit parameter) to the 3S 1N/O fit [110]. Based on the EXAFS data, it is not clear if a S-donor is replaced by a N/O-donor when Zn(II) bind to the low affinity site. Similarly, changes in the low affinity site, the G-domain, were observed when Ni(II) binds to the high affinity site. XANES analysis showed that the geometry of the zinc site is not altered when Ni(II) binds as it is four-coordinate tetrahedral regardless of Ni(II) being present in the high affinity site [110]. When Ni(II) binds to the high affinity site the coordination environment of Zn(II) in the low affinity site changes from Zn(S)$_{2.5}$N/OZn(N$_{Im}$)$_{0.5}$ to Zn(S)$_2$N/OZn(N$_{Im}$). EXAFS analysis revealed that in the absence of Ni(II), the zinc site contains 2 Zn(II) ions with an average of 2.5 S-donors at 2.29 Å, 1 N/O-donor at 2.02 Å, 0.5 imidazole nitrogen ligand at 1.98 Å, and a Zn–Zn interaction at 4.01 Å [110]. When Ni(II) is bound to the high affinity site, the zinc site still contains 2 Zn(II) ions with an average of 2 S-donors at 2.29 Å, 1 N/O-donor at 2.01 Å, 1 imidazole nitrogen ligand at 1.95 Å, and a Zn–Zn interaction at 3.98 Å [110]. The structural changes in the metal sites when another metal binds at the alternate site in HypB are subtle but the XAS studies suggest an allosteric recognition mechanism may be present in HypB similar to that seen in HypA.

H. pylori HypB binds Ni(II) ions at the G-domain with a 1:1 stoichiometry and a micromolar affinity determined by Ni(II) ion titrations monitored by UV–vis [106]. Competition metal binding studies involving the metal chelator Mag-fura 2 determined a K_d for Ni(II) of 150 nM and 1.2 nM for Zn(II) [111]. Similar to *E. coli* HypB, *H. pylori* HypB binds zinc tighter than nickel in the G-domain [109,111]. Mutagenesis studies involving the conserved residues Cys106 and His107 determined that these residues are important for Ni(II) and Zn(II) binding [111]. The C106A and H107A mutations disrupted Ni(II) binding and weakened Zn(II) binding by 2 orders of magnitude [111]. These results suggest that the coordination spheres of Ni(II) and Zn(II) are not identical but some of the ligands used to bind nickel also bind zinc [111]. *E. coli* HypB GTP hydrolysis is slow [114]. Ni(II) binding to HypB promotes dimerization and has no effect on GTPase activity while zinc binding does not promote dimerization and inhibits GTPase activity [111]. The crystal structure of *H. pylori* HypB with Ni(II) and GDP bound shows that the nickel is ligated by four cysteines in a square planar geometry and that this metal binding site bridges the HypB dimer (Figure 10) [114]. Studies done by Sydor et al. demonstrate that the nucleotide bound state of the protein influences the coordination environment of the nickel ion [114]. These studies suggest that H107 is a nickel ligand in the absence of nucleotide binding to the protein [114]. Additionally, Cys142 was identified as an important residue for coupling metal binding to *H. pylori* HypB with GTPase activity [114].

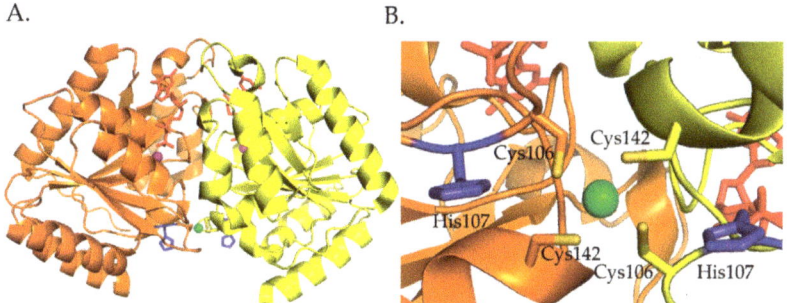

Figure 10. (**A**) Crystal structure of *H. pylori* HypB (PDB ID 4LPS) [114]. Prior to crystallization, the protein was incubated with Guanosine 5'-O-(3-thiotriphosphate) (GTPγS) and magnesium chloride before adding nickel chloride [114]. The protein monomers are colored orange and yellow, guanosine diphosphate (GDP) molecules are shown as red spheres, the magnesium ions are shown as purple spheres, and the nickel ion is shown as a green sphere, (**B**) Close-up of the bridging G-domain metal site with a nickel ion bound. (This figure was adapted from Sydor et al [114]. The research was originally published in the Journal of Biological Chemistry. Sydor, A.M.; Lebrette, H.; Ariyakumaran, R.; Cavazza, C.; Zamble, D.B. Relationship between Ni(II) and Zn(II) coordination and nucleotide binding by the *Helicobacter pylori* [NiFe]-hydrogenase and urease maturation factor HypB. *J. Biol. Chem.* **2014**, *289*, 3828–3841. © the American Society for Biochemistry and Molecular Biology *or* © the Author(s)).

HypB proteins have also been crystallized from *Methanocaldococcus jannaschii* (*M. jannaschii*) with Zn(II)) and (GTPγS) bound [113]. The protein was purified and incubated in buffer containing zinc chloride and crystallized from a solution containing GTPγS [113]. *T. kodakarensis* with ADP bound [112], and *Archeoglobus fulgidus* (*A. fulgidus*) in the nucleotide free apo form [119]. *T. kodakarensis* hypB is the only HypB protein discussed in this review is an ATPase. *T. kodakarensis* HypB is a member of the Mrp/MinD family of ATPase-type HypB [112]. All four HypB crystal structures show that in HypB there is a homodimer nucleotide-binding site near the dimer interface. Surprisingly, a nickel binding site was not identified for *T. kodakarensis* HypB, which suggest that it functions different from the other HypBs that have been characterized [112]. The HypB crystal structure from *M. jannaschii* revealed that there are two metal binding sites located in the G domain [113]. Like *H. pylori* HypB, the metal binding site bridges the dimer interface [113,114]. Additionally, the location of the nucleotide binding site in *M. jannaschii* is almost identical to that of *H. pylori* HypB [114]. In the crystal structure two zinc ions are bound in two unique sites that are bridged by a cysteine residue (Cys95). [113] One zinc ion is coordinated by three cysteines from one monomer (Cys95 and Cys127), a cysteine from another monomer (Cys95) and a water molecule. [113]. The second zinc ion is coordinated by two cysteine residues (Cys95 and Cys127), histidine (His96) and a water molecule [113]. The structure also revealed that GTPγS binds at the dimer interface [113].

Studies involving *A. fulgidus* and *E. coli* HypB determined that the dimerization of HypB proteins is important for [NiFe]-hydrogenase activity [119,120]. Studies by Cai et al. on *E. coli* HypB determined that Leu242 and Leu246 are important for dimerization of the protein [120]. The residues correspond to Leu171 and Val175 in *M. janaschii* HypB. The crystal structure of *M. Jannaschii* showed that Leu171 and Val175 are located in a hydrophobic patch at the dimer interface [113]. Mutant Leu242A and Leu246A *E. coli* HypB proteins could still bind metal and hydrolyze GTP, however, they could not dimerize resulting in a decrease [NiFe]-hydrogenase activity (half of the {NiFe}-hydrogenase activity that was observed with the wild-type HypB protein) [120]. This result is also supported by work done on *A. fulgidus*. *A. fulgidus* HypB is a monomer in the apo and the GDP-bound forms but the GTP bound form of the protein is a dimer [119]. This dimer was not observed in a Lys148A (Lys224 in *E. coli*) mutant *A. fulgidus* hypB though the protein could still bind GTP. It was shown that a K22A mutant HypB protein could not recover [NiFe]-hydrogenase activity in a Δ*hypB E. coli* strain [119].

HypB proteins with an N-terminal histidine rich region that are capable of binding several nickel ions have been characterized from *Bradyrhizobium japonicum* (*B. japonium*) and *R. leguminosarum* [121–123]. It has been suggested that these histidine rich tails play a role in nickel storage and buffering [122,124]. *B. japonicum* HypB has 24 histidine residues located near the N-terminus. The protein binds 9 Ni(II) per monomer with a K_d of 2.3 µM [121]. Other metal ions, including Zn(II), Cu(II), Co(II), Cd(II), and Mn(II) can bind to the protein [121]. *R. leguminosarum* HypB has seventeen histidine residues near the N-terminus and binds four Ni(II) ions per monomer [123]. Other metal ions like Co(II), Cu(II), and Zn(II) can compete with Ni(II) for binding to the protein [123]. The significance of these histidine rich regions in some HypB proteins is unknown but it has been suggested that they may be involved in nickel storage [17].

3.3. SlyD

SlyD is a member of the FK506-binding protein (FKBP) family of peptidyl-prolyl cis/trans isomerases [125]. The NMR structures of SlyD from *E.coli* show that the protein is a two-domain protein: a peptidyl-prolyl isomerase (PPIase) domain and the "inserted in the flap" (IF) domain (Figure 11) [126,127]. Prolyl isomerases are involved in numerous biological functions including catalyzing the cis-trans isomerization of proline peptide bonds, which is a crucial step in the folding pathway of some proteins [125,128]. However, in *E. coli* the isomerase function of SlyD is nonessential in [NiFe]-hydrogenase maturation [129]. In *E.coli* it was determined that SlyD is important for [NiFe]-hydrogenase activity as a *slyD* mutation resulted in a 50% reduction in H_2 production in cultures during exponential phase growth [16]. *E. coli* SlyD has an unstructured C-terminal metal-binding tail (Figure 11), which is rich in metal binding residues: histidines, cysteines, aspartates, and glutamates [17]. The C-terminal tail is variable among SlyDs [126,127,130–132]. *E. coli* SlyD can bind up to 7 Ni(II) per protein with submicromolar ($K_d < 1.8$ µM) affinity [133], which reversibly inhibits PPIase activity [134]. A precise model of the Ni(II) site in SlyD could not be determined by XAS studies as there are multiple nickel sites in the protein that had different geometries and coordination numbers [133]. Based on the ability of SlyD to bind multiple metal ions, it is thought that SlyD contributes to Ni(II) storage and is a possible source for Ni(II) ions during [NiFe]-hydrogenase assembly in *E. coli* [15,135].

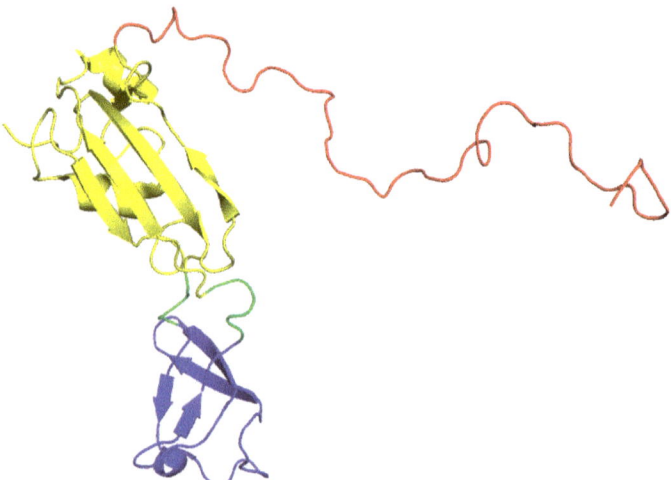

Figure 11. NMR structure of *E. coli* SlyD (PDB ID 2KFW) [126]. The IF domain is colored blue, the PPIase domain is colored yellow, and the unstructured C-terminal metal-binding tail is colored red.

H. pylori SlyD also has a histidine and cysteine rich C-terminal tail that binds 2.4 Ni(II) per protein or 3.3 Zn(II) ions per protein with K_{ds} of 2.74 and 3.79 µM, respectively [132]. A SlyD homolog from *Thermus thermophilius* (*T. thermophilius*) that does not contain the extended cysteine and histidine rich C-terminal tail seen for *E. coli* and *H. pylori* SlyD has a metal binding site composed of three histidines [130]. The crystal structure shows that either Ni(II) or Zn(II) is bound by the conserved sequence (HGHXaaH) [130]. A truncated *E. coli* SlyD protein, SlyD155, was created to resemble the *T. thermophilius* SlyD protein. SlyD155 was able to bind a single nickel ion with a K_d of 65 nM and an octahedral geometry with 2–3 imidazole ligands from His149, His151, and His153, and 3–4 N/O-donors, but could not activate [NiFe]-hydrogenase activity in vivo [129].

3.4. UreE

UreE delivers nickel to the UreABC-UreDFG complex (Figure 2) [28]. UreE from *K. aerogenes* is a homodimer that binds six nickels per homodimer with a K_d of ~9.6 µM [136]. XAS studies determined that UreE binds nickel in different sites. The average Ni(II) site determined for a Ni(II)UreE sample prepared by mixing UreE with approximately three equivalents of Ni(II) per dimer is pseudo-octahedral with six N/O donors of which three to five are histidine ligands [136]. Most of the nickel is bound at the histidine rich C-terminal tail (HGHHHAHHDHHAHSH) [28,137]. *H. pylori* and *Sporosarcina pasteurii* (*S. pasteurii* formerly *Bacillus pasteurii* [138]) lack this histidine rich tail. The *H. pylori* UreE is a homodimer that binds one Ni(II) or Zn(II) per dimer with a K_d of 0.15 and 0.49 µM, respectively [139,140]. The nickel site is six-coordinate and features four histidine ligands while the zinc site is five-coordinate with two or three histidine ligands [141]. *S. pasteurii* UreE binds two Ni(II) per dimer with an overall K_d of 35 µM [142]. The nickel site is also six-coordinate and features each Ni(II) ion bound by an average of two histidine and 4 N/O residues [142,143]. Studies done by Brayman, Hausinger, and Colpas show that the histidine rich C-terminal tail of *K aerogenes* UreE is not necessary for UreE to deliver nickel to urease [144,145]. A H144* mutant UreE from *K. aerogenes* was generated by removing the last fifteen residues, which includes ten histidine residues [144]. Like the wild-type protein, the H144* UreE is a homodimer, but it bound two Ni(II) per dimer versus the six Ni(II) per dimer, which was observed for the wild-type protein [146]. Urease activity in cells containing H144* UreE was similar to that of cells with the wild-type protein [144].

Cd(II), Co(II), Zn(II) and Cu(II) ions are able to compete with Ni(II) ions for the nickel binding site in H144* UreE [144]. These metals bind to the protein with stoichiometry of two metal ions per dimer using different ligands and with different geometries [146]. The two Ni(II) ions bind with peudo-octahedral geometry and the two Ni(II) sites are spectroscopically distinguishable as one Ni(II) site binds one less histidine residue than the other [146,147]. Equilibrium dialysis experiments determined that Ni(II) bound with K_ds of 47 µM and 1.45 µM [147]. ITC experiments determined that the two Ni(II) ions bind with a K_d of 1.6 nM but do not distinguish between the two Ni(II) binding to UreE [148]. Similar to *E. coli* NikR, a difference in the K_ds determined using two different techniques is observed. The two Cu(II) ions bind to H144* UreE with tetragonal geometry with each Cu(II) being coordinated by two histidine ligands and one Cu(II) ion is bound by a cysteine residue, Cys79 [146,147]. Like Ni(II), Co(II) ions adopt pseudo-octahedral geometry, and the two cobalt sites differ in the number of histidine ligands that coordinate to each Co(II) ion [146]. The metal binding ligands were determined using a combination of mutagenesis and urease activity studies coupled with metal binding studies involving equilibrium dialysis, and UV-visible, EPR, and NMR spectroscopies [147]. It was suggested that one Ni(II) or Co(II) is bound by His96, His112, and 1 N/O-donors from each monomer forming a six-coordinate metal site. The second Ni(II) or Co(II) is bound by His110, and two N/O-donors from each monomer forming a six-coordinate metal site [147]. Proton NMR studies suggest that the nickel and cobalt sites differ in the number of histidine ligands at each site. The first nickel site has one histidine and the second has two histidines; the number of histidine residues are reversed for the cobalt site [146]. It is important to note that the metal binding ligand His96 is conserved among UreE proteins [147] and was shown to be important for urease activation both in vitro [149] and in vivo [147]. Proton NMR studies suggest that the nickel and cobalt sites differ in the number of histidine ligands at each site. The first nickel site has one histidine and the second has two histidines, the number of histidine residues are reversed for the cobalt site [146].

The *K. aerogenes* H144* crystal structure showed that the protein binds three Cu(II) ions per dimer [150]. This crystal structure was solved from apo H144* crystals that were soaked with copper(II) sulphate [150]. One copper is bound in between the two monomers by two His96 residues, one from each monomer. The other two coppers are each coordinated by His110 and His112 within each monomer. The differences in the metal binding stoichiometries of Cu(II) binding to H144* UreE determined by Colpas et al. [146,147] in the crystal structure are explained by ITC experiments. These studies determined that the number of Ni(II) and Cu(II) ions bound per dimer is dependent on the concentration of the protein. At low concentrations (<10 μM), the dimeric H144* UreE binds 2 Ni(II), or Cu(II) ions per dimer but at higher concentrations, 25 μM, binds 3 Ni(II) or Cu(II) per dimer [148]. The ITC data also showed evidence of the formation of a tetramer (dimer of dimers) at higher concentrations [148]. Additionally, Cu(II) binds tighter to *K. aerogenes* than Ni(II) but Ni(II) binding is enthalpically favored.

The first crystal structure of UreE from *S. pasteurii* shows that Zn(II) binds at the interface of two dimers and is bound by four histidine residues (His 100 equivalent to His96 from *K. aerogenes*) from each monomer [151]. The protein crystals were obtained from UreE protein that copurified with Zn(II) ions [151]. A second set of crystal structures of *S. pasteurii* UreE bound to Ni(II) or Zn(II) show that the Ni(II) is coordinated by His100 in a site that is consistent with octahedral geometry while the Zn(II) site adopts pseudo-tetrahedral geometry and is bound by His9 and Asp12 from two different dimers [138]. In this study, the protein crystals were obtained from apo UreE that was incubated with nickel sulphate or zinc sulphate [138]. Crystallographic studies coupled with ITC suggest that there is a high affinity and a low affinity nickel site located at the C-terminus of the protein. The high affinity site features His100 as Ni(II) ligands and the low affinity site features His145 or His147 [138,142]. It has been suggested that the low affinity site funnels nickel to the high affinity site, which can bind either Ni(II) or Zn(II), and could be involved in UreE–UreG interactions [138].

H. pylori UreE has been crystallized in the apo, Ni(II) bound, Cu(II) bound, and Zn(II) bound forms [141,152]. The structures determined by Shi et al. show the apo protein as a dimer and the metal bound Cu(II) and Ni(II) forms of the protein as a tetramer, a dimer of dimers [152]. It is unclear what the exact source of the Cu(II) and Ni(II) ions is in these structures [152]. The Cu(II) and Ni(II) UreE crystals resulted from an attempt crystallize the UreE–HypA and the UreE–UreG complex, respectively [152]. However, in the structures obtained by Banaszak et al. the protein exists as a dimer in the apo, Zn(II) and Ni(II) forms [141]. These ZnUreE and NiUreE structures were obtained from apo UreE that was incubated with either zinc sulfate or nickel sulphate and crystallized [141]. Both sets of structure show that His102 (equivalent to His96 and His100 in *K. aerogenes* and *S. pasteurii, respectively*) is involved in binding the metal ion [141,152]. *H. pylori* UreE has a single His residue, His152, on the C-terminus. Mutagenesis studies coupled with ITC experiments determined that His152 is a ligand for Zn(II) and not Ni(II) [140]. It was also determined by ITC experiments that His102 is a ligand for both Ni(II) and Zn(II) [140]. The crystal structure and the XAS data revealed that the *H. pylori* Ni(II) site is pseudo-octahedral and the Zn(II) site is tetrahedral and both metals are bound by His102 and His152 [141]. Although there is a difference in the ligands bound to Ni(II) in the H152A UreE compared to wildtype, Banaszak et al. suggests that mutating His152, which may be a weakly coordinated residue, results in a rearrangement in the metal site [141].

4. Protein–Protein Interactions

The maturation of [NiFe]-hydrogenase and urease is a highly choreographed process that involves several accessory proteins. Protein–protein interactions between the metallochaperones HypA, HypB and SlyD have been observed experimentally, however, a ternary complex has not been detected [153]. Work done by Khorasani-Motlagh et al. suggests that HypB is the central component of nickel delivery to [NiFe]-hydrogenase as it interacts with both SlyD and HypA, individually [153]. Based on cyclic voltammetry (CV) and electrochemical impedance spectroscopy (EIS) measurements, the relative affinities of the HypB complexes are on the order of HypB–SlyD > HypB–HypA > HypA–SlyD [153].

HypB and SlyD form a complex in both *E. coli* [15,129,154] and *H. pylori* [155]. In both bacteria, SlyD enhances the GTPase activity of HypB [135,156]. *E. coli* SlyD also forms a complex with a Strep-tagII variant of the large subunit of [NiFe]-hydrogenase 3, HycE [157]. Additionally, it was determined that the C-terminal tail of SlyD is important for stimulating nickel release from HypB [129]. HypB also forms a heterodimer with HypA in *H. pylori* [95,106], and *E.coli* (Figure 12) [94]. Nickel is transferred from the G-domain of HypB to HypA and the rate of transfer increases significantly in the presence of GDP [102]. The transfer of nickel is also more efficient when HypA and HypB from a complex [102]. Once HypA is loaded with nickel the HypAB protein complex dissociates (Figure 12) [117]. In *E. coli*, HypA can form a complex with HycE in the absence of HypB or SlyD [158]. It is thought that HypA and HypB preassemble before reaching HycE as both proteins can interact in the absence of HycE [102,106,158]. Deletion of the *hypA* gene prevents HypB from interacting with HycE, which suggests that HypA serves as a scaffold for HypB to dock to the large [NiFe]-hydrogenase precursor protein [107,158].

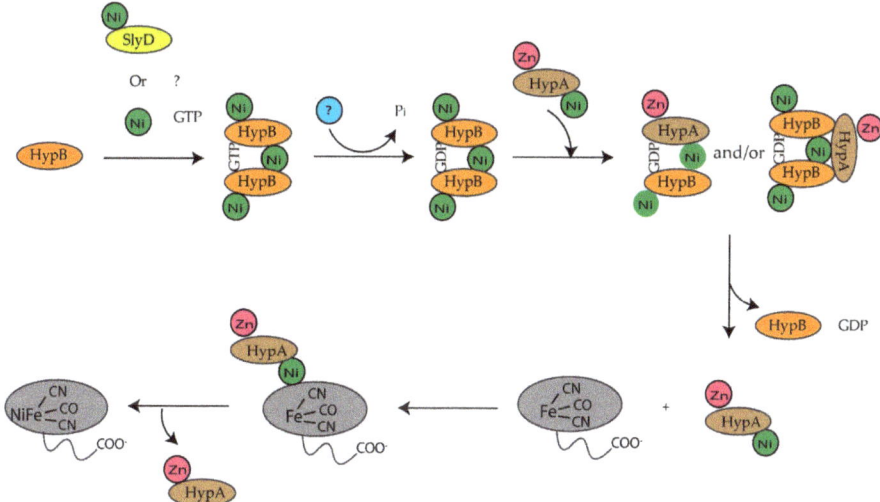

Figure 12. Model of the protein complexes required for the insertion of nickel into the large [NiFe]-hydrogenase precursor protein in *E. coli* [17,117,135]. Monomeric HypB dimerizes when it binds to GTP and then binds to nickel from SlyD or some other source. GTP hydrolysis occurs, which weakens the affinity of HypB for nickel and promotes the formation of a complex between NiGDPHypB and ZnHypA. Nickel is transferred from NiGDPHypB to ZnHypA. NiZnHypA then delivers nickel to the large [NiFe]-hydrogenase precursor protein.

In *T. kodakarensis*, HypA enhances the ATPase activity of HypB threefold [101]. Similar to *E. coli* HypAB interactions, HypAB interactions in *T. kodakarensis* are regulated by nucleotide hydrolysis. HypA and HypB form a heterotetrametric structure containing two HypA and two HypB in the presence of ATP (Figure 13) [101]. The HypAB complex binds Ni(II) with nanomolar affinity [101]. The crystal structure shows that the two HypA molecules are bound to the opposite surface of the ATP-binding site of the HypB dimer (Figure 13) [101]. Complex formation between HypA and HypB results in conformational changes that create a new nickel binding site with nanomolar affinity as determined by ITC (Figure 13) [101]. In the HypAB complex, His98 from HypA moves closer to the N-terminus of the protein forming a new nickel site involving His98 and the N-terminal MHE motif (Figure 13) [101]. The Ni(II) ion is bound in a four-coordinate, distorted square planar geometry and is coordinated by the N-terminal amine, the amide nitrogen and side chain of His2, and the side chain of His98 [101].

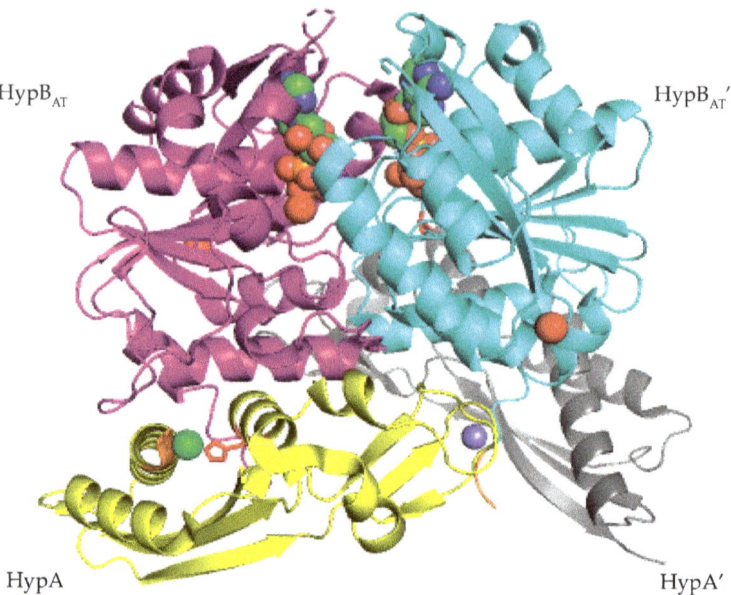

Figure 13. Crystal structure of the transient HypAB complex from *T. kodakarensis* (PDB ID 5AUN) [101]. The protein was crystallized from a solution containing equal molar concentrations of *T. kodakarensis* HypA and HypB in the presence of ATPγS and nickel chloride. The HypA monomers (shown in yellow and gray) bind on either site of the HypB dimer (shown in purple and cyan). The nickel ion is shown as a green sphere and is coordinated by the MHE motif (colored orange) and His98 (colored red) of HypA. The zinc ion is shown as a slate sphere, the magnesium ion as a red sphere, and the ADP molecules are shown as red, green, and blue spheres. (This figure is adapted from Watanabe, S.; Kawashima, T.; Nishitani, Y.; Kanai, T.; Wada, T.; Inaba, K.; Atomi, H.; Imanaka, T.; Miki, K. Structural basis of a Ni acquisition cycle for [NiFe] hydrogenase by Ni-metallochaperone HypA and its enhancer. *Proc. Natl. Acad. Sci. USA* **2015**, *112*, 7701–7706. Copyright (2015) National Academy of Sciences).

HypA has also been shown to form a complex with UreE that results in the formation of a new nickel site [96,103,159]. In *H. pylori*, HypA–UreE interactions are essential for urease maturation [31]. Crosslinking, static light scattering, and ITC studies show that the UreE dimer binds HypA using residues located in the C-terminus of the protein to form a hetero-complex, HypA–UreE (Figure 14) [103,159]. The dissociation constant for apo ZnHypA or NiZnHypA to apo-UreE$_2$ is 1μM at pH 7.2 [159]. Similarly, the dissociation constant of ZnHypA to ZnUreE$_2$ is also 1 μM [159]. However, the interactions between HypA and UreE$_2$ were weakened between Zn-HypA or NiZnHypA and Ni-UreE$_2$ at pH 7.2 [159].

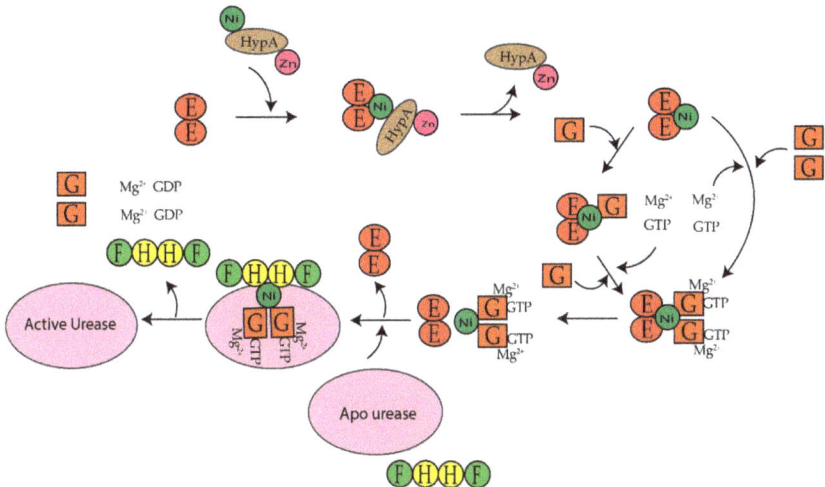

Figure 14. Model of the protein complexes required for the insertion of nickel into *H. pylori* urease [103, 160,161]. NiZnHypA forms a complex with UreE$_2$ and transfers nickel to UreE$_2$. NiUreE$_2$ will form a NiUreE$_2$–UreG$_2$ complex either with UreG$_2$ or with 2 UreG in the presence of magnesium ions and GTP. A NiUreE$_2$–UreG$_2$ complex with magnesium ions, GTP and UreG$_2$ is formed and nickel is transformed from UreE$_2$ to UreG$_2$. UreG$_2$ forms a complex with UreF–UreH and apo urease. GTP is hydrolyzed by UreG and nickel is inserted into apo urease in the presence of potassium bicarbonate (not shown) and ammonium bicarbonate (not shown). UreF, UreH, and UreG dissociate from the Ni-urease enzyme.

At both pH 7.2 and pH 6.3 *H. pylori* UreE$_2$ binds Ni(II) one order of magnitude tighter than *H. pylori* HypA [96,97,103,140]. It has been determined that Ni(II) is transferred from HypA to UreE$_2$ (Figure 14) [159]. ITC studies and fluorometric studies revealed that Ni(II) binding to ZnHypA·ApoUreE$_2$ results in at least two distinct isotherms with micromolar and nanomolar affinity at both pH values [103]. The micromolar binding is similar to that seen for Ni(II) binding to apoUreE$_2$ [103]. Nanomolar binding was also observed when NiZnHypA was titrated into apoUreE$_2$ suggesting that a new high affinity nickel binding site is formed when Ni(II) is added to ZnHypA·UreE$_2$ complex [103]. This result was further supported by metal binding experiments conducted using L2*HypA mutant protein. It was determined that ZnL2*HypA binds apoUreE$_2$ similar to wild-type HypA with micromolar affinity and that ZnL2*HypA has a weaker affinity for Ni(II) than the wild-type HypA protein [96,103]. Additionally, titrating ZnL2*HypA·apoUreE$_2$ with Ni(II) at pH 7.2 resulted in a single isotherm [103].

In *K. aerogenes* the urease accessory proteins UreD, UreF, UreG, and UreE bind sequentially to the urease enzyme (Figure 1). First, UreD forms a complex with apo-urease, UreABC–UreD [28]. In the presence of UreD, UreF forms a UreABC–UreDF complex [28,162] and UreG forms a complex with UreABC–UreDF [28,163]. Additionally, the UreDFG complex can bind directly to UreABC [28]. Finally, UreE delivers Ni(II) to the UreABC–UreDFG [149]. In *H. pylori* two different UreE–UreG complexes have been observed (Figure 14). The first complex is composed of two monomers of *H. pylori* UreG bound to one *H. pylori* UreE dimer (UreE$_2$–UreG$_2$) with a K_d of 4.0 µM [140,160]. A second UreE–UreG complex has been observed that features a UreE dimer with a UreG monomer (UreE$_2$–UreG) (Figure 14) [160]. It was determined that UreE accepts Ni(II) from HypA and the UreE$_2$–UreG$_2$ complex facilitates Ni(II) transfer from UreE to UreG and enhances GTP binding [160].

Several studies have been done on individual proteins and protein–protein interactions between the Ni(II) chaperones and accessory proteins. These studies have led to models for nickel insertion into *E. coli* [NiFe]-hydrogenase (Figure 12) and *H. pylori urease* (Figure 14). For nickel insertion into

E. coli [NiFe]-hydrogenase involves the dimerization of HypB when it binds to GTP and nickel. GTP is hydrolyzed to GDP, and NiGDPHypB forms a complex with ZnHypA. The stoichiometry of the HypA–HypB complex is unknown [117]. Nickel is then transferred from the G-domain nickel site of HypB to N-terminal nickel site of ZnHypA, and NiZnHypA forms a complex with and transfers nickel to the large [NiFe]-hydrogenase precursor protein. For nickel insertion into *H. pylori* urease activation involves NiZnHypA delivering nickel to $UreE_2$. $UreE_2$ binds to $UreG_2$ and transfers nickel to $UreG_2$. $NiUreG_2$ form a complex with $UreH_2F_2$ and apo urease. GTP hydrolysis occurs and nickel is delivered to urease followed by the dissociation of UreF, UreE, and UreG from the active urease enzyme.

5. Conclusions

Over the past 20 years significant progress has been made towards elucidating the role that nickel plays in the proper functioning of metalloregulators and chaperones. Multiple themes have emerged from the studies conducted on nickel metalloregulators and chaperones regarding nickel binding and protein allostery. The proteins discussed in this review bind nickel in one of three different geometries: four-coordinate square planar, four-coordinate tetrahedral, or six-coordinate octahedral geometry. Four of the proteins, RcnR, NmtR, HypA, and *E. coli* HypB, utilize the N-terminal amine to coordinate the nickel ion. The Ni(II)-responsive metalloregulators: RcnR, DmeR, and KmtR are also Co(II)-responsive.

Many of the metalloregulators discussed in this review can bind other first row transition metals but only cognate metal, nickel and in some cases cobalt, binding results in the allosteric regulation of DNA binding. NikR proteins have been studied extensively using a combination of NMR, X-ray crystallography, XAS, and metal binding studies. These studies determined that NikR can bind to the first-row transition metals, Co(II), Ni(II), Cu(II), and Zn(II) with various geometries [47,57,59]. The binding affinities clearly follow the Irving Williams series, yet NikR is a Ni(II) responsive metalloregulator [59]. These results suggest that only when nickel binds to NikR forming a four coordinate square planar complex using 3 $N_{Imidazole}$ ligands and 1 S-dnonors does the correct structural changes occur in the protein that favors DNA binding [57]. These results are corroborated by studies done in RcnR, NmtR, and InrS where nickel binding disfavors DNA binding. Collectively, these results support the theme that metal responsiveness is most closely linked to coordination number and ligand selection [36,164]. Protein allostery is also observed when Ni(II) binds to *H. pylori* HypA [97,104] and *E. coli* HypB [110].

Another theme that emerges from the work discussed in this review is that sequence similarities between the metalloregulators are not sufficient to determine the metal selectivity of metalloregulators. For example InrS possesses all the ligands that bind Cu(I) in *M. tuberculosis* CsoR, however, InrS uses these ligands plus an additional His ligand to bind nickel [63,77]. Interestingly, both RcnR and InrS belong to the same family of metalloregulators and thus have a similar protein fold but they coordinate nickel with a different ligand set and geometry. *E. coli* RcnR coordinates nickel in a six-coordinate octahedral site [70,72,73]. InrS utilizes a different ligand set to coordinate nickel in four-coordinate square planar geometry [79]. Why has nature designed two metalloregulators that carryout similar functions with the same fold but different metal binding geometries and ligand sets? It will be interesting to see how DmeR binds nickel. A similar situation exists for the two Ni(II) and Co(II) regulators from *M. tuberculosis*, NmtR and KmtR [84–86,89].

HypA, HypB, and SlyD proteins have been characterized from various organisms including *E. coli* and *H. pylori* and the metal site structures have been elucidated. Variations in the number of metal binding sites and metal ligands among homologous proteins from different organisms have been observed. The reasons for these variations are unknown but it may be linked to differences in intracellular nickel concentrations or the nickel requirements of the organism [17] as well as the various protein–protein interactions that occur in the shuttling of nickel from one protein to another. All HypA proteins have two metal binding sites, an N-terminal nickel site and a zinc site in a loop region with 2 CXXC motifs. However, both the flanking His residues, His79 and His95, in *H. pylori* HypA are not conserved in all HypA proteins, some have one or none [97]. Mutating any of the Cys or His residues resulted in a HypA protein that no longer sensed nickel binding or pH changes [97]. The variations in the number of flanking His residues in HypA proteins suggest that if nickel binding and pH changes are communicated in other HypA proteins, they do not result in similar structural changes observed for *H. pylori* HypA. A regulatory role for the *H. pylori* zinc site has been proposed [33,97,104,105]. It has been suggested that changes in the zinc site may affect interactions between HypA and other nickel proteins [33,97] and may affect the delivery of nickel to [NiFe]-hydrogenase and urease [33,105]. Studies involving mutant *hypA* strains have demonstrated that all the cysteine residues in the zinc site are important for [NiFe]-hydrogenase and urease activity in *H. pylori* [33,105]. The His residues may play a role in protein–protein interaction as they are not required for [NiFe]-hydrogenase and urease activity [33,105].

The number of metal binding sites varies in HypB proteins. All GTPase HypB proteins have a nickel site located in the G domain that binds either nickel or zinc. This site has a lower affinity for nickel than zinc but is involved in transferring nickel to HypA [109,111,117]. A N-terminal high affinity site has been characterized for *E. coli* HypB [109]. The exact role of this site still needs to be elucidated [109]. Leach et al. have suggested that that the N-terminal nickel site plays a structural and/or regulatory role [109]. They have also suggested that it could be a source of nickel for [NiFe]-hydrogenase when nickel concentrations in the cell are limited [109]. Some HypB proteins have histidine rich regions that bind multiple nickel ions that may be involved in nickel storage. Studies done by Sydor et al. on *H. pylori* HypB showed that there is a link between nucleotide binding and the ligands used to coordinate nickel. When GDP is bound to the protein, *H. pylori* HypB binds nickel with 4 S-donors from four cysteine residues, however, in the absence of nucleotide bound to the protein His107 ligates the nickel [114]. The importance of this change is not fully understood but it may play a role in hydrogenase and urease maturation [114].

SlyD is important for [NiFe] hydrogenase maturation [16] and is known to form a complex with HypB, [15,129,154,155]. However, the details of the interaction between SlyD and HypB still need to be determined. The importance of histidine rich C-terminal tails vary in [NiFe]-hydrogenase and urease maturation. Both *H. pylori* SlyD [132] and *K. aerogenes* UreE [136] have histidine rich C-terminal tails. It was shown that the histidine rich tail in SlyD is essential for [NiFe]-hydrogenase maturation [129], but the C-terminal tail of UreE is not necessary for urease maturation [144]. All UreE proteins characterized use a single conserved His residue and other residues to bind nickel in octahedral or pseudo octahedral geometry at the dimer interface. *S. pasteurii* UreE, like *E. coli* NikR and *E. coli* HypB has a high and low affinity nickel binding site [138,142]. It is thought that the low affinity site funnels nickel to the high affinity site, and the high affinity site is involved in UreE–UreG interactions [138].

The bioinorganic chemistry of nickel metalloregulators and chaperones is fascinating. Numerous studies have been conducted on individual chaperones and accessory proteins involved in [NiFe]-hydrogenase and urease maturation but some details regarding the sequential protein–protein interactions that are necessary to deliver nickel to [NiFe]-hydrogenase and urease still need to be elucidated. How does HypB get nickel? What is the exact role of SlyD in the delivery of nickel to hydrogenase? What role does HypB play in urease maturation? In *E. coli* what is the oligomeric state of the HypA–HypB complex? What favors HypA delivering nickel to [NiFe]-hydrogenase versus UreE and vice versa? What ligands compose the HypA–UreE metal binding site? How does UreG deliver nickel to urease? What is the role of UreD? How does UreF stimulate the GTPase activity of UreG?

For *H. pylori*, studies involving animal models have determined that the pathogenicity of *H. pylori* is dependent on both [NiFe]-hydrogenase and urease [26,165,166]. Additional studies done by Blum et al. determined that urease and not [NiFe] hydrogenase is responsible for acid resistance in *H. pylori* [105]. To date there are no mammalian enzymes that utilize nickel [1], therefore, drug therapies targeting proteins involved in nickel ion homeostasis could be used to treat *H. pylori* infections [24]. Prior to developing such therapies, an understanding of how nickel is bound by the various proteins in the cells and the details of protein–protein interactions involved in delivering nickel to [NiFe]-hydrogenase and urease in *H. pylori* is crucial.

Supplementary Materials: The following are available online at http://www.mdpi.com/2304-6740/7/8/104/s1, Table S1: The metal binding properties of the Ni(II)-responsive metalloregulators, Table S2: The metal binding properties of the Ni(II) chaperones.

Funding: This research was funded by the Rhode Island Institutional Development Award (IDeA) Network of Biomedical Research Excellence from the National Institute of General Medical Sciences of the National Institutes of Health under grant number P20GM103430.

Conflicts of Interest: The author declares no conflict of interest.

References

1. Maroney, M.J.; Ciurli, S. Nonredox nickel enzymes. *Chem. Rev.* **2014**, *114*, 4206–4228. [CrossRef] [PubMed]
2. Maier, R.J. Use of molecular hydrogen as an energy substrate by human pathogenic bacteria. *Biochem. Soc. Trans.* **2005**, *33*, 83–85. [CrossRef] [PubMed]
3. Mulrooney, S.B.; Hausinger, R.P. Nickel uptake and utilization by microorganisms. *FEMS Microbiol. Rev.* **2003**, *27*, 239–261. [CrossRef]
4. Macomber, L.; Hausinger, R.P. Mechanisms of nickel toxicity in microorganisms. *Metallomics* **2011**, *3*, 1153–1162. [CrossRef] [PubMed]
5. Higgins, K.A.; Carr, C.E.; Maroney, M.J. Specific metal recognition in nickel trafficking. *Biochemistry* **2012**, *51*, 7816–7832. [CrossRef]
6. De Pina, K.; Desjardin, V.; Mandrand-Berthelot, M.A.; Giordano, G.; Wu, L.F. Isolation and characterization of the *nikR* gene encoding a nickel-responsive regulator in *Escherichia coli*. *J. Bacteriol.* **1999**, *181*, 670–674. [PubMed]
7. Blériot, C.; Effantin, G.; Lagarde, F.; Mandrand-Berthelot, M.A.; Rodrigue, A. RcnB is a periplasmic protein essential for maintaining intracellular Ni and Co concentrations in *Escherichia coli*. *J. Bacteriol.* **2011**, *193*, 3785–3793. [CrossRef]
8. Iwig, J.S.; Rowe, J.L.; Chivers, P.T. Nickel homeostasis in *Escherichia coli* - the *rcnR-rcnA* efflux pathway and its linkage to NikR function. *Mol. Microbiol.* **2006**, *62*, 252–262. [CrossRef]
9. Böck, A.; King, P.W.; Blokesch, M.; Posewitz, M.C. Maturation of hydrogenases. *Adv. Microb. Physiol.* **2006**, *51*, 1–71.
10. Forzi, L.; Sawers, R.G. Maturation of [NiFe]-hydrogenases in *Escherichia coli*. *Biometals* **2007**, *20*, 565–578. [CrossRef]
11. Fontecilla-Camps, J.C.; Amara, P.; Cavazza, C.; Nicolet, Y.; Volbeda, A. Structure–function relationships of anaerobic gas-processing metalloenzymes. *Nature* **2009**, *460*, 814–822. [CrossRef]
12. Peters, J.W.; Schut, G.J.; Boyd, E.S.; Mulder, D.W.; Shepard, E.M.; Broderick, J.B.; King, P.W.; Adams, M.W.W. [FeFe]- and [NiFe]-hydrogenase diversity, mechanism, and maturation. *Biochim. Biophys. Acta-Mol. Cell Res.* **2015**, *1853*, 1350–1369. [CrossRef] [PubMed]
13. Watanabe, S.; Sasaki, D.; Tominaga, T.; Miki, K. Structural basis of [NiFe] hydrogenase maturation by Hyp proteins. *Biol. Chem.* **2012**, *393*, 1089–1100. [CrossRef] [PubMed]
14. Leach, M.R.; Zamble, D.B. Metallocenter assembly of the hydrogenase enzymes. *Curr. Opin. Chem. Biol.* **2007**, *11*, 159–165. [CrossRef] [PubMed]
15. Zhang, J.W.; Butland, G.; Greenblatt, J.F.; Emili, A.; Zamble, D.B. A role for SlyD in the *Escherichia coli* hydrogenase biosynthetic pathway. *J. Biol. Chem.* **2005**, *280*, 4360–4366. [CrossRef] [PubMed]
16. Pinske, C.; Sargent, F.; Sawers, R.G. SlyD-dependent nickel delivery limits maturation of [NiFe]-hydrogenases in late-stationary phase *Escherichia coli* cells. *Metallomics* **2015**, *7*, 683–690. [CrossRef] [PubMed]
17. Lacasse, M.J.; Zamble, D.B. [NiFe]-Hydrogenase maturation. *Biochemistry* **2016**, *55*, 1689–1701. [CrossRef] [PubMed]
18. Hube, M.; Blokesch, M.; Böck, A. Network of hydrogenase maturation in *Escherichia coli*: Role of accessory proteins HypA and HybF. *J. Bacteriol.* **2002**, *184*, 3879–3885. [CrossRef] [PubMed]
19. Reissmann, S.; Hochleitner, E.; Wang, H.; Paschos, A.; Lottspeich, F.; Glass, R.S.; Böck, A. Taming of a poison: Biosynthesis of the NiFe-hydrogenase cyanide ligands. *Science* **2003**, *299*, 1067–1070. [CrossRef]
20. Li, Y.; Zamble, D.B. Nickel homeostasis and nickel regulation: An overview. *Chem. Rev.* **2009**, *109*, 4617–4643. [CrossRef]

21. Lubitz, W.; Ogata, H.; Rüdiger, O.; Reijerse, E. Hydrogenases. *Chem. Rev.* **2014**, *114*, 4081–4148. [CrossRef] [PubMed]
22. Dunn, B.E.; Cohen, H.; Blaser, M.J. Helicobacter pylori. *Clin. Microbiol. Rev.* **1997**, *10*, 720–741. [CrossRef] [PubMed]
23. Kusters, J.G.; Van Vliet, A.H.M.; Ernst, J.; Kuipers, E.J. Pathogenesis of *Helicobacter pylori* infection. *Clin. Microbiol. Rev.* **2006**, *19*, 449–490. [CrossRef] [PubMed]
24. Kao, C.-Y.; Sheu, B.S.; Wu, J.J. *Helicobacter pylori* infection: An overview of bacterial virulence factors and pathogenesis. *Biomed. J.* **2016**, *39*, 14–23. [CrossRef] [PubMed]
25. Yamaoka, Y. Mechanisms of disease: *Helicobacter pylori* virulence factors. *Nat. Rev. Gastroenterol. Hepatol.* **2010**, *7*, 629–641. [CrossRef]
26. Olson, J.W. Molecular hydrogen as an energy source for *Helicobacter pylori*. *Science* **2002**, *298*, 1788–1790. [CrossRef] [PubMed]
27. Farrugia, M.A.; Macomber, L.; Hausinger, R.P. Biosynthesis of the urease metallocenter. *J. Biol. Chem.* **2013**, *288*, 13178–13185. [CrossRef]
28. Carter, E.L.; Flugga, N.; Boer, J.L.; Mulrooney, S.B.; Hausinger, R.P. Interplay of metal ions and urease. *Metallomics* **2009**, *1*, 207–221. [CrossRef]
29. Zambelli, B.; Musiani, F.; Benini, S.; Ciurli, S. Chemistry of Ni^{2+} in urease: Sensing, trafficking, and catalysis. *Acc. Chem. Res.* **2011**, *44*, 520–530. [CrossRef]
30. Zeer-Wanklyn, C.J.; Zamble, D.B. Microbial nickel: Cellular uptake and delivery to enzyme centers. *Curr. Opin. Chem. Biol.* **2017**, *37*, 80–88. [CrossRef]
31. Benoit, S.L.; Mehta, N.; Weinberg, M.V.; Maier, C.; Maier, R.J. Interaction between the *Helicobacter pylori* accessory proteins HypA and UreE is needed for urease maturation. *Microbiology* **2007**, *153*, 1474–1482. [CrossRef]
32. Olson, J.W.; Mehta, N.S.; Maier, R.J. Requirement of nickel metabolism proteins HypA and HypB for full activity of both hydrogenase and urease in *Helicobacter pylori*. *Mol. Microbiol.* **2001**, *39*, 176–182. [CrossRef]
33. Johnson, R.C.; Hu, H.Q.; Merrell, D.S.; Maroney, M.J. Dynamic HypA zinc site is essential for acid viability and proper urease maturation in *Helicobacter pylori*. *Metallomics* **2015**, *7*, 674–682. [CrossRef]
34. Jones, M.D.; Li, Y.; Zamble, D.B. Acid-responsive activity of the *Helicobacter pylori* metalloregulator NikR. *Proc. Natl. Acad. Sci. USA* **2018**, *115*, 8966–8971. [CrossRef]
35. Wolfram, L.; Bauerfeind, P. Activities of urease and nickel uptake of *Helicobacter pylori* proteins are media- and host-dependent. *Helicobacter* **2009**, *14*, 264–270. [CrossRef] [PubMed]
36. Higgins, K.A.; Giedroc, D.P. Metal specificity of metallosensors. In *Encyclopedia of Inorganic and Bioinorganic Chemistry*; John Wiley & Sons, Ltd.: Chichester, UK, 2013; pp. 209–224.
37. Musiani, F.; Zambelli, B.; Bazzani, M.; Mazzei, L.; Ciurli, S. Nickel-responsive transcriptional regulators. *Metallomics* **2015**, *7*, 1305–1318. [CrossRef]
38. Chivers, P.T.; Sauer, R.T. NikR repressor: High-affinity nickel binding to the C-terminal domain regulates binding to operator DNA. *Chem. Biol.* **2002**, *9*, 1141–1148. [CrossRef]
39. Bloom, S.L.; Zamble, D.B. Metal-selective DNA-binding response of *Escherichia coli* NikR. *Biochemistry* **2004**, *43*, 10029–10038. [CrossRef]
40. Chivers, P.T.; Sauer, R.T. Regulation of high affinity nickel uptake in bacteria. Ni^{2+}-dependent interaction of NikR with wild-type and mutant operator sites. *J. Biol. Chem.* **2000**, *275*, 19735–19741. [CrossRef]
41. Budnick, J.A.; Prado-Sanchez, E.; Caswell, C.C. Defining the regulatory mechanism of NikR, a nickel-responsive transcriptional regulator, in *Brucella abortus*. *Microbiology* **2018**, *164*, 1320–1325. [CrossRef]
42. Fabini, E.; Zambelli, B.; Mazzei, L.; Ciurli, S.; Bertucci, C. Surface plasmon resonance and isothermal titration calorimetry to monitor the Ni(II)-dependent binding of *Helicobacter pylori* NikR to DNA. *Anal. Bioanal. Chem.* **2016**, *408*, 7971–7980. [CrossRef]
43. Contreras, M.; Thiberge, J.M.; Mandrand-Berthelot, M.A.; Labigne, A. Characterization of the roles of NikR, a nickel-responsive pleiotropic autoregulator of *Helicobacter pylori*. *Mol. Microbiol.* **2003**, *49*, 947–963. [CrossRef]
44. Vannini, A.; Pinatel, E.; Costantini, P.E.; Pelliciari, S.; Roncarati, D.; Puccio, S.; De Bellis, G.; Peano, C.; Danielli, A. Comprehensive mapping of the *Helicobacter pylori* NikR regulon provides new insights in bacterial nickel responses. *Sci. Rep.* **2017**, *7*, 45458. [CrossRef]
45. Schreiter, E.R.; Sintchak, M.D.; Guo, Y.; Chivers, P.T.; Sauer, R.T.; Drennan, C.L. Crystal structure of the nickel-responsive transcription factor NikR. *Nat. Struct. Biol.* **2003**, *10*, 794–799. [CrossRef]

46. Schreiter, E.R.; Wang, S.C.; Zamble, D.B.; Drennan, C.L. NikR-operator complex structure and the mechanism of repressor activation by metal ions. *Proc. Natl. Acad. Sci. USA* **2006**, *103*, 13676–13681. [CrossRef]
47. Phillips, C.M.; Schreiter, E.R.; Guo, Y.; Wang, S.C.; Zamble, D.B.; Drennan, C.L. Structural basis of the metal specificity for nickel regulatory protein NikR. *Biochemistry* **2008**, *47*, 1938–1946. [CrossRef]
48. Phillips, C.M.; Schreiter, E.R.; Stultz, C.M.; Drennan, C.L. Structural basis of low-affinity nickel binding to the nickel-responsive transcription factor NikR from *Escherichia coli*. *Biochemistry* **2010**, *49*, 7830–7838. [CrossRef]
49. Chivers, P.T.; Tahirov, T.H. Structure of *Pyrococcus horikoshii* NikR: Nickel sensing and implications for the regulation of DNA recognition. *J. Mol. Biol.* **2005**, *348*, 597–607. [CrossRef]
50. Bahlawane, C.; Dian, C.; Muller, C.; Round, A.; Fauquant, C.; Schauer, K.; de Reuse, H.; Terradot, L.; Michaud-Soret, I. Structural and mechanistic insights into *Helicobacter pylori* NikR activation. *Nucleic Acids Res.* **2010**, *38*, 3106–3118. [CrossRef]
51. Benini, S.; Cianci, M.; Ciurli, S. Holo-Ni^{2+} *Helicobacter pylori* NikR contains four square-planar nickel-binding sites at physiological pH. *Dalt. Trans.* **2011**, *40*, 7831–7833. [CrossRef]
52. West, A.L.; St John, F.; Lopes, P.E.; MacKerell, A.D., Jr.; Pozharski, E.; Michel, S.L. Holo-Ni(II)HpNikR is an asymmetric tetramer containing two different nickel-binding sites. *J. Am. Chem. Soc.* **2010**, *132*, 14447–14456. [CrossRef]
53. West, A.L.; Evans, S.E.; Gonzalez, J.M.; Carter, L.G.; Tsuruta, H.; Pozharski, E.; Michel, S.L.J. Ni(II) coordination to mixed sites modulates DNA binding of HpNikR via a long-range effect. *Proc. Natl. Acad. Sci. USA* **2012**, *109*, 5633–5638. [CrossRef]
54. Dian, C.; Schauer, K.; Kapp, U.; McSweeney, S.M.; Labigne, A.; Terradot, L. Structural basis of the nickel response in Helicobacter pylori: Crystal structures of HpNikR in apo and nickel-bound states. *J. Mol. Biol.* **2006**, *361*, 715–730. [CrossRef]
55. Chivers, P.T.; Sauer, R.T. NikR is a ribbon-helix-helix DNA-binding protein. *Protein Sci.* **1999**, *8*, 2494–2500. [CrossRef]
56. Musiani, F.; Bertosa, B.; Magistrato, A.; Zambelli, B.; Turano, P.; Losasso, V.; Micheletti, C.; Ciurli, S.; Carloni, P. Computational study of the DNA-binding protein *Helicobacter pylori* NikR: The role of Ni^{2+}. *J. Chem. Theory Comput.* **2010**, *6*, 3503–3515. [CrossRef]
57. Leitch, S.; Bradley, M.J.; Rowe, J.L.; Chivers, P.T.; Maroney, M.J. Nickel-specific response in the transcriptional regulator, *Escherichia coli* NikR. *J. Am. Chem. Soc.* **2007**, *129*, 5085–5095. [CrossRef]
58. Zambelli, B.; Bellucci, M.; Danielli, A.; Scarlato, V.; Ciurli, S. The Ni^{2+} binding properties of *Helicobacter pylori* NikR. *Chem. Commun.* **2007**, *35*, 3649–3651. [CrossRef]
59. Wang, S.C.; Dias, A.V.; Bloom, S.L.; Zamble, D.B. Selectivity of metal binding and metal-induced stability of *Escherichia coli* NikR. *Biochemistry* **2004**, *43*, 10018–10028. [CrossRef]
60. Ahn, B.E.; Cha, J.; Lee, E.J.; Han, A.R.; Thompson, C.J.; Roe, J.H. Nur, a nickel-responsive regulator of the Fur family, regulates superoxide dismutases and nickel transport in *Streptomyces coelicolor*. *Mol. Microbiol.* **2006**, *59*, 1848–1858. [CrossRef]
61. Kim, H.M.; Shin, J.-H.; Cho, Y.-B.; Roe, J.-H. Inverse regulation of Fe- and Ni-containing SOD genes by a Fur family regulator Nur through small RNA processed from 3′UTR of the sodF mRNA. *Nucleic Acids Res.* **2014**, *42*, 2003–2014. [CrossRef]
62. An, Y.J.; Ahn, B.E.; Han, A.R.; Kim, H.M.; Chung, K.M.; Shin, J.H.; Cho, Y.B.; Roe, J.H.; Cha, S.S. Structural basis for the specialization of Nur, a nickel-specific Fur homolog, in metal sensing and DNA recognition. *Nucleic Acids Res.* **2009**, *37*, 3442–3451. [CrossRef]
63. Liu, T.; Ramesh, A.; Ma, Z.; Ward, S.K.; Zhang, L.; George, G.N.; Talaat, A.M.; Sacchettini, J.C.; Giedroc, D.P. CsoR is a novel *Mycobacterium tuberculosis* copper-sensing transcriptional regulator. *Nat. Chem. Biol.* **2007**, *3*, 60–68. [CrossRef]
64. Foster, A.W.; Pernil, R.; Patterson, C.J.; Scott, A.J.P.; Pålsson, L.O.; Pal, R.; Cummins, I.; Chivers, P.T.; Pohl, E.; Robinson, N.J. A tight tunable range for Ni(II) sensing and buffering in cells. *Nat. Chem. Biol.* **2017**, *13*, 409–414. [CrossRef]
65. Ma, Z.; Cowart, D.M.; Ward, B.P.; Arnold, R.J.; DiMarchi, R.D.; Zhang, L.; George, G.N.; Scott, R.A.; Giedroc, D.P. Unnatural amino acid substitution as a probe of the allosteric coupling pathway in a mycobacterial Cu(I) sensor. *J. Am. Chem. Soc.* **2009**, *131*, 18044–18045. [CrossRef]
66. Rodrigue, A.; Effantin, G.; Mandrand-Berthelot, M.A. Identification of rcnA (yohM), a nickel and cobalt resistance gene in *Escherichia coli*. *J. Bacteriol.* **2005**, *187*, 2912–2916. [CrossRef]

67. Blaha, D.; Arous, S.; Bleriot, C.; Dorel, C.; Mandrand-Berthelot, M.A.; Rodrigue, A. The *Escherichia coli* metallo-regulator RcnR represses rcnA and rcnR transcription through binding on a shared operator site: Insights into regulatory specificity towards nickel and cobalt. *Biochimie* **2011**, *93*, 434–439. [CrossRef]
68. Iwig, J.S.; Chivers, P.T. DNA recognition and wrapping by *Escherichia coli* RcnR. *J. Mol. Biol.* **2009**, *393*, 514–526. [CrossRef]
69. Iwig, J.S.; Leitch, S.; Herbst, R.W.; Maroney, M.J.; Chivers, P.T. Ni(II) and Co(II) sensing by *Escherichia coli* RcnR. *J. Am. Chem. Soc.* **2008**, *130*, 7592–7606. [CrossRef]
70. Higgins, K.A.; Chivers, P.T.; Maroney, M.J. Role of the N-terminus in determining metal-specific responses in the *E. coli* Ni- and Co-responsive metalloregulator, RcnR. *J. Am. Chem. Soc.* **2012**, *134*, 7081–7093. [CrossRef]
71. Madeira, F.; Park, Y.M.; Lee, J.; Buso, N.; Gur, T.; Madhusoodanan, N.; Basutkar, P.; Tivey, A.; Potter, S.C.; Lopez, R. The EMBL-EBI search and sequence analysis tools APIs in 2019. *Nucleic Acids Res.* **2019**, *47*, W636–W641. [CrossRef]
72. Higgins, K.A.; Hu, H.Q.; Chivers, P.T.; Maroney, M.J. Effects of select histidine to cysteine mutations on transcriptional regulation by *Escherichia coli* RcnR. *Biochemistry* **2013**, *52*, 84–97. [CrossRef]
73. Carr, C.E.; Musiani, F.; Huang, H.; Chivers, P.T.; Ciurli, S.; Maroney, M.J. Glutamate ligation in the Ni(II)- and Co(II)-responsive *Escherichia coli* transcriptional regulator, RcnR. *Inorg. Chem.* **2017**, *56*, 6459–6476. [CrossRef]
74. Chang, F.M.J.; Martin, J.E.; Giedroc, D.P. Electrostatic occlusion and quaternary structural ion pairing are key determinants of Cu(I)-mediated allostery in the copper-sensing operon repressor (CsoR). *Biochemistry* **2015**, *54*, 2463–2472. [CrossRef]
75. Denby, K.J.; Iwig, J.; Bisson, C.; Westwood, J.; Rolfe, M.D.; Sedelnikova, S.E.; Higgins, K.; Maroney, M.J.; Baker, P.J.; Chivers, P.T.; et al. The mechanism of a formaldehyde-sensing transcriptional regulator. *Sci. Rep.* **2016**, *6*, 38879. [CrossRef]
76. Huang, H.-T.; Bobst, C.E.; Iwig, J.S.; Chivers, P.T.; Kaltashov, I.A.; Maroney, M.J. Co(II) and Ni(II) binding of the *Escherichia coli* transcriptional repressor RcnR orders its N terminus, alters helix dynamics, and reduces DNA affinity. *J. Biol. Chem.* **2018**, *293*, 324–332. [CrossRef]
77. Foster, A.W.; Patterson, C.J.; Pernil, R.; Hess, C.R.; Robinson, N.J. Cytosolic Ni(II) sensor in cyanobacterium. *J. Biol. Chem.* **2012**, *287*, 12142–12151. [CrossRef]
78. Foster, A.W.; Pernil, R.; Patterson, C.J.; Robinson, N.J. Metal specificity of cyanobacterial nickel-responsive repressor InrS: Cells maintain zinc and copper below the detection threshold for InrS. *Mol. Microbiol.* **2014**, *92*, 797–812. [CrossRef]
79. Carr, C.E.; Foster, A.W.; Maroney, M.J. An XAS investigation of the nickel site structure in the transcriptional regulator InrS. *J. Inorg. Biochem.* **2017**, *177*, 352–358. [CrossRef]
80. Ma, Z.; Cowart, D.M.; Scott, R.A.; Giedroc, D.P. Molecular insights into the metal selectivity of the copper(I)-sensing repressor CsoR from *Bacillus subtilis*. *Biochemistry* **2009**, *48*, 3325–3334. [CrossRef]
81. Rubio-Sanz, L.; Prieto, R.I.; Imperial, J.; Palacios, J.M.; Brito, B. Functional and expression analysis of the metal-inducible *dmeRF* System from *Rhizobium leguminosarum* bv. viciae. *Appl. Environ. Microbiol.* **2013**, *79*, 6414–6422. [CrossRef]
82. Dokpikul, T.; Chaoprasid, P.; Saninjuk, K.; Sirirakphaisarn, S.; Johnrod, J.; Nookabkaew, S.; Sukchawalit, R.; Mongkolsuk, S. Regulation of the cobalt/nickel efflux operon *dmeRF* in *Agrobacterium tumefaciens* and a Link between the iron-Sensing regulator RirA and cobalt/nickel resistance. *Appl. Environ. Microbiol.* **2016**, *82*, 4732–4742. [CrossRef]
83. Li, Z.; Song, X.; Wang, J.; Bai, X.; Gao, E.; Wei, G. Nickel and cobalt resistance properties of *Sinorhizobium meliloti* isolated from *Medicago lupulina* growing in gold mine tailing. *PeerJ* **2018**, *6*, e5202. [CrossRef]
84. Cavet, J.S.; Meng, W.; Pennella, M.A.; Appelhoff, R.J.; Giedroc, D.P.; Robinson, N.J. A nickel-cobalt-sensing ArsR-SmtB family repressor. Contributions of cytosol and effector binding sites to metal selectivity. *J. Biol. Chem.* **2002**, *277*, 38441–38448. [CrossRef]
85. Pennella, M.A.; Shokes, J.E.; Cosper, N.J.; Scott, R.A.; Giedroc, D.P. Structural elements of metal selectivity in metal sensor proteins. *Proc. Natl. Acad. Sci. USA* **2003**, *100*, 3713–3718. [CrossRef]
86. Reyes-Caballero, H.; Lee, C.W.; Giedroc, D.P. *Mycobacterium tuberculosis* NmtR harbors a nickel sensing site with parallels to *Escherichia coli* RcnR. *Biochemistry* **2011**, *50*, 7941–7952. [CrossRef]
87. Lee, C.W.; Chakravorty, D.K.; Chang, F.-M.J.; Reyes-Caballero, H.; Ye, Y.; Merz, K.M.; Giedroc, D.P. Solution structure of *Mycobacterium tuberculosis* NmtR in the apo state: Insights into Ni(II)-mediated allostery. *Biochemistry* **2012**, *51*, 2619–2629. [CrossRef]

88. Kim, J.-S.; Kang, S.-O.; Lee, J.K. The protein complex composed of nickel-binding SrnQ and DNA binding motif-bearing SrnR of *Streptomyces griseus* represses *sodF* transcription in the presence of nickel. *J. Biol. Chem.* **2003**, *278*, 18455–18463. [CrossRef]
89. Campbell, D.R.; Chapman, K.E.; Waldron, K.J.; Tottey, S.; Kendall, S.; Cavallaro, G.; Andreini, C.; Hinds, J.; Stoker, N.G.; Robinson, N.J.; et al. Mycobacterial cells have dual nickel-cobalt sensors: Sequence relationships and metal sites of metal-responsive repressors are not congruent. *J. Biol. Chem.* **2007**, *282*, 32298–32310. [CrossRef]
90. Zhu, T.; Tian, J.; Zhang, S.; Wu, N.; Fan, Y. Identification of the transcriptional regulator NcrB in the nickel resistance determinant of *Leptospirillum ferriphilum UBK03*. *PLoS ONE* **2011**, *6*, e17367. [CrossRef]
91. Benoit, S.L.; Maier, R.J. Mua (HP0868) is a nickel-binding protein that modulates urease activity in *Helicobacter pylori*. *MBio* **2011**, *2*, e00039-11. [CrossRef]
92. Kidd, S.P.; Djoko, K.Y.; Ng, J.; Argente, M.P.; Jennings, M.P.; McEwan, A.G. A novel nickel responsive MerR-like regulator, NimR, from *Haemophilus influenzae*. *Metallomics* **2011**, *3*, 1009–1018. [CrossRef]
93. Blokesch, M.; Rohrmoser, M.; Rode, S.; Böck, A. HybF, a zinc-containing protein involved in NiFe hydrogenase maturation. *J. Bacteriol.* **2004**, *186*, 2603–2611. [CrossRef]
94. Atanassova, A.; Zamble, D.B. *Escherichia coli* HypA is a zinc metalloprotein with a weak affinity for nickel. *J. Bacteriol.* **2005**, *187*, 4689–4697. [CrossRef]
95. Mehta, N.; Olson, J.W.; Maier, R.J. Characterization of *Helicobacter pylori* nickel metabolism accessory proteins needed for maturation of both urease and hydrogenase. *J. Bacteriol.* **2003**, *185*, 726–734. [CrossRef]
96. Hu, H.Q.; Johnson, R.C.; Merrell, D.S.; Maroney, M.J. Nickel ligation of the N-terminal amine of HypA is required for urease maturation in *Helicobacter pylori*. *Biochemistry* **2017**, *56*, 1105–1116. [CrossRef]
97. Herbst, R.W.; Perovic, I.; Martin-Diaconescu, V.; O'Brien, K.; Chivers, P.T.; Pochapsky, S.S.; Pochapsky, T.C.; Maroney, M.J. Communication between the zinc and nickel sites in dimeric HypA: Metal recognition and pH sensing. *J. Am. Chem. Soc.* **2010**, *132*, 10338–10351. [CrossRef]
98. Xia, W.; Li, H.; Sze, K.-H.; Sun, H. Structure of a nickel chaperone, HypA, from *Helicobacter pylori* reveals two distinct metal binding sites. *J. Am. Chem. Soc.* **2009**, *131*, 10031–10040. [CrossRef]
99. Watanabe, S.; Arai, T.; Matsumi, R.; Atomi, H.; Imanaka, T.; Miki, K. Crystal structure of HypA, a nickel-binding metallochaperone for [NiFe] hydrogenase maturation. *J. Mol. Biol.* **2009**, *394*, 448–459. [CrossRef]
100. Spronk, C.A.E.M.; Żerko, S.; Górka, M.; Koźmiński, W.; Bardiaux, B.; Zambelli, B.; Musiani, F.; Piccioli, M.; Basak, P.; Blum, F.C.; et al. Structure and dynamics of *Helicobacter pylori* nickel-chaperone HypA: An integrated approach using NMR spectroscopy, functional assays and computational tools. *J. Biol. Inorg. Chem.* **2018**, *23*, 1309–1330. [CrossRef]
101. Watanabe, S.; Kawashima, T.; Nishitani, Y.; Kanai, T.; Wada, T.; Inaba, K.; Atomi, H.; Imanaka, T.; Miki, K. Structural basis of a Ni acquisition cycle for [NiFe] hydrogenase by Ni-metallochaperone HypA and its enhancer. *Proc. Natl. Acad. Sci. USA* **2015**, *112*, 7701–7706. [CrossRef]
102. Douglas, C.D.; Ngu, T.T.; Kaluarachchi, H.; Zamble, D.B. Metal Transfer within the *Escherichia coli* HypB–HypA Complex of Hydrogenase Accessory Proteins. *Biochemistry* **2013**, *52*, 6030–6039. [CrossRef]
103. Hu, H.Q.; Huang, H.; Maroney, M.J. The *Helicobacter pylori* HypA·UreE2 complex contains a novel high-affinity Ni(II)-binding site. *Biochemistry* **2018**, *57*, 2932–2942. [CrossRef]
104. Kennedy, D.C.; Herbst, R.W.; Iwig, J.S.; Chivers, P.T.; Maroney, M.J. A dynamic Zn site in *Helicobacter pylori* HypA: A potential mechanism for metal-specific protein activity. *J. Am. Chem. Soc.* **2007**, *129*, 16–17. [CrossRef]
105. Blum, F.C.; Hu, H.Q.; Servetas, S.L.; Benoit, S.L.; Maier, R.J.; Maroney, M.J.; Merrell, D.S. Structure-function analyses of metal-binding sites of HypA reveal residues important for hydrogenase maturation in *Helicobacter pylori*. *PLoS ONE* **2017**, *12*, e0183260. [CrossRef]
106. Xia, W.; Li, H.; Yang, X.; Wong, K.B.; Sun, H. Metallo-GTPase HypB from *Helicobacter pylori* and its interaction with nickel chaperone protein HypA. *J. Biol. Chem.* **2012**, *287*, 6753–6763. [CrossRef]
107. Kwon, S.; Watanabe, S.; Nishitani, Y.; Kawashima, T.; Kanai, T.; Atomi, H.; Miki, K. Crystal structures of a [NiFe] hydrogenase large subunit HyhL in an immature state in complex with a Ni chaperone HypA. *Proc. Natl. Acad. Sci. USA* **2018**, *115*, 7045–7050. [CrossRef]
108. Leipe, D.D.; Wolf, Y.I.; Koonin, E.V.; Aravind, L. Classification and evolution of P-loop GTPases and related ATPases. *J. Mol. Biol.* **2002**, *317*, 41–72. [CrossRef]
109. Leach, M.R.; Sandal, S.; Sun, H.; Zamble, D.B. Metal binding activity of the *Escherichia coli* hydrogenase maturation factor HypB. *Biochemistry* **2005**, *44*, 12229–12238. [CrossRef]

110. Dias, A.V.; Mulvihill, C.M.; Leach, M.R.; Pickering, I.J.; George, G.N.; Zamble, D.B. Structural and biological analysis of the metal sites of *Escherichia coli* hydrogenase accessory protein HypB. *Biochemistry* **2008**, *47*, 11981–11991. [CrossRef]
111. Sydor, A.M.; Liu, J.; Zamble, D.B. Effects of metal on the biochemical properties of *Helicobacter pylori* HypB, a maturation factor of [NiFe]-hydrogenase and urease. *J. Bacteriol.* **2011**, *193*, 1359–1368. [CrossRef]
112. Sasaki, D.; Watanabe, S.; Matsumi, R.; Shoji, T.; Yasukochi, A.; Tagashira, K.; Fukuda, W.; Kanai, T.; Atomi, H.; Imanaka, T.; et al. Identification and structure of a novel archaeal HypB for [NiFe] hydrogenase maturation. *J. Mol. Biol.* **2013**, *425*, 1627–1640. [CrossRef]
113. Gasper, R.; Scrima, A.; Wittinghofer, A. Structural insights into HypB, a GTP-binding protein that regulates metal binding. *J. Biol. Chem.* **2006**, *281*, 27492–27502. [CrossRef]
114. Sydor, A.M.; Lebrette, H.; Ariyakumaran, R.; Cavazza, C.; Zamble, D.B. Relationship between Ni(II) and Zn(II) coordination and nucleotide binding by the *Helicobacter pylori* [NiFe]-hydrogenase and urease maturation factor HypB. *J. Biol. Chem.* **2014**, *289*, 3828–3841. [CrossRef]
115. Maier, T.; Lottspeich, F.; Böck, A. GTP hydrolysis by HypB is essential for nickel insertion into hydrogenases of *Escherichia coli*. *Eur. J. Biochem.* **1995**, *230*, 133–138. [CrossRef]
116. Maier, T.; Jacobi, A.; Sauter, M.; Böck, A. The product of the *hypB* gene, which is required for nickel incorporation into hydrogenases, is a novel guanine nucleotide-binding protein. *J. Bacteriol.* **1993**, *175*, 630–635. [CrossRef]
117. Lacasse, M.J.; Douglas, C.D.; Zamble, D.B. Mechanism of selective nickel transfer from HypB to HypA, *Escherichia coli* [NiFe]-hydrogenase accessory proteins. *Biochemistry* **2016**, *55*, 6821–6831. [CrossRef]
118. Chan Chung, K.C.; Cao, L.; Dias, A.V.; Pickering, I.J.; George, G.N.; Zamble, D.B. A high-affinity metal-binding peptide from Escherichia coli HypB. *J. Am. Chem. Soc.* **2008**, *130*, 14056–14057. [CrossRef]
119. Chan, K.-H.; Li, T.; Wong, C.-O.; Wong, K.-B. Structural basis for GTP-dependent dimerization of hydrogenase maturation factor HypB. *PLoS ONE* **2012**, *7*, e30547. [CrossRef]
120. Cai, F.; Ngu, T.T.; Kaluarachchi, H.; Zamble, D.B. Relationship between the GTPase, metal-binding, and dimerization activities of *E. coli* HypB. *J. Biol. Inorg. Chem.* **2011**, *16*, 857–868. [CrossRef]
121. Fu, C.; Olson, J.W.; Maier, R.J. HypB protein of *Bradyrhizobium japonicum* is a metal-binding GTPase capable of binding 18 divalent nickel ions per dimer. *Proc. Natl. Acad. Sci. USA* **1995**, *92*, 2333–2337. [CrossRef]
122. Olson, J.W.; Fu, C.; Maier, R.J. The HypB protein from *Bradyrhizobium japonicum* can store nickel and is required for the nickel-dependent transcriptional regulation of hydrogenase. *Mol. Microbiol.* **1997**, *24*, 119–128. [CrossRef]
123. Rey, L.; Imperial, J.; Palacios, J.M.; Ruiz-Argüeso, T. Purification of *Rhizobium leguminosarum* HypB, a nickel-binding protein required for hydrogenase synthesis. *J. Bacteriol.* **1994**, *176*, 6066–6073. [CrossRef]
124. Olson, J.W.; Maier, R.J. Dual roles of *Bradyrhizobium japonicum* nickel in protein in nickel storage and GTP-dependent Ni mobilization. *J. Bacteriol.* **2000**, *182*, 1702–1705. [CrossRef]
125. Kovermann, M.; Schmid, F.X.; Balbach, J. Molecular function of the prolyl cis/trans isomerase and metallochaperone SlyD. *Biol. Chem.* **2013**, *394*, 965–975. [CrossRef]
126. Martino, L.; He, Y.; Hands-Taylor, K.L.; Valentine, E.R.; Kelly, G.; Giancola, C.; Conte, M.R. The interaction of the *Escherichia coli* protein SlyD with nickel ions illuminates the mechanism of regulation of its peptidyl-prolyl isomerase activity. *FEBS J.* **2009**, *276*, 4529–4544. [CrossRef]
127. Weininger, U.; Haupt, C.; Schweimer, K.; Graubner, W.; Kovermann, M.; Bruser, T.; Scholz, C.; Schaarschmidt, P.; Zoldak, G.; Schmid, F.X.; et al. NMR solution structure of SlyD from *Escherichia coli*: Spatial separation of prolyl isomerase and chaperone function. *J. Mol. Biol.* **2009**, *387*, 295–305. [CrossRef]
128. Schmidpeter, P.A.; Ries, L.K.; Theer, T.; Schmid, F.X. Prolyl isomerization and its catalysis in protein folding and protein function. *J. Mol. Biol.* **2015**, *427*, 1609–1631. [CrossRef]
129. Kaluarachchi, H.; Altenstein, M.; Sugumar, S.R.; Balbach, J.; Zamble, D.B.; Haupt, C. Nickel binding and [NiFe]-hydrogenase maturation by the metallochaperone SlyD with a single metal-binding site in *Escherichia coli*. *J. Mol. Biol.* **2012**, *417*, 28–35. [CrossRef]
130. Löw, C.; Neumann, P.; Tidow, H.; Weininger, U.; Haupt, C.; Friedrich-Epler, B.; Scholz, C.; Stubbs, M.T.; Balbach, J. Crystal structure determination and functional characterization of the metallochaperone SlyD from *Thermus thermophilus*. *J. Mol. Biol.* **2010**, *398*, 375–390. [CrossRef]

131. Quistgaard, E.M.; Weininger, U.; Ural-Blimke, Y.; Modig, K.; Nordlund, P.; Akke, M.; Löw, C. Molecular insights into substrate recognition and catalytic mechanism of the chaperone and FKBP peptidyl-prolyl isomerase SlyD. *BMC Biol.* **2016**, *14*, 82. [CrossRef]
132. Cheng, T.; Li, H.; Xia, W.; Sun, H. Multifaceted SlyD from *Helicobacter pylori*: Implication in [NiFe] hydrogenase maturation. *J. Biol. Inorg. Chem.* **2012**, *17*, 331–343. [CrossRef]
133. Kaluarachchi, H.; Sutherland, D.E.; Young, A.; Pickering, I.J.; Stillman, M.J.; Zamble, D.B. The Ni(II)-binding properties of the metallochaperone SlyD. *J. Am. Chem. Soc.* **2009**, *131*, 18489–18500. [CrossRef]
134. Hottenrott, S.; Schumann, T.; Plückthun, A.; Fischer, G.; Rahfeld, J.U. The *Escherichia coli* SlyD is a metal ion-regulated peptidyl-prolyl cis/trans-isomerase. *J. Biol. Chem.* **1997**, *272*, 15697–15701. [CrossRef]
135. Kaluarachchi, H.; Zhang, J.W.; Zamble, D.B. Escherichia coli SlyD, more than a Ni(II) reservoir. *Biochemistry* **2011**, *50*, 10761–10763. [CrossRef]
136. Lee, M.H.; Pankratz, H.S.; Wang, S.; Scott, R.A.; Finnegan, M.G.; Johnson, M.K.; Ippolito, J.A.; Christianson, D.W.; Hausinger, R.P. Purification and characterization of *Klebsiella aerogenes* UreE protein: A nickel-binding protein that functions in urease metallocenter assembly. *Protein Sci.* **1993**, *2*, 1042–1052. [CrossRef]
137. Mulrooney, S.B.; Hausinger, R.P. Sequence of the *Klebsiella aerogenes* urease genes and evidence for accessory proteins facilitating nickel incorporation. *J. Bacteriol.* **1990**, *172*, 5837–5843. [CrossRef]
138. Zambelli, B.; Banaszak, K.; Merloni, A.; Kiliszek, A.; Rypniewski, W.; Ciurli, S. Selectivity of Ni(II) and Zn(II) binding to *Sporosarcina pasteurii* UreE, a metallochaperone in the urease assembly: A calorimetric and crystallographic study. *J. Biol. Inorg. Chem.* **2013**, *18*, 1005–1017. [CrossRef]
139. Benoit, S.; Maier, R.J. Dependence of *Helicobacter pylori* urease activity on the nickel-sequestering ability of the UreE accessory protein. *J. Bacteriol.* **2003**, *185*, 4787–4795. [CrossRef]
140. Bellucci, M.; Zambelli, B.; Musiani, F.; Turano, P.; Ciurli, S. *Helicobacter pylori* UreE, a urease accessory protein: Specific Ni^{2+}- and Zn^{2+}-binding properties and interaction with its cognate UreG. *Biochem. J.* **2009**, *422*, 91–100. [CrossRef]
141. Banaszak, K.; Martin-Diaconescu, V.; Bellucci, M.; Zambelli, B.; Rypniewski, W.; Maroney, M.J.; Ciurli, S. Crystallographic and X-ray absorption spectroscopic characterization of *Helicobacter pylori* UreE bound to Ni^{2+} and Zn^{2+} reveals a role for the disordered C-terminal arm in metal trafficking. *Biochem. J.* **2012**, *441*, 1017–1035. [CrossRef]
142. Stola, M.; Musiani, F.; Mangani, S.; Turano, P.; Safarov, N.; Zambelli, B.; Ciurli, S. The nickel site of *Bacillus pasteurii* UreE, a urease metallo-chaperone, as revealed by metal-binding studies and X-ray absorption spectroscopy. *Biochemistry* **2006**, *45*, 6495–6509. [CrossRef]
143. Ciurli, S.; Safarov, N.; Miletti, S.; Dikiy, A.; Christensen, S.K.; Kornetzky, K.; Bryant, D.A.; Vandenberghe, I.; Devreese, B.; Samyn, B.; et al. Molecular characterization of *Bacillus pasteurii* UreE, a metal-binding chaperone for the assembly of the urease active site. *J. Biol. Inorg. Chem.* **2002**, *7*, 623–631. [CrossRef]
144. Brayman, T.G.; Hausinger, R.P. Purification, characterization, and functional analysis of a truncated *Klebsiella aerogenes* UreE urease accessory protein lacking the histidine-rich carboxyl terminus. *J. Bacteriol.* **1996**, *178*, 5410–5416. [CrossRef]
145. Colpas, G.J.; Hausinger, R.P. In vivo and in vitro kinetics of metal transfer by the *Klebsiella aerogenes* urease nickel metallochaperone, UreE. *J. Biol. Chem.* **2000**, *275*, 10731–10737. [CrossRef]
146. Colpas, G.J.; Brayman, T.G.; McCracken, J.; Pressler, M.A.; Babcock, G.T.; Ming, L.J.; Colangelo, C.M.; Scott, R.A.; Hausinger, R.P. Spectroscopic characterization of metal binding by *Klebsiella aerogenes* UreE urease accessory protein. *J. Biol. Inorg. Chem.* **1998**, *3*, 150–160. [CrossRef]
147. Colpas, G.J.; Brayman, T.G.; Ming, L.J.; Hausinger, R.P. Identification of metal-binding residues in the *Klebsiella aerogenes* urease nickel metallochaperone, UreE. *Biochemistry* **1999**, *38*, 4078–4088. [CrossRef]
148. Grossoehme, N.E.; Mulrooney, S.B.; Hausinger, R.P.; Wilcox, D.E. Thermodynamics of Ni^{2+}, Cu^{2+}, and Zn^{2+} binding to the urease metallochaperone UreE. *Biochemistry* **2007**, *46*, 10506–10516. [CrossRef]
149. Soriano, A.; Colpas, G.J.; Hausinger, R.P. UreE stimulation of GTP-dependent urease activation in the UreD-UreF-UreG-urease apoprotein complex. *Biochemistry* **2000**, *39*, 12435–12440. [CrossRef]
150. Song, H.K.; Mulrooney, S.B.; Huber, R.; Hausinger, R.P. Crystal structure of *Klebsiella aerogenes* UreE, a bickel-binding metallochaperone for urease activation. *J. Biol. Chem.* **2001**, *276*, 49359–49364. [CrossRef]
151. Remaut, H.; Safarov, N.; Ciurli, S.; Van Beeumen, J. Structural Basis for Ni^{2+} transport and assembly of the urease active site by the metallochaperone UreE from *Bacillus pasteurii*. *J. Biol. Chem.* **2001**, *276*, 49365–49370. [CrossRef]

152. Shi, R.; Munger, C.; Asinas, A.; Benoit, S.L.; Miller, E.; Matte, A.; Maier, R.J.; Cygler, M. Crystal structures of apo and metal-bound forms of the UreE protein from *Helicobacter pylori*: Role of multiple metal binding sites. *Biochemistry* **2010**, *49*, 7080–7088. [CrossRef]
153. Khorasani-Motlagh, M.; Noroozifar, M.; Kerman, K.; Zamble, D.B. Complex formation between the *Escherichia coli* [NiFe]-hydrogenase nickel maturation factors. *BioMetals* **2019**, *32*, 521–532. [CrossRef]
154. Leach, M.R.; Zhang, J.W.; Zamble, D.B. The role of complex formation between the *Escherichia coli* hydrogenase accessory factors HypB and SlyD. *J. Biol. Chem.* **2007**, *282*, 16177–16186. [CrossRef]
155. Stingl, K.; Schauer, K.; Ecobichon, C.; Labigne, A.; Lenormand, P.; Rousselle, J.-C.; Namane, A.; de Reuse, H. In vivo interactome of *Helicobacter pylori* urease revealed by tandem affinity purification. *Mol. Cell. Proteomics* **2008**, *7*, 2429–2441. [CrossRef]
156. Cheng, T.; Li, H.; Yang, X.; Xia, W.; Sun, H. Interaction of SlyD with HypB of *Helicobacter pylori* facilitates nickel trafficking. *Metallomics* **2013**, *5*, 804–807. [CrossRef]
157. Chan Chung, K.C.; Zamble, D.B. The Escherichia coli metal-binding chaperone SlyD interacts with the large subunit of [NiFe]-hydrogenase 3. *FEBS Lett.* **2011**, *585*, 291–294. [CrossRef]
158. Chan Chung, K.C.; Zamble, D.B. Protein Interactions and localization of the Escherichia coli accessory protein HypA during nickel insertion to [NiFe] hydrogenase. *J. Biol. Chem.* **2011**, *286*, 43081–44390. [CrossRef]
159. Yang, X.; Li, H.; Cheng, T.; Xia, W.; Lai, Y.-T.; Sun, H. Nickel translocation between metallochaperones HypA and UreE in *Helicobacter pylori*. *Metallomics* **2014**, *6*, 1731–1736. [CrossRef]
160. Yang, X.; Li, H.; Lai, T.P.; Sun, H. UreE-UreG complex facilitates nickel transfer and preactivates GTPase of UreG in *Helicobacter pylori*. *J. Biol. Chem.* **2015**, *290*, 12474–12485. [CrossRef]
161. Fong, Y.H.; Wong, H.C.; Yuen, M.H.; Lau, P.H.; Chen, Y.W.; Wong, K.-B. Structure of UreG/UreF/UreH complex reveals how urease accessory proteins facilitate maturation of *Helicobacter pylori* Urease. *PLoS Biol.* **2013**, *11*, e1001678. [CrossRef]
162. Moncrief, M.B.; Hausinger, R.P. Purification and activation properties of UreD-UreF-urease apoprotein complexes. *J. Bacteriol.* **1996**, *178*, 5417–5421. [CrossRef]
163. Soriano, A.; Hausinger, R.P. GTP-dependent activation of urease apoprotein in complex with the UreD, UreF, and UreG accessory proteins. *Proc. Natl. Acad. Sci. USA* **1999**, *96*, 11140–11144. [CrossRef]
164. Giedroc, D.P.; Arunkumar, A.I. Metal sensor proteins: Nature's metalloregulated allosteric switches. *Dalt. Trans.* **2007**, *29*, 3107–3120. [CrossRef]
165. Eaton, K.A.; Krakowka, S. Effect of gastric pH on urease-dependent colonization of gnotobiotic piglets by *Helicobacter pylori*. *Infect. Immun.* **1994**, *62*, 3604–3607.
166. Eaton, K.A.; Brooks, C.L.; Morgan, D.R.; Krakowka, S. Essential role of urease in pathogenesis of gastritis induced by *Helicobacter pylori* in gnotobiotic piglets. *Infect. Immun.* **1991**, *59*, 2470–2475.

© 2019 by the author. Licensee MDPI, Basel, Switzerland. This article is an open access article distributed under the terms and conditions of the Creative Commons Attribution (CC BY) license (http://creativecommons.org/licenses/by/4.0/).

Review

The Maturation Pathway of Nickel Urease

Yap Shing Nim and Kam-Bo Wong *

School of Life Sciences, Centre for Protein Science and Crystallography, State Key Laboratory of Agrobiotechnology, The Chinese University of Hong Kong, Hong Kong 999077, China
* Correspondence: kbwong@cuhk.edu.hk

Received: 7 June 2019; Accepted: 4 July 2019; Published: 6 July 2019

Abstract: Maturation of urease involves post-translational insertion of nickel ions to form an active site with a carbamylated lysine ligand and is assisted by urease accessory proteins UreD, UreE, UreF and UreG. Here, we review our current understandings on how these urease accessory proteins facilitate the urease maturation. The urease maturation pathway involves the transfer of Ni^{2+} from UreE → UreG → UreF/UreD → urease. To avoid the release of the toxic metal to the cytoplasm, Ni^{2+} is transferred from one urease accessory protein to another through specific protein–protein interactions. One central theme depicts the role of guanosine triphosphate (GTP) binding/hydrolysis in regulating the binding/release of nickel ions and the formation of the protein complexes. The urease and [NiFe]-hydrogenase maturation pathways cross-talk with each other as UreE receives Ni^{2+} from hydrogenase maturation factor HypA. Finally, the druggability of the urease maturation pathway is reviewed.

Keywords: urease maturation; metallochaperone; nickel; G-protein; conformational change

1. Introduction

Urease catalyzes the hydrolysis of urea into carbon dioxide and carbamate, which spontaneously decomposes into ammonia and another carbon dioxide. The enzyme is involved in nitrogen metabolism that is found in bacteria, archaea, fungi, plants, and some invertebrates [1–3]. While the Jack bean (*Canavanlia ensiformis*) urease was the first nickel enzyme identified [4] and the first enzyme to be crystallized [5], the mechanism of urease maturation is the most well studied in bacteria, in particular *Klebsiella aerogenes* and *Helicobacter pylori*. Urease has been implicated in the pathogenesis of bacterial infections. For example, *H. pylori* can colonize in the acidic stomach due to the ureolytic activity of urease [6]. *H. pylori* infection increases the risk of peptic ulcer and gastric cancer [7,8]. Most of the ureases discovered so far are nickel enzymes. One notable exception is an iron urease found in *Helicobacter mustelae* in ferret stomach [9].

The concentration of free nickel ions is tightly regulated in cells because Ni^{2+} can inactivate enzymes by displacing weaker ions such as Mg^{2+} in the active sites [10,11]. To avoid cytotoxicity, cells have to evolve a mechanism to deliver nickel from one protein to another without releasing the toxic metal to the cytoplasm. In the urease maturation pathway, there are four urease accessory proteins, UreD, UreE, UreF and UreG, involved in the nickel delivery. In this article, we review how these metallochaperones interact with each other and with urease to facilitate the transfer of nickel ions in the urease maturation pathway. Correct metalation is ensured by specific protein–protein interactions that are allosterically regulated by binding/hydrolysis of guanosine triphosphate (GTP). We also review the cross-talking between the maturation pathways of ureases and [NiFe]-hydrogenases, urease maturation in plants and the potential of the urease maturation pathway in antibacterial drug discovery. Finally, some unanswered questions on the urease maturation pathway are discussed.

2. Structures of Urease

The urease sequence is highly conserved. Most of the bacterial ureases are comprised of three subunits [12]. For example, *K. aerogenes* urease has three subunits: α-subunit (UreC), β-subunit (UreB) and γ-subunit (UreA) (Figure 1A) [13]. In *H. pylori*, the two smaller β and γ subunits are fused to form the UreA, and the large α-subunit is renamed as UreB [14]. In plants, all three subunits are fused into one polypeptide chain as UreA [15]. In *K. aerogenes* [16,17] and *Sporosarcina pasteurii* [18], the urease contains three catalytic sites constituted by the α, β and γ subunits. Three copies of the subunits UreABC form a trimeric disc-like structure (Figure 1A). In plant urease, the disc-like structure dimerizes to form a hexameric urease (UreA$_3$)$_2$ (Figure 1B) [19]. *H. pylori* UreA contains a 27-residue C-terminal extension that is responsible for the formation of a dodecameric quaternary structure (UreAB$_3$)$_4$ [20].

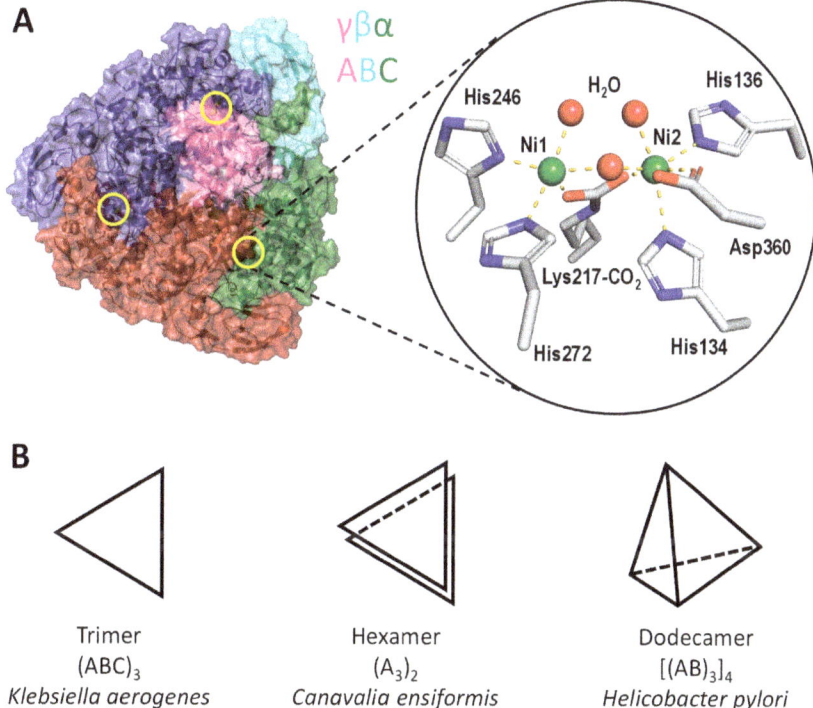

Figure 1. Structures of ureases. (**A**) Crystal structure of *K. aerogenes* urease (PDB: 1FWJ [16]). The basic catalytic unit of ureases consists of α, β and γ subunits encoding by *ureC*, *ureB* and *ureA*, respectively. In *K. aerogenes* and *S. pasteurii*, three copies of UreABC form a trimeric disk-like structure. The positions of three actives sites are circled in yellow and zoomed in at the right. The two nickel ions are coordinated by a carbamylated lysine residue, four histidine, one aspartate, and three water molecules. (**B**) The trimeric urease, schematically represented as triangles, forms the basic repeating unit of ureases to form more complex quaternary structures. In plant ureases, such as *Canavalia ensiformis* (PDB: 3LA4 [19]), two urease trimers are stacked together in a C2 symmetry to form a hexameric quaternary structure. In *H. pylori*, four trimers assemble in a tetrahedral symmetry to form a dodecameric urease (PDB: 1E9Z [20]).

The urease active site is located in the α-subunit and is highly conserved. It contains two Ni^{2+} ions [16,18–20] bridged by the carboxyl group of a carbamylated lysine [17,20,21], and chelated by four histidine residues and one aspartate residue (Figure 1A). Purified *K. aerogenes* urease can be activated in vitro by incubation with high concentration of bicarbonate and Ni^{2+} [22]. The carbamylation is likely to come from carbon dioxide, instead of bicarbonate as suggested by the result of pH jump experiment [22]. Mutagenesis studies suggest that the carbamylated lysine residues are essential to urease maturation [23]. Urease activity can be inhibited by Zn^{2+}, Cu^{2+}, Co^{2+} and Mn^{2+} [24].

3. Genetic Studies Showed the Importance of Urease Accessory Proteins

In *K. aerogenes*, the urease operon is arranged as *ureDABCEFG* [25,26] (Figure 2). *ureA*, *ureB* and *ureC* respectively encode the γ, β and α subunits of the urease structural genes. *ureD*, *ureE*, *ureF* and *ureG* encode for the urease accessory proteins essential for the maturation of urease. In *H. pylori*, the operon is arranged as *ureABIEFGH* [27,28]. There is a unique *ureI* encoding for an acid-gated urea channel for colonization in the acidic stomach [29,30]. *ureH* is an orthologue of *ureD*. For simplicity, UreD is used in this article to denote the protein product of *ureH* in *H. pylori* or *ureD* in other species.

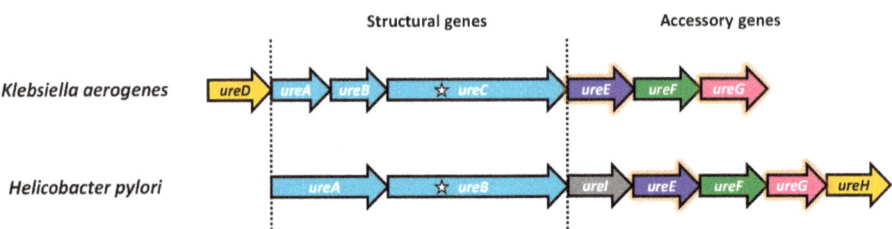

Figure 2. Comparison of *K. aerogenes* and *H. pylori* urease operons. Orthologous genes are indicated using the same color. The locations of the active-site lysine residue involved in carbamylation and nickel binding are indicated as stars. *H. pylori* contains an extra *ureI* gene encoding for an acid-gated urea channel.

Bacterial cells harboring the urease operon showed nickel-dependent activation of urease [26,31,32]. Urease purified from cells containing the intact urease operon was active, whereas the urease purified from cells containing only the urease structural genes was inactive [33,34]. This observation suggests that the urease maturation is assisted by the urease accessory proteins. Deletion [26], knockout [35] and transposon mutagenesis [28] of individual accessory genes, *ureG*, *ureF* and *ureD*, abolished urease activity. The activity could be partially [26] or fully [32] recovered by gene complementation. Deletion of *ureE* either lowered [26] or abolished the urease activity [32,35,36], which was partially regenerated by adding nickel [32]. These results suggested that UreG, UreF and UreD are absolutely required for urease maturation, while UreE facilitates the process.

4. The Formation of UreGFD Complex

Urease apoprotein was shown to interact with UreD [37,38], UreF/UreD [39,40], or UreG/UreF/UreD [38,40–42]. It has been known that UreD and UreF can form a UreFD complex, which can then recruit UreG to form a UreGFD complex [43]. The UreGFD complex can form an activation complex [43,44] with urease apoprotein and activate urease in a GTP-dependent manner [43,44]. Substitution of the P-loop residues of UreG abolished its ability to activate urease [45,46].

The structural studies provided insights into how UreD, UreF and UreG interact with each other to form the protein complexes required for urease maturation. UreD forms a complex with UreF [35,38,40,47]. Crystal structure of the *H. pylori* UreFD complex revealed that it is a 2:2 heterodimer with UreF at the middle providing the dimerization interface (Figure 3A) [48]. UreD bound to both ends of the UreF dimer, forming a rod-shape head-to-head dimer of heterodimer UreFD (Figure 3A). UreF contains highly conserved residues at its C-terminal tail, which is unstructured and susceptible to proteolytic cleavage when expressed alone [48,49]. The C-terminal tail of UreF was shown to be essential for the interaction with UreD [40,48], the assembly of an activation complex [40], and urease activation [40,48]. Upon binding to UreD, these C-terminal residues become structured and form an extra helix-10 and a loop structure stabilized by hydrogen bonds involving a conserved Arg-250 residue [48]. These conformational changes were shown to be important for recruiting UreG to form the UreGFD complex by mutagenesis studies. For example, the R250A variant abolished the formation of UreGFD complex and urease activation [43].

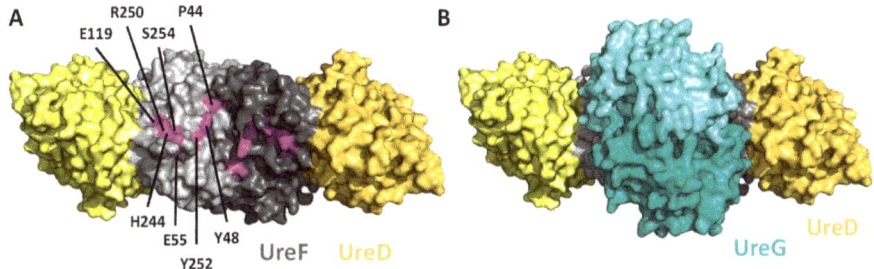

Figure 3. Dimerization of UreF creates a saddle-like structure for recruiting UreG to the UreGFD complex. (**A**) Crystal structure of the UreFD complex (PDB: 3SF5 [48]). UreF (grey) interacts with UreD (yellow) to form a 2:2 heterodimer in a rod-shape topological arrangement of D:F:F:D. A saddle-like structure is formed at the UreF dimer surface. Invariant residues of UreF that have been shown to be essential to UreG binding and urease activation are highlighted in purple and are numbered according to the *H. pylori* sequence [48,50]. (**B**) Crystal structure of the UreGFD complex showing that UreG dimer (cyan) binds to the saddle-like structure of UreF (PDB: 4HI0 [43]).

The crystal structure of the *H. pylori* UreGFD complex (PDB: 4HI0) revealed that it is a dimer of the heterotrimers of UreGFD [43] (Figure 3B). The complex contains a guanosine diphosphate (GDP)-bound UreG dimer sitting on the UreF wherein the dimerization axes of UreG and UreFD are almost perpendicular to each other. UreF-mediated dimerization is required to provide a complete UreG binding site. Substitutions that broke the dimerization of UreF abolished the recruitment of UreG to the UreGFD complex and urease maturation [40,43]. UreG binds to a saddle-like structure of UreF consisting of clusters of conserved residues [43]—some of them have been identified to be important for recruiting UreG to the activation complex and for activating urease [50] (Figure 3A).

5. UreG Dissociates from the UreGFD Complex and Forms a Dimer in the Presence of Ni/GTP

That UreG undergoes Ni/GTP-dependent dimerization was identified when Ni^{2+} and GTP were added to the GDP-bound UreGFD complex, which was then dissociated into a UreG dimer and the UreFD complex [43]. UreG is a nickel chaperone [43,51,52] and a SIMIBI (after Signal recognition particle, MinD and BioD) class GTPase [53]. UreG remains as a monomer in the absence of GTP and it binds Ni^{2+} with lower affinity [54,55]. UreG only dimerizes when both GTP and Ni^{2+} were present, and the UreG dimer binds one Ni^{2+} per dimer with a K_d of 0.36 µM [43,52]. Zn^{2+} can also induce dimerization of UreG [54,56,57], but the Zn/UreG dimer was not stable and dissociated to monomer when the excess Zn^{2+} was removed by gel filtration [43]. Moreover, this Zn/UreG dimer is inactive in GTP hydrolysis [52]. Dimerization is required for GTP hydrolysis. The cysteine and histidine in the conserved CPH motif (Cys-Pro-His) are important for UreG dimerization and nickel binding [43,54], as well as urease activation [43,46,58]. After GTP hydrolysis, the UreG dimer dissociates back to monomer and releases one Ni^{2+}, providing a plausible mechanism for coupling GTP hydrolysis to nickel delivery [43]. GTP-dependent dimerization [59–63] and conformational changes [64] were also observed in HypB, another SIMIBI GTPase involved in the [NiFe]-hydrogenase maturation pathway.

Structural insights into how Ni^{2+} and GTP induce dimerization of UreG were provided by the crystal structure of the UreG dimer in complex with Ni^{2+} and GMPPNP, a nonhydrolyzable analogue of GTP [51]. The GTP binding pocket is sandwiched between two UreG chains and a Ni^{2+} ion is coordinated by the conserved CPH motif from each chain [51]. The structure of Ni/GMPPNP-bound

UreG dimer is compared to that of the GDP-bound UreG in the UreGFD complex (Figure 4). Upon GTP binding, the γ-phosphate of GTP introduces a charge–charge repulsion on Asp37 in the G2 switch, initiating a swinging motion of helix-2, and Glu42 forms a hydrogen bond with Arg130 of the opposite chain. Consequently, the zip-up motion of β2 and β3 propagates the conformational changes to the CPH motif, where Cys66 and His68 reorientate towards the dimeric interface to chelate a Ni^{2+} ion in a square-planar geometry (Figure 4B) [51]. The structural changes observed also explain why UreG dissociates from the UreGFD complex upon GTP binding as residue Tyr39 in the G2 region swings outward and makes steric clashes with UreF [48].

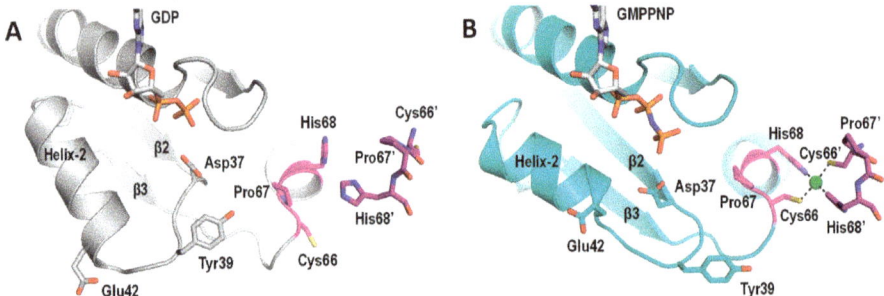

Figure 4. Guanosine triphosphate (GTP)-dependent conformational changes of UreG. (**A**) The structures of GDP-bound *H. pylori* UreG (PDB: 4HI0 [43]) is compared to the (**B**) Ni/GMPPNP-bound *K. pneumoniae* UreG (PDB: 5XKT [51]). Invariant residues involved in the conformational changes and nickel binding are indicated and numbered according to the *H. pylori* sequence. Residues in the opposite chain are numbered with apostrophes. Noteworthy, Cys66 and His68 of the CPH motif (magenta) are pointing away from each other in the GDP-bound UreG. Upon binding of GMPPNP, they reorient to form a square-planar nickel binding site at the dimeric interface. Moreover, the swinging motion of Tyr39 creates steric clashes that induce dissociation of UreG from the UreGFD complex.

6. UreG–UreE Interaction Is GTP-Dependent

UreG interacts with UreE in 2:2 ratio to form a $UreE_2G_2$ complex in the presence of GTP and Mg^{2+}. The formation of $UreE_2G_2$ is independent of Zn^{2+} and Ni^{2+} [52]. In the presence of GDP, the complex dissociates into the $UreE_2G$ complex and a UreG monomer [52]. Interestingly, addition of Ni^{2+} and GTPγS to the UreGFD complex and UreE promotes UreG to switch its protein binding partner from UreFD to UreE, resulting in the formation of the $UreE_2G_2$ and UreFD complexes [51]. The D37A/E42A variant of UreG failed to dissociate from the UreGFD complex and formed the Ni/GTP-bound UreG dimer, presumably disrupting the conformational changes induced by GTP binding [51]. Interestingly, the D37A/E42A variant also failed to form the $UreE_2G_2$ complex in the presence of Ni/GTP, further supporting that the conformational changes in UreG are important for the UreE–UreG interaction [51]. The formation of the $UreE_2G_2$ complex facilitates the transfer of Ni^{2+} from UreE to UreG. By monitoring the thiolate-to-Ni^{2+} transition at 337 nm, it has been shown that Ni^{2+} was transferred from UreE to UreG within the $UreE_2G_2$ complex in its GTP-bound state, but not in the $UreE_2G$ complex in the presence of GDP [52].

How UreE interacts with UreG is not known. Crystal structures of UreE from *S. pasteurii* [65], *K. aerogenes* [66] and *H. pylori* [67,68] have been solved, and share high structural homology. UreE exists as a dimer in solution [68–71]. High protein concentration and the presence of Zn^{2+}, Cu^{2+} or Ni^{2+} can induce UreE tetramerization [65,72–74], but mutagenesis studies showed that the formation of the tetramer was not essential to urease activation [68]. The C-terminal domain of two UreE chains interact with each other to form a dimer. At the dimeric interface, UreE has one conserved metal-binding site formed by His102 (numbered according to the *H. pylori* sequence) in a GNXH motif from each of the UreE chain [65–68]. This central histidine is essential to urease maturation [68,75,76].

The variable C-terminus histidine-rich tail of UreE [76] has been shown to bind Ni^{2+} (residue 143–157 in *K. aerogenes* [70,72,77]; residue 137–147 in *S. pasteurii* [73]). In the crystal structure of *H. pylori* UreE, an extra histidine (His152) of this C-terminal tail was shown to bind a Ni^{2+} or Zn^{2+} at the dimeric interface (Figure 5B) [67,68]. Neither the H102A substitution nor truncation of the variable C-terminal tail (residue 158–170 of *H. pylori* UreE) affect the formation of the $UreE_2G_2$ complex [52]. Truncation of the C-terminal tail of *K. aerogenes* UreE showed a 25–60% decrease in urease activation [77,78]. The functional role of the C-terminal tail is unclear. However, it has been proposed that the C-terminal tail may play a role in regulating the binding and release of Ni^{2+} [65,67]. Based on charge and shape complementarity, models of how UreE interacts with UreG have been proposed [71,79,80]. The models predict that the nickel binding sites of UreE and UreG should point towards each other, which is supported by the mutagenesis studies that show R101A UreE or C66A UreG destabilized the $UreE_2G_2$ complex [52].

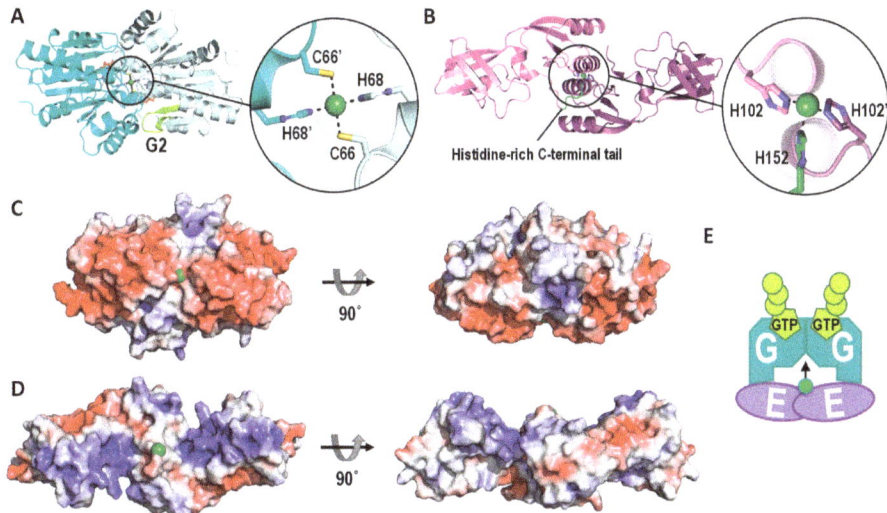

Figure 5. Structures of UreE and UreG. Cartoon representation of *K. pneumoniae* Ni/GMPPNP-bound UreG (**A**, PDB: 5XKT [51]) and *H. pylori* Ni-bound UreE (**B**, PDB: 3TJ8 [67]). Both UreG and UreE bind a nickel ion at the dimeric interface. The surface electrostatic potentials of (**C**) UreG and (**D**) UreE were calculated using the APBS program [81] and are color coded (red, −5 kT/e; blue, +5 kT/e). (**E**) It has been suggested that UreG and UreE are complementary in charge and shape and are likely to form a $UreE_2G_2$ complex with their nickel binding sites pointing towards each other [71,80].

7. How Urease Accessory Proteins Facilitate Urease Maturation

GTP-dependent conformational changes of UreG provide a mechanism where GTP binding/hydrolysis facilitates the delivery of nickel along the urease maturation pathway (Figure 6). It has been shown that the Ni/UreE dimer, providing the sole nickel source, can activate urease in the presence of UreGFD and GTP [51]. Binding of GTP induces UreG to dissociate from the UreGFD complex and bind with UreE to form the $UreE_2G_2$ complex. The $UreE_2G_2$ complex, which can also activate urease in the presence of UreFD complex [51], facilitates the transfer of nickel from UreE to UreG [52]. Direct protein–protein interactions among the urease accessory proteins are required as separating Ni/UreE and UreGFD/urease by a dialysis membrane abolished the urease activation in vitro [51]. After UreG gets its Ni^{2+}, it can interact with UreFD and urease apoprotein to form the activation complex [43]. Mutagenesis studies suggested that UreG binds to UreFD in the activation complex [46,50]. That Ni/GTP-bound UreG, which dissociates from the UreFD complex, can interact with UreFD in the activation complex suggests that UreFD may undergo conformational changes upon binding of urease apoprotein to accommodate Ni/GTP-bound UreG in the activation complex. Whether UreE is involved in this activation complex is unclear. It has been shown that UreG can pull down UreE, UreD, UreF and urease [46,82]. Given that UreG can also interact with UreE, this observation did not conclusively demonstrate the interaction of UreE to the activation complex. Furthermore, it has been shown that urease can be activated in the presence of Ni/UreG, UreFD and urease apoprotein without UreE [51], suggesting that UreE is not essential in the formation of the activation complex.

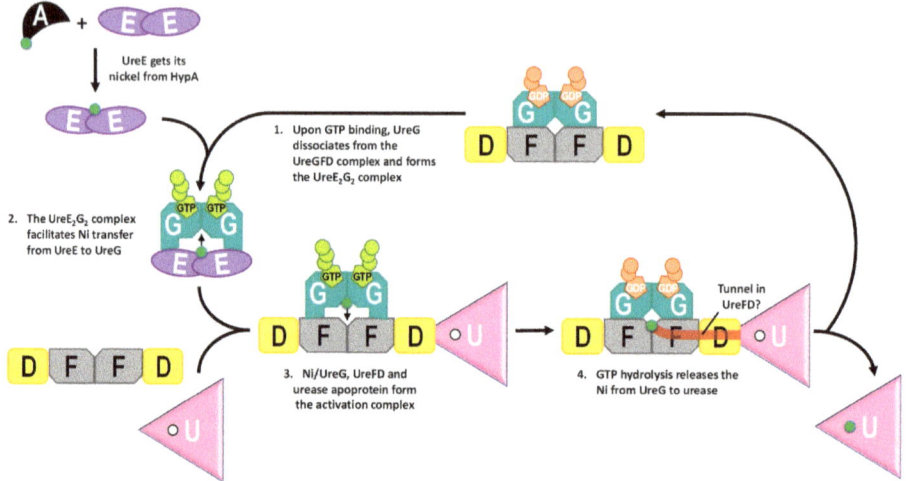

Figure 6. The urease maturation pathway.

It is well established that GTP hydrolysis is essential to urease maturation [41,83]. Substituting the conserved Lys-20 and Thr-21 residues to alanine in the P-loop region of K. aerogenes UreG abolished its ability to bind ATP-linked agarose, the formation of the activation complex and urease maturation [45]. As the Ni/UreG dimer will dissociate into monomers and release the bound Ni^{2+} upon GTP hydrolysis [43], the GTPase activity of UreG must be tightly regulated to prevent nonproductive release of Ni^{2+} outside the activation complex. It has been shown that the GTPase activity is only detectable by addition of bicarbonate [43]. One hypothesis is that bicarbonate can serve as one of the substrates of UreG that results in the formation of an intermediate of a carboxy-phosphate [41]. The effect of bicarbonate on the UreG activity exhibited a classical Michaelis–Menten kinetics ([52], Figure 7). Using a saturated concentration of GTP, we estimated the K_m value of bicarbonate was 13 ± 4 mM. In addition to bicarbonate, GTPase activity of UreG can also be increased by ~2–3 fold by addition of UreE, NH_4^+ or K^+ [52]. Even under the most favorable conditions, the activity of UreG is still very low. In our hands, the turnover number (k_{cat}) of H. pylori UreG is ~0.006 s^{-1} at 37 °C in the presence of K^+ and saturated concentration of GTP (Figure 7). The low intrinsic activity of UreG should be of advantage as it prevents premature hydrolysis that releases the bound Ni^{2+} to the cytoplasm. It is unclear how and when the GTPase activity of UreG is activated. One possibility is that the GTPase activity of UreG is stimulated by the conformational changes induced during the formation of the activation complex. Conformational changes in the formation of the UreFD/urease complex have been suggested by cross-linking experiments [84].

GTP hydrolysis changes the conformation of the CPH motif in such a way that the square-planer coordination by Cys66 and His68 is disrupted (Figure 4), promoting the release of Ni^{2+} in the activation complex. Interestingly, it has been shown that interaction between UreG and UreF is essential to the urease maturation [46,48,50]. Mutations that broke the UreG–UreF interaction also abolished urease maturation, suggesting that UreG is likely bound to the activation complex via the UreFD complex.

As the nickel binding site of UreG is far away from the active site of urease, it is not fully understood how the Ni^{2+} is transferred from UreG to urease after GTP hydrolysis. One intriguing hypothesis suggested that the Ni^{2+} reaches urease through a tunnel within the core of the UreFD complex [82,85,86]. Large cavities have been identified in the structures of *H. pylori* UreFD complex [82,85,86]. The cavities form a tunnel, which is wide enough for Ni^{2+} to pass through, connecting the nickel binding site of UreG to an exit at UreD. The tunnel hypothesis is supported by: (1) Both UreF and UreD were able to bind Ni^{2+} [38,86]; (2) Tunnel-disrupting variants of UreD were shown to greatly reduce the urease maturation without affecting the UreD–urease interaction or the formation of the UreGFD complex [82,86]. As UreD is responsible for binding urease in the activation complex, the tunnel within the UreFD complex should provide a mechanism for the transfer of Ni^{2+} from UreG to the urease within the activation complex.

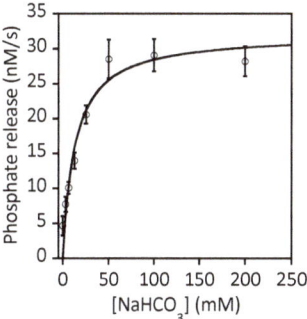

Figure 7. Dependency of GTPase activity of UreG on bicarbonate concentration. GTPase activity of *H. pylori* UreG (5 µM) was measured at 37 °C in 0–200 mM $NaHCO_3$ at saturated concentration of GTP (2 mM) in 200 mM KCl, 2.5 µM $NiSO_4$, 2 mM $MgSO_4$, 1 mM TCEP, 20 mM HEPES pH 7.5 by following the concentration of phosphate released as described previously [87]. Fitting the data to the Michaelis–Mention equation yielded a K_m value of 13 ± 4 mM and a k_{cat} value of $6.4 \pm 0.6 \times 10^{-3}$ s^{-1}.

8. UreE Gets Its Nickel from Cross-Talking to the Hydrogenase Maturation Pathway

Hydrogenase maturation factors HypA and HypB, which are responsible for the delivery of nickel ions to the active site of [NiFe]-hydrogenase [88–94], are also essential for urease maturation [31,90]. Genes of hydrogenase maturation factors were first identified in *Escherichia coli* [94,95], where *hypA* and *hypB* are encoded in the *hyp* operon. In *H. pylori*, the two genes are located in different locations in the chromosome [96]. Urease activity was greatly reduced in *hypA* or *hypB* knockout strains of *H. pylori* [31,90]. The urease activity could be restored by addition of Ni^{2+} in the growth medium. These observations suggest that the hydrogenase maturation factors HypA and HypB are required for the urease maturation. Structures of HypA [97,98] revealed a mixed α/β topology containing a nickel binding domain with micromolar affinity [99–102], and a zinc binding domain [98,100,102–105] with nanomolar affinity [102]. Residues coordinating the Zn^{2+} [100,105] and Ni^{2+} [36,99,101] are essential for urease maturation. HypA interacts with the C-terminal domain of UreE [52,106–109] to form a 1:2 complex [52,101,107,108]. The complex formation creates a unique nanomolar affinity nickel-binding site, which is not found in any of the individual proteins [107]. Ni^{2+} can be transferred from HypA to UreE in the complex [52,108]. It has been shown that HypA and UreG compete with each other for UreE [109]. In the presence of Mg^{2+} and GTP, UreE dissociates from the HypA/$UreE_2$ complex and interacts with UreG to form the $UreE_2G_2$ complex, suggesting the switching of the protein binding partners is GTP-dependent [52]. It has been demonstrated that Ni^{2+} can be transferred from HypA to UreE to UreG [52]. HypA, on the other hand, obtains its Ni^{2+} from HypB [60,110] with the help of SlyD [106,111–114] of the hydrogenase maturation pathway, which has been reviewed in [115,116].

9. Urease Maturation Pathway in Plants

In plants, urease and its accessory genes contain introns and are scattered on different chromosomes. The urease structural genes and homologues of UreD, UreF and UreG have been identified in plants [117,118]. Similar to the maturation of bacterial ureases, the formation of the UreGFD/urease activation complexes was shown to be essential to the maturation of ureases in rice (*Oryza sativa*) [119,120] and *Arabidopsis* [120,121]. Plant UreG, previously named as p32, is encoded by *Eu3* [122]. It has a GTPase domain homologous to bacterial UreG and an histidine-rich N-terminal extension and a plant-specific region containing two HXH motifs [120]. The two HXH motifs, but not the histidine-rich extension, were found to be essential for urease activation [120]. Interestingly, no UreE homologue was found in plants. It was postulated that the plant UreG, with the plant specific HXH motifs, may combine the function of bacterial UreE and UreG into one protein [56,120–122].

10. Urease Maturation Pathway Is Druggable

Urease is a microbial virulence factor for gastric infection by *H. pylori*, urinary tract infection by *Proteus mirabilis* and *Klebsiella* strains, and lung infection by fungus *Cryptococcus neoformans* [123]. The first line treatment of *H. pylori* infection is combining proton-pump inhibitors with antibiotics in the triple therapy [124]. However, the emerging antibiotic resistance make the development of new drugs pressing. Recent research findings suggest that the urease maturation pathway could become a new druggable target [65,123]. Bismuth is a heavy metal that is effective for *H. pylori* eradication [124]. Recent study showed that colloidal bismuth subcitrate inhibits the biosynthesis of active *H. pylori* urease by displacing Ni^{2+} in UreG [125]. Virtual screening has identified two compounds that inhibit the GTPase activity of UreG, reduce urease activity with higher potency than clinically using acetohydroxamic acid, and suppress *H. pylori* infection in a mammalian cell model [125]. These results demonstrate that the urease maturation pathway is a novel druggable target for treatment of *H. pylori* infection.

11. Future Perspectives

While major progress has been made on how urease accessory proteins interplay to deliver Ni^{2+} to ureases, some interesting questions remain unanswered. (1) The structure of the activation complex is not known despite several structural models that have been proposed previously [126,127]. A high-resolution structure of the activation complex should provide novel structural insights into how GTPase activity of UreG is stimulated and the mechanism of nickel transfer within the complex. (2) The mechanism of carbamylation of the active site lysine is relatively understudied. Does it take place before or after the insertion of Ni^{2+}? Is the carbamylation reaction catalyzed by other factors? (3) Urease requires two Ni^{2+} to function. Since only one Ni^{2+} ion is carried per UreG dimer, it takes two round trips of UreG to activate urease. MBP-UreD and MBP-UreFD were shown to dissociate from urease after activation [37,39]. It is unclear what triggers the dissociation of the activation complex after urease activation. (4) What could be the driving force for the transfer of Ni^{2+} from UreG to the urease through the tunnel? How is the free energy change of GTP hydrolysis coupled to the nickel transfer? (5) Unlike the release of Ni^{2+} from UreG that can be explained by GTP hydrolysis, it is not known why nickel is transferred unidirectionally from HypA to UreE and from UreE to UreG. Structure determination of the HypA/$UreE_2$ and $UreE_2G_2$ complexes should help to provide structural insights to this question.

12. Conclusions

The urease maturation pathway represents the most well-studied paradigm on how nickel ions are delivered from one protein to another through specific protein–protein interactions. One central theme depicts the binding/hydrolysis of GTP allosterically regulate the switching of protein-binding partners and the transfer of nickel ions from UreE → UreG → UreF/UreD → urease. This mechanism ensures that the nickel ions are always protein-bound to avoid leaking of the toxic metal to the cytoplasm. UreE also receives its nickel ions from interacting with HypA of the hydrogenase maturation pathway. A better understanding of the urease maturation pathway allow us to develop drugs against the *H. pylori* infection.

Funding: This work was funded by the grants from the Research Grants Council of Hong Kong (14117314, AoE/M-05/12, and AoE/M-403/16) and from the Research Committee of The Chinese University of Hong Kong (3132814 and 3132815).

Conflicts of Interest: The authors declare no conflicts of interest.

References

1. Krajewska, B. Ureases I. Functional, Catalytic and Kinetic Properties: A Review. *J. Mol. Catal. B Enzym.* **2009**, *59*, 9–21. [CrossRef]
2. Hanlon, D.P. The Distribution of Arginase and Urease in Marine Invertebrates. *Comp. Biochem. Physiol. B Biochem. Mol. Biol.* **1975**, *52*, 261–264. [CrossRef]
3. Alonso-Saez, L.; Waller, A.S.; Mende, D.R.; Bakker, K.; Farnelid, H.; Yager, P.L.; Lovejoy, C.; Tremblay, J.-E.; Potvin, M.; Heinrich, F.; et al. Role for Urea in Nitrification by Polar Marine Archaea. *Proc. Natl. Acad. Sci. USA* **2012**, *109*, 17989–17994. [CrossRef] [PubMed]
4. Dixon, N.E.; Gazzola, C.; Watters, J.J.; Blakeley, R.L.; Zerner, B. Jack Bean Urease (EC 3.5.1.5). A Metalloenzyme. A Simple Biological Role for Nickel? *J. Am. Chem. Soc.* **1975**, *97*, 4131–4133. [CrossRef] [PubMed]
5. Sumner, J.B. The Isolation and Crystallization of the Enzyme Urease. *J. Biol. Chem.* **1926**, *69*, 435–442.
6. Kao, C.-Y.; Sheu, B.-S.; Wu, J.-J. *Helicobacter pylori* infection: An overview of bacterial virulence facztors and pathogenesis. *Biomed. J.* **2016**, *39*, 14–23. [CrossRef] [PubMed]
7. Ishaq, S.; Nunn, L. *Helicobacter pylori* and Gastric Cancer: A State of the Art Review. *Gastroenterol. Hepatol. Bed Bench* **2015**, *8*, S6–S14. [PubMed]
8. Narayanan, M.; Reddy, K.M.; Marsicano, E. Peptic Ulcer Disease and *Helicobacter pylori* Infection. *Mo. Med.* **2018**, *115*, 219–224. [PubMed]
9. Carter, E.L.; Tronrud, D.E.; Taber, S.R.; Karplus, P.A.; Hausinger, R.P. Iron-Containing Urease in a Pathogenic Bacterium. *Proc. Natl. Acad. Sci. USA* **2011**, *108*, 13095–13099. [CrossRef]
10. Capdevila, D.A.; Edmonds, K.A.; Giedroc, D.P. Metallochaperones and Metalloregulation in Bacteria. *Essays Biochem.* **2017**, *61*, 177–200. [CrossRef]
11. Waldron, K.J.; Robinson, N.J. How Do Bacterial Cells Ensure That Metalloproteins Get the Correct Metal? *Nat. Rev. Microbiol.* **2009**, *7*, 25–35. [CrossRef] [PubMed]
12. Kappaun, K.; Piovesan, A.R.; Carlini, C.R.; Ligabue-Braun, R. Ureases: Historical Aspects, Catalytic, and Non-Catalytic Properties—A Review. *J. Adv. Res.* **2018**, *13*, 3–17. [CrossRef] [PubMed]
13. Todd, M.J.; Hausinger, R.P. Purification and Characterization of the Nickel-Containing Multicomponent Urease from *Klebsiella aerogenes*. *J. Biol. Chem.* **1987**, *262*, 5963–5967. [PubMed]
14. Hu, L.T.; Mobley, H.L. Purification and N-Terminal Analysis of Urease from *Helicobacter pylori*. *Infect. Immun.* **1990**, *58*, 992–998. [PubMed]
15. Blakeley, R.L.; Zerner, B. Jack Bean Urease: The First Nickel Enzyme. *J. Mol. Catal.* **1984**, *23*, 263–292. [CrossRef]
16. Pearson, M.A.; Michel, L.O.; Hausinger, R.P.; Karplus, P.A. Structures of Cys319 Variants and Acetohydroxamate-Inhibited *Klebsiella aerogenes* Urease. *Biochemistry* **1997**, *36*, 8164–8172. [CrossRef] [PubMed]
17. Jabri, E.; Carr, M.B.; Hausinger, R.P.; Karplus, P.A. The Crystal Structure of Urease from *Klebsiella aerogenes*. *Science* **1995**, *268*, 998–1004. [CrossRef]
18. Benini, S.; Rypniewski, W.R.; Wilson, K.S.; Miletti, S.; Ciurli, S.; Mangani, S. A New Proposal for Urease Mechanism Based on the Crystal Structures of the Native and Inhibited Enzyme from *Bacillus pasteurii*: Why Urea Hydrolysis Costs Two Nickels. *Structure* **1999**, *7*, 205–216. [CrossRef]
19. Balasubramanian, A.; Ponnuraj, K. Crystal Structure of the First Plant Urease from Jack Bean: 83 Years of Journey from Its First Crystal to Molecular Structure. *J. Mol. Biol.* **2010**, *400*, 274–283. [CrossRef]
20. Ha, N.; Oh, S.; Sung, J.Y.; Cha, K.A.; Lee, M.H.; Oh, B. Supramolecular Assembly and Acid Resistance of *Helicobacter pylori* Urease. *Nat. Struct. Mol. Biol.* **2001**, *8*, 505–509. [CrossRef]

21. Ciurli, S.; Benini, S.; Rypniewski, W.R.; Wilson, K.S.; Miletti, S.; Mangani, S. Structural Properties of the Nickel Ions in Urease: Novel Insights into the Catalytic and Inhibition Mechanisms. *Coord. Chem. Rev.* **1999**, *190–192*, 331–355. [CrossRef]
22. Park, I.S.; Hausinger, R.P. Requirement of Carbon Dioxide for in vitro Assembly of the Urease Nickel Metallocenter. *Science* **1995**, *267*, 1156–1158. [CrossRef] [PubMed]
23. Pearson, M.A.; Schaller, R.A.; Michel, L.O.; Karplus, P.A.; Hausinger, R.P. Chemical Rescue of *Klebsiella aerogenes* Urease Variants Lacking the Carbamylated-Lysine Nickel Ligand. *Biochemistry* **1998**, *37*, 6214–6220. [CrossRef] [PubMed]
24. Park, I.S.; Hausinger, R.P. Metal Ion Interactions with Urease and UreD-Urease Apoproteins. *Biochemistry* **1996**, *35*, 5345–5352. [CrossRef] [PubMed]
25. Mulrooney, S.B.; Hausinger, R.P. Sequence of the *Klebsiella aerogenes* Urease Genes and Evidence for Accessory Proteins Facilitating Nickel Incorporation. *J. Bacteriol.* **1990**, *172*, 5837–5843. [CrossRef] [PubMed]
26. Lee, M.H.; Mulrooney, S.B.; Renner, M.J.; Markowicz, Y.; Hausinger, R.P. *Klebsiella aerogenes* Urease Gene Cluster: Sequence of *UreD* and Demonstration That Four Accessory Genes (*UreD*, *UreE*, *UreF*, and *UreG*) Are Involved in Nickel Metallocenter Biosynthesis. *J. Bacteriol.* **1992**, *174*, 4324–4330. [CrossRef] [PubMed]
27. Akada, J.K.; Shirai, M.; Takeuchi, H.; Tsuda, M.; Nakazawa, T. Identification of the Urease Operon in *Helicobacter pylori* and Its Control by mRNA Decay in Response to pH. *Mol. Microbiol.* **2000**, *36*, 1071–1084. [CrossRef]
28. Cussac, V.; Ferrero, R.L.; Labigne, A. Expression of *Helicobacter pylori* Urease Genes in *Escherichia coli* Grown under Nitrogen-Limiting Conditions. *J. Bacteriol.* **1992**, *174*, 2466–2473. [CrossRef]
29. Weeks, D.L.; Eskandari, S.; Scott, D.R.; Sachs, G. A H^+-Gated Urea Channel: The Link between *Helicobacter pylori* Urease and Gastric Colonization. *Science* **2000**, *287*, 482–485. [CrossRef]
30. Rektorschek, M.; Buhmann, A.; Weeks, D.; Schwan, D.; Bensch, K.W.; Eskandari, S.; Scott, D.; Sachs, G.; Melchers, K. Acid Resistance of *Helicobacter pylori* Depends on the UreI Membrane Protein and an Inner Membrane Proton Barrier. *Mol. Microbiol.* **2000**, *36*, 141–152. [CrossRef]
31. Benoit, S.L.; Zbell, A.L.; Maier, R.J. Nickel Enzyme Maturation in *Helicobacter hepaticus*: Roles of Accessory Proteins in Hydrogenase and Urease Activities. *Microbiology* **2007**, *153*, 3748–3756. [CrossRef] [PubMed]
32. Sriwanthana, B.; Island, M.D.; Maneval, D.; Mobley, H.L.T. Single-Step Purification of *Proteus mirabilis* Urease Accessory Protein UreE, a Protein with a Naturally Occurring Histidine Tail, by Nickel Chelate Affinity Chromatography. *J. Bacteriol.* **1994**, *176*, 6836–6841. [CrossRef] [PubMed]
33. Hu, L.; Foxall, P.A.; Russell, R.; Mobley, H.L.T. Purification of Recombinant *Helicobacter pylori* Urease Apoenzyme Encoded by *UreA* and *UreB*. *Infect. Immun.* **1992**, *60*, 2657–2667. [PubMed]
34. Lee, M.H.; Mulrooney, S.B.; Hausinger, R.P. Purification, Characterization, and in vivo Reconstitution of *Klebsiella aerogenes* Urease Apoenzyme. *J. Bacteriol.* **1990**, *172*, 4427–4431. [CrossRef] [PubMed]
35. Voland, P.; Weeks, D.L.; Marcus, E.A.; Prinz, C.; Sachs, G.; Scott, D. Interactions among the Seven *Helicobacter pylori* Proteins Encoded by the Urease Gene Cluster. *Am. J. Physiol. Gastrointest. Liver Physiol.* **2003**, *284*, G96–G106. [CrossRef] [PubMed]
36. Benoit, S.L.; Mehta, N.; Weinberg, M.V.; Maier, C.; Maier, R.J. Interaction between the *Helicobacter pylori* Accessory Proteins HypA and UreE Is Needed for Urease Maturation. *Microbiology* **2007**, *153*, 1474–1482. [CrossRef] [PubMed]
37. Park, I.S.; Carr, M.B.; Hausinger, R.P. In vitro Activation of Urease Apoprotein and Role of UreD as a Chaperone Required for Nickel Metallocenter Assembly. *Proc. Natl. Acad. Sci. USA* **1994**, *91*, 3233–3237. [CrossRef] [PubMed]
38. Carter, E.L.; Hausinger, R.P. Characterization of the *Klebsiella aerogenes* Urease Accessory Protein UreD in Fusion with the Maltose Binding Protein. *J. Bacteriol.* **2010**, *192*, 2294–2304. [CrossRef]
39. Moncrief, M.B.C.; Hausinger, R.P. Purification and Activation Properties of UreD-UreF-Urease Apoprotein Complexes. *J. Bacteriol.* **1996**, *178*, 5417–5421. [CrossRef]
40. Kim, J.K.; Mulrooney, S.B.; Hausinger, R.P. The UreEF Fusion Protein Provides a Soluble and Functional Form of the UreF Urease Accessory Protein. *J. Bacteriol.* **2006**, *188*, 8413–8420. [CrossRef]
41. Soriano, A.; Hausinger, R.P. GTP-Dependent Activation of Urease Apoprotein in Complex with the UreD, UreF, and UreG Accessory Proteins. *Proc. Natl. Acad. Sci. USA* **1999**, *96*, 11140–11144. [CrossRef] [PubMed]

42. Park, I.; Hausinger, R.P. Evidence for the Presence of Urease Apoprotein Complexes Containing UreD, UreF, and UreG in Cells That Are Competent for in vivo Enzyme Activation. *J. Bacteriol.* **1995**, *177*, 1947–1951. [CrossRef] [PubMed]
43. Fong, Y.H.; Wong, H.C.; Yuen, M.H.; Lau, P.H.; Chen, Y.W.; Wong, K.-B. Structure of UreG/UreF/UreH Complex Reveals How Urease Accessory Proteins Facilitate Maturation of *Helicobacter pylori* Urease. *PLoS Biol.* **2013**, *11*, e1001678. [CrossRef] [PubMed]
44. Farrugia, M.A.; Han, L.; Zhong, Y.; Boer, J.L.; Ruotolo, B.T.; Hausinger, R.P. Analysis of a Soluble (UreD:UreF:UreG)$_2$ Accessory Protein Complex and Its Interactions with *Klebsiella aerogenes* Urease by Mass Spectrometry. *J. Am. Soc. Mass Spectrom.* **2013**, *24*, 1328–1337. [CrossRef] [PubMed]
45. Moncrief, M.B.C.; Hausinger, R.P. Characterization of UreG, Identification of a UreD-UreF-UreG Complex, and Evidence Suggesting That a Nucleotide-Binding Site in UreG Is Required for in vivo Metallocenter Assembly of *Klebsiella aerogenes* Urease. *J. Bacteriol.* **1997**, *179*, 4081–4086. [CrossRef]
46. Boer, J.L.; Quiroz-Valenzuela, S.; Anderson, K.L.; Hausinger, R.P. Mutagenesis of *Klebsiella aerogenes* UreG to Probe Nickel Binding and Interactions with Other Urease-Related Proteins. *Biochemistry* **2010**, *49*, 5859–5869. [CrossRef]
47. Rain, J.C.; Selig, L.; De Reuse, H.; Battaglia, V.; Reverdy, C.; Simon, S.; Lenzen, G.; Petel, F.; Wojcik, J.; Schächter, V.; et al. The Protein-Protein Interaction Map of *Helicobacter pylori*. *Nature* **2001**, *409*, 211–215. [CrossRef]
48. Fong, Y.H.; Wong, H.C.; Chuck, C.P.; Chen, Y.W.; Sun, H.; Wong, K.-B. Assembly of Preactivation Complex for Urease Maturation in *Helicobacter pylori*: Crystal Structure of UreF-UreH Protein Complex. *J. Biol. Chem.* **2011**, *286*, 43241–43249. [CrossRef]
49. Lam, R.; Romanov, V.; Johns, K.; Battaile, K.P.; Wu-Brown, J.; Guthrie, J.L.; Hausinger, R.P.; Pai, E.F.; Chirgadze, N.Y. Crystal Structure of a Truncated Urease Accessory Protein UreF from *Helicobacter pylori*. *Proteins* **2010**, *78*, 2839–2848. [CrossRef]
50. Boer, J.L.; Hausinger, R.P. *Klebsiella aerogenes* UreF: Identification of the UreG Binding Site and Role in Enhancing the Fidelity of Urease Activation. *Biochemistry* **2012**, *51*, 2298–2308. [CrossRef]
51. Yuen, M.H.; Fong, Y.H.; Nim, Y.S.; Lau, P.H.; Wong, K.-B. Structural Insights into how GTP-Dependent Conformational Changes in a Metallochaperone UreG Facilitate Urease Maturation. *Proc. Natl. Acad. Sci. USA* **2017**, *114*, E10890–E10898. [CrossRef] [PubMed]
52. Yang, X.; Li, H.; Lai, T.-P.; Sun, H. UreE-UreG Complex Facilitates Nickel Transfer and Preactivates GTPase of UreG in *Helicobacter pylori*. *J. Biol. Chem.* **2015**, *290*, 12474–12485. [CrossRef] [PubMed]
53. Leipe, D.D.; Wolf, Y.I.; Koonin, E.V.; Aravind, L. Classification and Evolution of P-Loop GTPases and Related ATPases. *J. Mol. Biol.* **2002**, *317*, 41–72. [CrossRef] [PubMed]
54. Zambelli, B.; Turano, P.; Musiani, F.; Neyroz, P.; Ciurli, S. Zn^{2+}-Linked Dimerization of UreG from *Helicobacter pylori*, a Chaperone Involved in Nickel Trafficking and Urease Activation. *Proteins Struct. Funct. Bioinform.* **2009**, *74*, 222–239. [CrossRef] [PubMed]
55. Zambelli, B.; Stola, M.; Musiani, F.; De Vriendt, K.; Samyn, B.; Devreese, B.; Van Beeumen, J.; Turano, P.; Dikiy, A.; Bryant, D.A.; et al. UreG, a Chaperone in the Urease Process, Is an Intrinsically Unstructured Protein Binding a Single Zn^{2+} Ion. *J. Biol. Chem.* **2005**, *280*, 4684–4695. [CrossRef] [PubMed]
56. Real-Guerra, R.; Staniscuaski, F.; Zambelli, B.; Musiani, F.; Ciurli, S.; Carlini, C.R. Biochemical and Structural Studies on Native and Recombinant *Glycine max* UreG: A Detailed Characterization of a Plant Urease Accessory Protein. *Plant Mol. Biol.* **2012**, *78*, 461–475. [CrossRef] [PubMed]
57. Miraula, M.; Ciurli, S.; Zambelli, B. Intrinsic Disorder and Metal Binding in UreG Proteins from Archae Hyperthermophiles: GTPase Enzymes Involved in the Activation of Ni(II) Dependent Urease. *J. Biol. Inorg. Chem.* **2015**, *20*, 739–755. [CrossRef] [PubMed]
58. Martin-Diaconescu, V.; Bellucci, M.; Musiani, F.; Ciurli, S.; Maroney, M.J. Unraveling the *Helicobacter pylori* UreG Zinc Binding Site Using X-Ray Absorption Spectroscopy (XAS) and Structural Modeling. *J. Biol. Inorg. Chem.* **2012**, *17*, 353–361. [CrossRef]
59. Xia, W.; Li, H.; Yang, X.; Wong, K.-B.; Sun, H. Metallo-GTPase HypB from *Helicobacter pylori* and Its Interaction with Nickel Chaperone Protein HypA. *J. Biol. Chem.* **2012**, *287*, 6753–6763. [CrossRef] [PubMed]
60. Lacasse, M.J.; Douglas, C.D.; Zamble, D.B. Mechanism of Selective Nickel Transfer from HypB to HypA, *Escherichia coli* [NiFe]-Hydrogenase Accessory Proteins. *Biochemistry* **2016**, *55*, 6821–6831. [CrossRef] [PubMed]

61. Gasper, R.; Scrima, A.; Wittinghofer, A. Structural Insights into HypB, a GTP-Binding Protein That Regulates Metal Binding. *J. Biol. Chem.* **2006**, *281*, 27492–27502. [CrossRef] [PubMed]
62. Cai, F.; Ngu, T.T.; Kaluarachchi, H.; Zamble, D.B. Relationship between the GTPase, Metal-Binding, and Dimerization Activities of *E. Coli* HypB. *J. Inorg. Biochem.* **2011**, *16*, 857–868. [CrossRef] [PubMed]
63. Chan, K.H.; Lee, K.M.; Wong, K.B. Interaction between Hydrogenase Maturation Factors HypA and HypB Is Required for [NiFe]-Hydrogenase Maturation. *PLoS ONE* **2012**, *7*, e32592. [CrossRef] [PubMed]
64. Chan, K.H.; Li, T.; Wong, C.O.; Wong, K.B. Structural Basis for GTP-Dependent Dimerization of Hydrogenase Maturation Factor HypB. *PLoS ONE* **2012**, *7*, e30547. [CrossRef] [PubMed]
65. Remaut, H.; Safarov, N.; Ciurli, S.; Van Beeumen, J. Structural Basis for Ni^{2+} Transport and Assembly of the Urease Active Site by the Metallochaperone UreE from *Bacillus pasteurii*. *J. Biol. Chem.* **2001**, *276*, 49365–49370. [CrossRef]
66. Song, H.K.; Mulrooney, S.B.; Huber, R.; Hausinger, R.P. Crystal Structure of *Klebsiella aerogenes* UreE, a Nickel-Binding Metallochaperone for Urease Activation. *J. Biol. Chem.* **2001**, *276*, 49359–49364. [CrossRef]
67. Banaszak, K.; Martin-Diaconescu, V.; Bellucci, M.; Zambelli, B.; Rypniewski, W.; Maroney, M.J.; Ciurli, S. Crystallographic and X-Ray Absorption Spectroscopic Characterization of *Helicobacter pylori* UreE Bound to Ni^{2+} and Zn^{2+} Reveals a Role for the Disordered C-Terminal Arm in Metal Trafficking. *Biochem. J.* **2012**, *441*, 1017–1026. [CrossRef]
68. Shi, R.; Munger, C.; Asinas, A.; Benoit, S.L.; Miller, E.; Matte, A.; Maier, R.J.; Cygler, M. Crystal Structures of apo and Metal-Bound Forms of the UreE Protein from *Helicobacter pylori*: Role of Multiple Metal Binding Sites. *Biochemistry* **2010**, *49*, 7080–7088. [CrossRef]
69. Stola, M.; Musiani, F.; Mangani, S.; Turano, P.; Safarov, N.; Zambelli, B.; Ciurli, S. The Nickel Site of *Bacillus pasteurii* UreE, a Urease Metallo-Chaperone, as Revealed by Metal-Binding Studies and X-Ray Absorption Spectroscopy. *Biochemistry* **2006**, *45*, 6495–6509. [CrossRef]
70. Lee, M.H.; Pankratz, H.S.; Wang, S.; Scott, R.A.; Finnegan, M.G.; Johnson, M.K.; Ippolito, J.A.; Christianson, D.W.; Hausinger, R.P. Purification and Characterization of *Klebsiella aerogenes* UreE Protein: A Nickel-binding Protein That Functions in Urease Metallocenter Assembly. *Protein Sci.* **1993**, *2*, 1042–1052. [CrossRef]
71. Bellucci, M.; Zambelli, B.; Musiani, F.; Turano, P.; Ciurli, S. *Helicobacter pylori* UreE, a Urease Accessory Protein: Specific Ni^{2+}- and Zn^{2+}-Binding Properties and Interaction with Its Cognate UreG. *Biochem. J.* **2009**, *422*, 91–100. [CrossRef] [PubMed]
72. Grossoehme, N.E.; Mulrooney, S.B.; Hausinger, R.P.; Wilcox, D.E. Thermodynamics of Ni^{2+}, Cu^{2+}, and Zn^{2+} Binding to the Urease Metallochaperone. *Biochemistry* **2007**, *46*, 10506–10516. [CrossRef] [PubMed]
73. Won, H.; Lee, Y.; Kim, J.; Shin, I.S.; Lee, M.H.; Lee, B. Structural Characterization of the Nickel-Binding Properties of *Bacillus pasteurii* Urease Accessory Protein (Ure)E in Solution. *J. Biol. Chem.* **2004**, *279*, 17466–17472. [CrossRef] [PubMed]
74. Ciurli, S.; Safarov, N.; Miletti, S.; Dikiy, A.; Christensen, S.K.; Kornetzky, K.; Bryant, D.A.; Vandenberghe, I.; Devreese, B.; Samyn, B.; et al. Molecular Characterization of *Bacillus pasteurii* UreE, a Metal-Binding Chaperone for the Assembly of the Urease Active Site. *J. Biol. Inorg. Chem.* **2002**, *7*, 623–631. [CrossRef] [PubMed]
75. Colpas, G.J.; Hausinger, R.P. In vivo and in citro Kinetics of Metal Transfer by the *Klebsiella aerogenes* Urease Nickel Metallochaperone, UreE. *J. Biol. Chem.* **2000**, *275*, 10731–10737. [CrossRef] [PubMed]
76. Colpas, G.J.; Brayman, T.G.; Ming, L.J.; Hausinger, R.P. Identification of Metal-Binding Residues in the *Klebsiella aerogenes* Urease Nickel Metallochaperone, UreE. *Biochemistry* **1999**, *38*, 4078–4088. [CrossRef] [PubMed]
77. Brayman, T.G.; Hausinger, R.P. Purification, Characterization, and Functional Analysis of a Truncated *Klebsiella aerogenes* UreE Urease Accessory Protein Lacking the Histidine-Rich Carboxyl Terminus. *J. Bacteriol.* **1996**, *178*, 5410–5416. [CrossRef] [PubMed]
78. Mulrooney, S.B.; Ward, S.K.; Hausinger, R.P. Purification and Properties of the *Klebsiella aerogenes* UreE Metal-Binding Domain, a Functional Metallochaperone of Urease. *J. Bacteriol.* **2005**, *187*, 3581–3585. [CrossRef]
79. Musiani, F.; Zambelli, B.; Stola, M.; Ciurli, S. Nickel Trafficking: Insights into the Fold and Function of UreE, a Urease Metallochaperone. *J. Inorg. Biochem.* **2004**, *98*, 803–813. [CrossRef]

80. Merloni, A.; Dobrovolska, O.; Zambelli, B.; Agostini, F.; Bazzani, M.; Musiani, F.; Ciurli, S. Molecular Landscape of the Interaction between the Urease Accessory Proteins UreE and UreG. *Biochim. Biophys. Acta Proteins Proteom.* **2014**, *1844*, 1662–1674. [CrossRef]
81. Jurrus, E.; Engel, D.; Star, K.; Monson, K.; Brandi, J.; Felberg, L.E.; Brookes, D.H.; Wilson, L.; Chen, J.; Liles, K.; et al. Improvements to the APBS Biomolecular Solvation Software Suite. *Protein Sci.* **2018**, *27*, 112–128. [CrossRef] [PubMed]
82. Farrugia, M.A.; Wang, B.; Feig, M.; Hausinger, R.P. Mutational and Computational Evidence That a Nickel-Transfer Tunnel in UreD Is Used for Activation of *Klebsiella aerogenes* Urease. *Biochemistry* **2015**, *54*, 6392–6401. [CrossRef] [PubMed]
83. Soriano, A.; Colpas, G.J.; Hausinger, R.P. UreE Stimulation of GTP-Dependent Urease Activation in the UreD-UreF-UreG-Urease Apoprotein Complex. *Biochemistry* **2000**, *39*, 12435–12440. [CrossRef] [PubMed]
84. Chang, Z.; Kuchar, J.; Hausinger, R.P. Chemical Cross-Linking and Mass Spectrometric Identification of Sites of Interaction for UreD, UreF, and Urease. *J. Biol. Chem.* **2004**, *279*, 15305–15313. [CrossRef]
85. Musiani, F.; Gioia, D.; Masetti, M.; Falchi, F.; Cavalli, A.; Recanatini, M.; Ciurli, S. Protein Tunnels: The Case of Urease Accessory Proteins. *J. Chem. Theory Comput.* **2017**, *13*, 2322–2331. [CrossRef]
86. Zambelli, B.; Berardi, A.; Martin-Diaconescu, V.; Mazzei, L.; Musiani, F.; Maroney, M.J.; Ciurli, S. Nickel Binding Properties of *Helicobacter pylori* UreF, an Accessory Protein in the Nickel-Based Activation of Urease. *J. Biol. Inorg. Chem.* **2014**, *19*, 319–334. [CrossRef] [PubMed]
87. Baykov, A.A.; Evtushenko, O.A.; Avaeva, S.M. A Malachite Green Procedure for Orthophosphate Determination and Its Use in Alkaline Phosphatase-Based Enzyme Immunoassay. *Anal. Biochem.* **1988**, *171*, 266–270. [CrossRef]
88. Maier, T.; Lottspeich, F.; Böck, A. GTP Hydrolysis by HypB Is Essential for Nickel Insertion into Hydrogenases of *Escherichia coli*. *Eur. J. Biochem.* **1995**, *230*, 133–138. [CrossRef]
89. Olson, J.W.; Fu, C.; Maier, R.J. The HypB Protein from *Bradyrhizobium japonicum* can Store Nickel and Is Required for the Nickel-dependent Transcriptional Regulation of Hydrogenase. *Mol. Microbiol.* **1997**, *24*, 119–128. [CrossRef]
90. Olson, J.W.; Mehta, N.S.; Maier, R.J. Requirement of Nickel Metabolism Proteins HypA and HypB for Full Activity of Both Hydrogenase and Urease in *Helicobacter pylori*. *Mol. Microbiol.* **2001**, *39*, 176–182. [CrossRef]
91. Hube, M.; Blokesch, M.; Böck, A. Network of Hydrogenase Maturation in *Escherichia coli*: Role of Accessory Proteins HypA and HybF. *J. Bacteriol.* **2002**, *184*, 3879–3885. [CrossRef] [PubMed]
92. Hoffmann, D.; Gutekunst, K.; Klissenbauer, M.; Schulz-Friedrich, R.; Appel, J. Mutagenesis of Hydrogenase Accessory Genes of *Synechocystis* sp. PCC 6803: Additional Homologues of *hypA* and *hypB* Are Not Active in Hydrogenase Maturation. *FEBS J.* **2006**, *273*, 4516–4527. [CrossRef] [PubMed]
93. Maier, T.; Jacobi, A.; Sauter, M.; Böck, A. The Product of the *HypB* Gene, Which Is Required for Nickel Incorporation into Hydrogenases, Is a Novel Guanine Nucleotide-Binding Protein. *J. Bacteriol.* **1993**, *175*, 630–635. [CrossRef] [PubMed]
94. Jacobi, A.; Rossmann, R.; Böck, A. The *hyp* Operon Gene Products Are Required for the Maturation of Catalytically Active Hydrogenase Isoenzymes in *Escherichia coli*. *Arch. Microbiol.* **1992**, *158*, 444–451. [CrossRef] [PubMed]
95. Lutz, S.; Jacobi, A.; Schlensog, V.; Böhm, R.; Sawers, G.; Böck, A. Molecular Characterization of an Operon (*hyp*) Necessary for the Activity of the Three Hydrogenase Isoenzymes in *Escherichia coli*. *Mol. Microbiol.* **1991**, *5*, 123–135. [CrossRef] [PubMed]
96. Tomb, J.F.; White, O.; Kerlavage, A.R.; Clayton, R.A.; Sutton, G.G.; Fleischmann, R.D.; Ketchum, K.A.; Klenk, H.P.; Gill, S.; Brian, A.D. The Complete Genome Sequence of the Gastric Pathogen *Helicobacter pylori*. *Nature* **1997**, *388*, 539–547. [CrossRef] [PubMed]
97. Spronk, C.A.E.M.; Żerko, S.; Górka, M.; Koźmiński, W.; Bardiaux, B.; Zambelli, B.; Musiani, F.; Piccioli, M.; Basak, P.; Blum, F.C.; et al. Structure and Dynamics of *Helicobacter pylori* Nickel-chaperone HypA: An Integrated Approach Using NMR Spectroscopy, Functional Assays and Computational Tools. *JBIC J. Biol. Inorg. Chem.* **2018**, *23*, 1309–1330. [CrossRef] [PubMed]
98. Xia, W.; Li, H.; Sze, K.H.; Sun, H. Structure of a Nickel Chaperone, HypA, from *Helicobacter pylori* Reveals Two Distinct Metal Binding Sites. *J. Am. Chem. Soc.* **2009**, *131*, 10031–10040. [CrossRef]
99. Mehta, N.; Olson, J.W.; Maier, R.J. Characterization of *Helicobacter pylori* Nickel Metabolism Accessory Proteins Needed for Maturation of Both Urease and Hydrogenase. *J. Bacteriol.* **2003**, *185*, 726–734. [CrossRef]

100. Herbst, R.W.; Perovic, I.; Martin-diaconescu, V.; Brien, K.O.; Chivers, P.T.; Pochapsky, S.S.; Pochapsky, T.C.; Maroney, M.J. Communication between the Zinc and Nickel Sites in Dimeric HypA: Metal Recognition and pH Sensing. *J. Am. Chem. Soc.* **2010**, *132*, 10338–10351. [CrossRef]
101. Hu, H.Q.; Johnson, R.C.; Merrell, D.S.; Maroney, M.J. Nickel Ligation of the N-Terminal Amine of HypA Is Required for Urease Maturation in *Helicobacter pylori*. *Biochemistry* **2017**, *56*, 1105–1116. [CrossRef] [PubMed]
102. Atanassova, A.; Zamble, D.B. *Escherichia coli* HypA Is a Zinc Metalloprotein with a Weak Affinity for Nickel. *J. Bacteriol.* **2005**, *187*, 4689–4697. [CrossRef] [PubMed]
103. Watanabe, S.; Arai, T.; Matsumi, R.; Atomi, H.; Imanaka, T.; Miki, K. Crystal Structure of HypA, a Nickel-Binding Metallochaperone for [NiFe] Hydrogenase Maturation. *J. Mol. Biol.* **2009**, *394*, 448–459. [CrossRef] [PubMed]
104. Kennedy, D.C.; Herbst, R.W.; Iwig, J.S.; Chivers, P.T.; Maroney, M.J. A Dynamic Zn Site in *Helicobacter pylori* HypA: A Potential Mechanism for Metal-Specific Protein Activity. *J. Am. Chem. Soc.* **2007**, *129*, 16–17. [CrossRef] [PubMed]
105. Johnson, R.C.; Hu, H.Q.; Merrell, D.S.; Maroney, M.J. Dynamic HypA Zinc Site Is Essential for Acid Viability and Proper Urease Maturation in *Helicobacter pylori*. *Metallomics* **2015**, *7*, 674–682. [CrossRef]
106. Stingl, K.; Schauer, K.; Ecobichon, C.; Labigne, A.; Lenormand, P.; Rousselle, J.-C.; Namane, A.; de Reuse, H. In vivo Interactome of *Helicobacter pylori* Urease Revealed by Tandem Affinity Purification. *Mol. Cell. Proteom.* **2008**, *7*, 2429–2441. [CrossRef]
107. Hu, H.Q.; Huang, H.T.; Maroney, M.J. The *Helicobacter pylori* HypA·UreE$_2$ Complex Contains a Novel High-Affinity Ni(II)-Binding Site. *Biochemistry* **2018**, *57*, 2932–2942. [CrossRef]
108. Yang, X.; Li, H.; Cheng, T.; Xia, W.; Lai, Y.-T.; Sun, H. Nickel Translocation between Metallochaperones HypA and UreE in *Helicobacter pylori*. *Metallomics* **2014**, *6*, 40–42. [CrossRef]
109. Benoit, S.L.; McMurry, J.L.; Hill, S.A.; Maier, R.J. *Helicobacter pylori* Hydrogenase Accessory Protein HypA and Urease Accessory Protein UreG Compete with Each Other for UreE Recognition. *Biochim. Biophys. Acta-Gen. Subj.* **2012**, *1820*, 1519–1525. [CrossRef]
110. Douglas, C.D.; Ngu, T.T.; Kaluarachchi, H.; Zamble, D.B. Metal Transfer within the *Escherichia coli* HypB–HypA Complex of Hydrogenase Accessory Proteins. *Biochemistry* **2013**, *52*, 6030–6039. [CrossRef]
111. Kaluarachchi, H.; Zhang, J.W.; Zamble, D.B. *Escherichia coli* SlyD, More Than a Ni(II) Reservoir. *Biochemistry* **2011**, *50*, 10761–10763. [CrossRef]
112. Cheng, T.; Li, H.; Yang, X.; Xia, W.; Sun, H. Interaction of SlyD with HypB of *Helicobacter pylori* Facilitates Nickel Trafficking. *Metallomics* **2013**, *5*, 804–807. [CrossRef]
113. Zhang, J.W.; Butland, G.; Greenblatt, J.F.; Emili, A.; Zamble, D.B. A Role for SlyD in the *Escherichia coli* Hydrogenase Biosynthetic Pathway. *J. Biol. Chem.* **2005**, *280*, 4360–4366. [CrossRef]
114. Leach, M.R.; Jie, W.Z.; Zamble, D.B. The Role of Complex Formation between the *Escherichia coli* Hydrogenase Accessory Factors HypB and SlyD. *J. Biol. Chem.* **2007**, *282*, 16177–16186. [CrossRef]
115. Lacasse, M.J.; Zamble, D.B. [NiFe]-Hydrogenase Maturation. *Biochemistry* **2016**, *55*, 1689–1701. [CrossRef]
116. Zeer-wanklyn, C.J.; Zamble, D.B. Microbial Nickel: Cellular Uptake and Delivery to Enzyme Centers. *Curr. Opin. Chem. Biol.* **2017**, *37*, 80–88. [CrossRef]
117. Witte, C.-P. Urea Metabolism in Plants. *Plant Sci.* **2011**, *180*, 431–438. [CrossRef]
118. Polacco, J.C.; Mazzafera, P.; Tezotto, T. Opinion—Nickel and Urease in Plants: Still Many Knowledge Gaps. *Plant Sci.* **2013**, *199–200*, 79–90. [CrossRef]
119. Cao, F.-Q.; Werner, A.K.; Dahncke, K.; Romeis, T.; Liu, L.-H.; Witte, C.-P. Identification and Characterization of Proteins Involved in Rice Urea and Arginine Catabolism. *Plant Physiol.* **2010**, *154*, 98–108. [CrossRef]
120. Myrach, T.; Zhu, A.; Witte, C.-P. The Assembly of the Plant Urease Activation Complex and the Essential Role of the Urease Accessory Protein G (UreG) in Delivery of Nickel to Urease. *J. Biol. Chem.* **2017**, *292*, 14556–14565. [CrossRef]
121. Witte, C.-P.; Rosso, M.G.; Romeis, T. Identification of Three Urease Accessory Proteins that are Required for Urease Activation in *Arabidopsis*. *Plant Physiol.* **2005**, *139*, 1155–1162. [CrossRef]
122. Freyermuth, S.K.; Bacanamwo, M.; Polacco, J.C. The Soybean *Eu3* Gene Encodes an Ni-Binding Protein Necessary for Urease Activity. *Plant J.* **2000**, *21*, 53–60. [CrossRef] [PubMed]
123. Rutherford, J.C. The Emerging Role of Urease as a General Microbial Virulence Factor. *PLoS Pathog.* **2014**, *10*, e1004062. [CrossRef] [PubMed]

124. Safavi, M.; Sabourian, R.; Foroumadi, A. Treatment of *Helicobacter pylori* Infection: Current and Future Insights. *World J. Clin. Cases* **2016**, *4*, 5–19. [CrossRef] [PubMed]
125. Yang, X.; Koohi-moghadam, M.; Wang, R.; Chang, Y.; Woo, P.C.Y.; Wang, J.; Li, H.; Sun, H. Metallochaperone UreG Serves as a New Target for Design of Urease Inhibitor: A Novel Strategy for Development of Antimicrobials. *PLoS Biol.* **2018**, *16*, e2003887. [CrossRef] [PubMed]
126. Eschweiler, J.D.; Farrugia, M.A.; Dixit, S.M.; Hausinger, R.P.; Ruotolo, B.T. A Structural Model of the Urease Activation Complex Derived from Ion Mobility-Mass Spectrometry and Integrative Modeling. *Structure* **2018**, *26*, 599–606.e3. [CrossRef] [PubMed]
127. Ligabue-Braun, R.; Real-Guerra, R.; Carlini, C.R.; Verli, H. Evidence-Based Docking of the Urease Activation Complex. *J. Biomol. Struct. Dyn.* **2013**, *31*, 854–861. [CrossRef] [PubMed]

© 2019 by the authors. Licensee MDPI, Basel, Switzerland. This article is an open access article distributed under the terms and conditions of the Creative Commons Attribution (CC BY) license (http://creativecommons.org/licenses/by/4.0/).

Article

Molecular Modelling of the Ni(II)-Responsive *Synechocystis* PCC 6803 Transcriptional Regulator InrS in the Metal Bound Form

Elia Barchi and Francesco Musiani *

Laboratory of Bioinorganic Chemistry, Department of Pharmacy and Biotechnology, University of Bologna, Viale G. Fanin 40, I-40127 Bologna, Italy; elia.barchi91@gmail.com
* Correspondence: francesco.musiani@unibo.it; Tel.: +39-051-209-6236

Received: 24 May 2019; Accepted: 20 June 2019; Published: 21 June 2019

Abstract: InrS (internal nickel-responsive sensor) is a transcriptional regulator found in cyanobacteria that represses the transcription of the nickel exporter NrsD in the apo form and de-represses expression of the exporter upon Ni(II) binding. Although a crystal structure of apo-InrS from *Synechocystis* PCC 6803 has been reported, no structure of the protein with metal ions bound is available. Here we report the results of a computational study aimed to reconstruct the metal binding site by taking advantage of recent X-ray absorption spectroscopy (XAS) data and to envisage the structural rearrangements occurring upon Ni(II) binding. The modelled Ni(II) binding site shows a square planar geometry consistent with experimental data. The structural details of the conformational changes occurring upon metal binding are also discussed in the framework of trying to rationalize the different affinity of the apo- and holo-forms of the protein for DNA.

Keywords: InrS; nickel-dependent transcriptional regulators; molecular modelling

1. Introduction

Nickel is an essential element for the metabolism of various pathogenic bacteria and is also used by plants, fungi, plants, and single-cell eukaryotic organisms, but not by higher animals such as humans [1]. This peculiarity can be exploited against the phenomenon of antibiotic resistance in pathogenic bacteria. Indeed, eight out of 12 dangerous microorganisms recently identified by the World Health Organization use nickel for their survival [2]. Consequently, the understanding of the molecular mechanisms underlying the use of this metal and regulating the genic expression of nickel-metabolism-related proteins is of fundamental importance.

In cyanobacteria, nickel ions are directly bound to urease and hydrogenases or through the tetrapyrrole ring of coenzyme F_{430}; several other nickel-binding proteins are deputed to nickel import, trafficking and export [3]. Among these proteins, the internal nickel-responsive sensor (InrS) is a transcriptional regulator exclusively found in cyanobacteria [4–6]. It belongs to the resistance to cobalt and nickel repressor (RcnR)/copper-sensitive operon repressor (CsoR) family of transcriptional regulators [7], consisting in disc-shaped all α-helical homo dimer (αβ) of dimers (αβ/α′β′) formed by three-helix bundles monomers (Figure S1) [4,8–10]. With the exception of the formaldehyde-sensing transcriptional regulator (FrmR) [11,12] and of the CsoR-like sulphur transferase repressor (CstR, where the binding to the operator is inhibited by the oxidation of an inter-subunit cysteine pair by sulphite) [13,14], all the characterized members of this family are metal-ion sensors (Figure 1 and Table S1) [5,7,10].

In the case of InrS, the function of the protein is to repress the expression of a nickel exporter (NrsD) and to enhance the expression of proteins involved in nickel import when the concentration of intracellular Ni(II) is too low by forming a stable protein-DNA complex [5,6,15]. Competition experiments conducted on InrS from *Synechocystis* PCC 6803 (*Sy*InrS), using the Ni(II) chelating ligands NTA, EGTA and EDTA, estimated a dissociation constant of *Sy*InrS for Ni(II) in the sub-picomolar range ($K_d = 2 \times 10^{-14}$ M) [4]. Calorimetric experiments conducted on the same protein, but in the absence of possible artefacts due to the presence of the exogenous competitors, revealed instead dissociation constants for Ni(II) in the nanomolar/micromolar range. In particular, a 2 + 2 binding mode for *Sy*InrS was observed, with the two binding events featuring dissociation constants of 70 nM and 4.5 mM, respectively [6]. The formation of a Ni(II)-*Sy*InrS complex decreases the DNA affinity of the protein and consequently de-represses the expression of the *nrsD* gene [16]. However, in the absence of a structural comparison involving the apo- and holo-forms of InrS, a rationalization for the function of this Ni-sensor cannot be achieved.

An initial study of the Ni(II)-binding site in *Sy*InrS performed by using electronic absorption spectroscopy revealed that Ni(II) ions bind in a four-coordinate planar geometry comprising at least one cysteine ligand [4]. The alignment between the *Sy*InrS sequence and the other CsoR/RcnR family members revealed that the canonical W–X–Y–Z fingerprint, used to characterize the first coordination shell residues [8,9], corresponds to His21, Cys53, His78, and Cys82 in *Sy*InrS (Figure 1). A subsequent mutagenesis study confirmed the roles of the latter three residues in Ni(II) binding [16], while His21 has been recently confirmed as the fourth Ni(II) ligand both through mutagenesis [17] and X-ray absorption spectroscopy (XAS) data [18]. The latter study, which investigated both Ni(II) and Cu(II) bound *Sy*InrS, also confirmed the presence of a planar four-coordinate [Ni(His)$_2$(Cys)$_2$] site and provided Ni(II)–ligands interatomic distances [18].

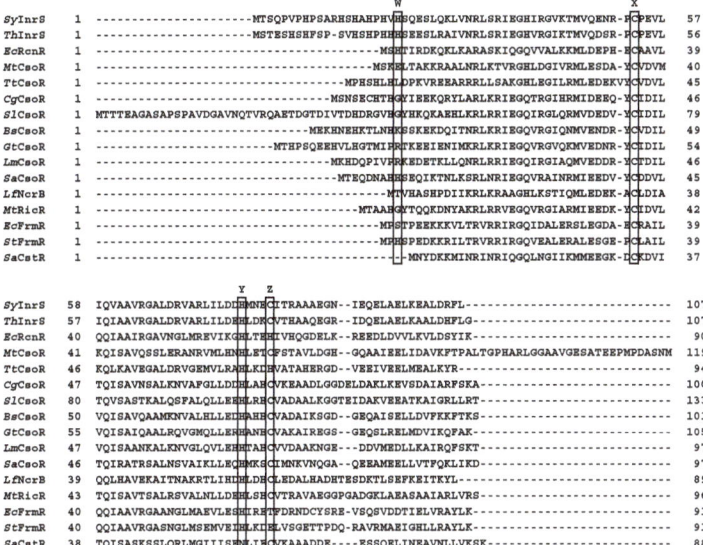

Figure 1. PROMALS3D [19] multiple sequence alignment of the known members of the RcnR/CsoR family. *Synechocystis* PCC 6803 InrS, *Sy*InrS; *Thermosynechococcus* sp. NK55a InrS, *Th*InrS; *Escherichia coli* RcnR, *Ec*RcnR; *Mycobacterium tuberculosis* CsoR, *Mt*CsoR; *Thermus thermophilus* CsoR, *Tt*CsoR; *Corynebacterium glutamicum* CsoR, *Cg*CsoR; *Streptomyces lividans* CsoR, *Sl*CsoR; *Bacillus subtilis* CsoR, *Bs*CsoR; *Geobacillus thermodenitrificans* CsoR, *Gt*CsoR; *Listeria monocytogenes* CsoR, *Lm*CsoR; *Staphylococcus aureus* CsoR, *Sa*CsoR; *Leptospirillum ferriphilum* NcrB, *Lf*NcrB; *Mycobacterium tuberculosis* RicR, *Mt*RicR; *Escherichia coli* FrmR, *Ec*FrmR; *Salmonella typhimurium* (strain SL1344) FrmR, *St*FrmR; *Staphylococcus aureus* NCTC 8325, *Sa*CsrT. The CsoR/RcnR W–X–Y–Z fingerprint residues have been highlighted by black boxes.

The recently solved crystal structure of apo-SyInrS (Figure 2a) showed the typical fold of the RcnR/CsoR family and the proximity of Cys53, His78, and Cys82, strengthening the hypothesis of their role in the Ni(II) binding site [17]. The metal binding site (Figure 2b) should thus be formed by Cys53, located at the beginning of helix α2 from one monomer and located in the vicinity of His78 and Cys82 (at the C-term of helix α2) from the adjacent monomer. On the other hand, His21 is located on an unstructured region at the beginning of helix α1, apparently far from the other three residues, but it can be envisioned that it can complement them through a conformational rearrangement of the N-terminal region.

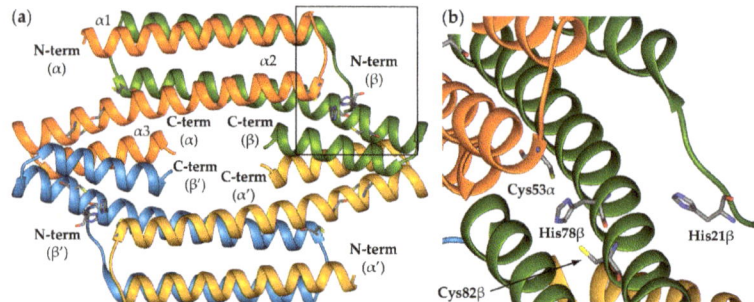

Figure 2. (a) Ribbon diagram of SyInrS crystal structure (PDB id: 5FMN [17]). Chains α, β, α', and β' are coloured in orange, green, yellow, and light blue, respectively. (b) Detail of the putative metal binding region [black square in panel (a)]. Proposed Ni(II) binding residues are reported as sticks coloured accordingly to the atom type.

In addition to a lack of structural information for the Ni(II) bound SyInrS, the structural details of the protein-DNA interaction of proteins belonging to the CsoR/RcnR family are also missing. Nevertheless, *Streptomyces lividans* CsoR (SlCsoR) binds the cognate DNA operator with a 2 SlCsoR: 1 DNA stoichiometry [20] and indirect structural models for CsoR:DNA operator binding have been proposed in the last few years in the case of *Bacillus subtilis* CsoR (BsCroR) [21] and SlCsoR [20]. Of all the members of this family, both CsoR and FrmR have been crystallized in both the ligand-free and -bound forms (see Figure S1 in Supplementary Materials). A comparison between the apo- and holo-FrmR structures revealed a conformational rearrangement in response to formaldehyde binding, which involves a translational movement of α-helices α1 and α2 of one monomer, which slide across the equivalent helices on the other monomer by ca. 1.5 α-helical turns [12]. Similar conformational changes, involving α-helix movement and rotation within the homo-tetramer assembly, were also observed in the case of SlCsoR [22] and *Geobacillus thermodenitrificans* (GtCsoR) [23]. This rearrangement causes the structure to compact and a rearrangement of the charges on the surface of the tetramers. Therefore, a similar behaviour can also be inferred for SyInrS. The aim of the present work is to gain structural information on the Ni(II)-bound SyInrS through the use of molecular modelling techniques guided by all the available experimental data.

2. Results and Discussion

2.1. Reconstruction of Apo-SyInrS N-terminal Region and Modelling of Holo-SyInrS

In principle, Ni(II) binding to SyInrS could induce a conformational rearrangement similar to that observed for FrmR [12], even though it is not possible to exclude that only a negligible structural effect could be induced by metal binding. Thus we performed the modelling of the nickel binding site by starting from two SyInrS structures: (i) The crystal structure of tetrameric apo-SyInrS (PDB id: 5FMN [17]) and (ii) the model structure of tetrameric SyInrS in the putative Ni(II)-bound conformation. In the first case, the crystal structure of apo-SyInrS was used after the addition of three residues at the N-terminal of chains α and α' (namely His19, Val20 and the putative metal binding residue His21) that are missing in the crystal structure probably because of disorder. This reconstruction was carried out through homology modelling, using the atomic positions of the corresponding residues in chains β and β', which are instead structurally well-defined in the same PDB file. For the model of apo-SyInrS in the holo conformation, the homology modelling technique was also used, but in this case the available crystal structures of Cu(I)-bound CsoR from *Mycobacterium tuberculosis* and *Geobacillus thermodenitrificans* NG80-2 were used as templates (Figure S1, PDB id: 2HH7 [7] and 4M1P [8], respectively).

Figure 3 reports the results of the reconstruction of the missing protein portions of SyInrS in the apo conformation and of the modelling in the holo conformation (apo-SyInrS and holo-SyInrS hereafter, respectively, uniquely to the conformation of the protein and not referred to the metalation state). In the model of holo-SyInrS it was possible to include residues 13–18 at the N-terminal of each monomer. These residues were not present in the crystal structure of apo-SyInrS. In both cases the model structures were analysed by using the software PROCHECK [24] and QMEAN [25], with very satisfactory results (≥97.5% most favoured regions and ≤2.5% additional allowed regions). The general fold of the protein is the same regardless of the conformation, with each monomer made of three α-helices and each dimer being formed by the interaction of the N-terminal portion of helix α2 from each monomer. The dimer of dimers is then formed by the interaction of the C-terminal portion of α-helix α2 and the entire α-helix α3 from each monomer. In both conformations, the protein quaternary structure is disc-shaped with some subtle differences. The axis of inertia ellipsoid inscribing the structure of apo-SyInrS results in axes of 77.0 Å, 60.2 Å and 29.2 Å; similarly, values of 76.8 Å, 65.8 Å and 28.8 Å were determined for holo-SyInrS (Figure S2). This indicates a significant increase (by ca. 6 Å), upon nickel binding, of the medium axis. This is caused by a movement of α-helix α1 of each monomer, which slides along the facing helix on the interacting monomer by ca. 1.5 α-helical turns (coherently with the same conformational change observed in the case of FrmR [12]). Moreover, α-helices α1 and α2 rotate by ca. 21 and 13 degrees around the tetramer vertical axis, respectively (Figure S3). The rearrangement of each dimer causes a consequent re-organisation of the interactions between α-helices α3 at the dimer of dimers interface and a narrowing of the cavity at the centre of the structure. Finally, holo-SyInrS loses the biconcave shape observed in the apo form (see Figure 3, lower panels) and assumes the shape of a convex disk.

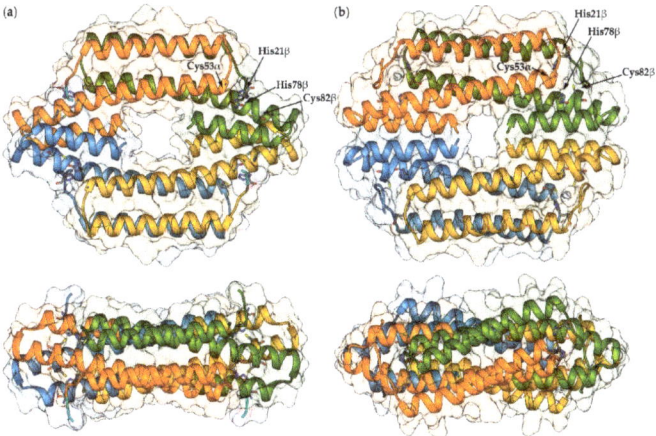

Figure 3. Ribbon diagram and molecular surface of apo-SyInrS (**a**) and holo-SyInrS (**b**). Chains α, β, α', and β' are coloured as in Figure 1. In panel (**a**), the reconstructed residues of chains α and α' are in red. Proposed Ni(II) binding residues are reported as sticks coloured accordingly to the atom type. The orientations in the bottom panels are rotated by 90° around the horizontal axis with respect to the orientations in the top panels.

The molecular surface of the SyInrS tetramer is characterized by two positively charged regions on each side of the disk-shaped protein (Figure 4). The presence of such regions is due to the presence of residues Lys29, Arg32, and Arg35 (α-helix α1) and Arg69 and Arg71 (α2) on the protein surface. The conformational change occurring upon transition from the apo to the holo form of SyIrnS causes a displacement of these positively charged regions toward the center of α-helices α1-2. Moreover, in the holo conformation there are two negatively charged regions on each side of the disk formed by the SyInrS tetramer that are much more reduced in the apo form.

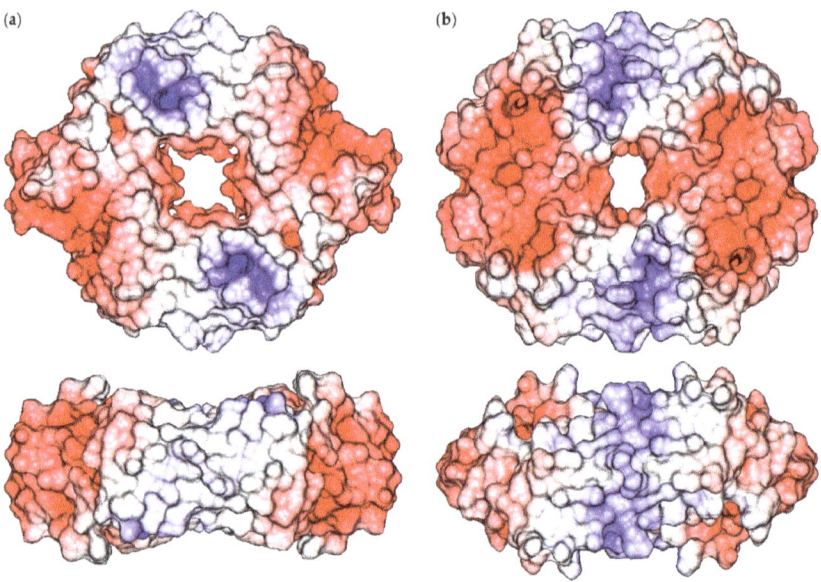

Figure 4. Molecular surface of apo-SyInrS (**a**) and holo-SyInrS (**b**). The surface is coloured according to the surface electrostatic potential calculated with APBS software [26]. The orientations in the bottom panels are rotated by 90° around the horizontal axis with respect to the orientations in the top panels.

The intermolecular interactions occurring among the four monomers that make up the protein structure were analysed by using the PDBsum server [27,28] and the results are reported in Table 1. From the data, it appears that the apo-to-holo transition significantly increases the number of non-bonded contacts both in the α-β dimer and at the dimer of dimers (α-β') interface. On the other hand, the interaction surface, as well as the number of involved residues, increases at the dimer interface and decreases at the dimer of dimers interface. The significance of the changes observed in the case of hydrogen bonds and salt bridges are difficult to interpret due to the intrinsic errors involved in the homology modelling procedure.

Table 1. PDBsum analysis of the interaction between the SyInrS protein chains.

Conformation	Interacting Monomers	Interface Residues	Interface Surface (Å^2)	Non-Bonded Contacts	Hydrogen Bonds	Salt Bridges
Apo	α-β (α'-β')	22	1490	87	8	6
	α-β' (α'-β)	17	952	55	5	2
Holo	α-β (α'-β')	30	1950	133	2	4
	α-β' (α'-β)	13	670	78	0	2

2.2. Modelling of Ni(II) Bound Forms of SyInrS

The inclusion of four Ni(II) ions (one per monomer) was performed through the application of a well-established protocol, already successfully used on another member of the RcnR/CsoR family: RcnR from *Escherichia coli* [29], and on UreG from *Helicobacter pylori*, a protein involved in the activation of the nickel-dependent enzyme urease [30]. In principle, each Ni(II) ion was constrained to the residues identified in previous studies [16] and by using the interatomic distances derived from the analysis of EXAFS data [18]. Additional angle and torsional restraints were added in order to ensure the correct orientation of the interacting residue with the Ni(II) ion (Table 2).

Table 2. Distances, angles, and dihedral constraints used in the modelling of Ni(II)-bound SyInrS. All constraints in the form mean ± 1 standard deviation.

	Constrained Atoms	Distance (Å)
	Ni(II)–His21(Nδ/ε)	1.9 ± 0.05
	Ni(II)–His78(Nδ/ε)	1.9 ± 0.05
	Ni(II)–Cys53(Sγ)	2.2 ± 0.05
	Ni(II)–His82(Sγ)	2.2 ± 0.05
Bonded Atoms	**Constrained Atoms**	**Angle (Degrees)**
Ni(II)–Cys(Sγ)	Ni(II)–Cys(Sγ)–Cys(Cβ)	109 ± 5
Ni(II)–His(Nδ)	Ni(II)–His(Nδ)–His(Cγ)	120 ± 10
	Ni(II)–His(Nδ)–His(Cε)	120 ± 10
Ni(II)–His(Nε)	Ni(II)–His(Nε)–His(Cδ)	120 ± 10
	Ni(II)–His(Nε)–His(Cε)	120 ± 10
Bonded Atoms	**Constrained Atoms**	**Dihedral (Degrees)**
Ni(II)–His(Nδ)	Ni(II)–His(Nδ)–His(Cε)–His(Nε)	180 ± 10
	Ni(II)–His(Nδ)–His(Cγ)–His(Cδ)	180 ± 10
Ni(II)–His(Nε)	Ni(II)–His(Nε)–His(Cε)–His(Nδ)	180 ± 10
	Ni(II)–His(Nε)–His(Cδ)–His(Cγ)	180 ± 10

Considering that two of the four involved residues are histidines (His21 and His78) and that it is not possible to evince from the fitting of the EXAFS data which imidazole N atom interacts with nickel, four modelling calculations were carried out by starting from each protein conformation and involving all the possible combinations: H21(Nε)–H78(Nδ), H21(Nδ)–H78(Nε), H21(Nδ)–H78(Nδ), and H21(Nε)–H78(Nε). Most importantly, no restraints were added to constrain the coordination geometry around the metal ions. This was done in order to verify the possibility of the two possible conformations to bind nickel ions without further conformational rearrangements.

Tables 3 and 4 report the analysis conducted on the obtained models. First, the coordination geometry of the four Ni(II) ions was inspected by using the tool Metal Geometry included in the software UCSF Chimera [31] (Table 3). In particular, the root mean square deviation (RMSD) of the ligation set and metal ion with respect to their idealized positions was analysed when four ligands are considered. In the case of the metal bound models generated starting from apo-SyInrS, none of the resulting metal binding sites were in the experimentally determined square planar coordination geometry (see Figure S4). However, a distorted tetrahedral geometry was found when four ligands were considered in the analysis. Moreover, in order to satisfy the constraints included for the modelling of the metal site, in two cases the imidazole ring of His78 resulted in a highly distorted geometry. Also, in the case of the H21(Nδ)–H78(Nε) and H21(Nδ)–H78(Nδ) models generated starting from holo-SyInrS, the resulting coordination geometry was tetrahedral with different degrees of distortion (Figure S5). Of the two remaining models generated starting from holo-SyInrS, the most probable four-ligands coordination geometry was square planar and the distortion of the metal binding site from the ideal geometry was nearly identical [0.29 and 0.27 Å for H21(Nε)–H78(Nδ) and H21(Nε)–H78(Nε), respectively] (Table 3 and Figure S5).

Table 3. Analysis of the metal binding sites in the *Sy*InrS Ni(II) bound models.

Initial Conformation Model	Apo-*Sy*InrS		Holo-*Sy*InrS	
	Best Coordination Geometry	RMSD (Å)	Best Conformation Geometry	RMSD (Å)
H21(Nε)–H78(Nδ)	Tetrahedral	0.54	Square planar	0.29
H21(Nδ)–H78(Nε)	Tetrahedral	0.23	Tetrahedral	0.53
H21(Nδ)–H78(Nδ)	Tetrahedral	0.65	Tetrahedral	0.36
H21(Nε)–H78(Nε)	Tetrahedral	0.39	Square planar	0.27

In order to possibly discriminate between the two models resulting in a square planar metal binding site, the model structures were inspected by using the software PROCHECK [24] (Table 4). The parameters selected for this analysis were the number of residues localized in the four regions of the Ramachandran plot and the G-factor, a measure of how "normal" the stereochemical properties of the residues composing the protein were. The standards of "normality" have been derived from an analysis of 163 non-homologous, high-resolution protein chains chosen from structures solved by X-ray crystallography to a resolution of 2.0 Å or better [24]. A low or a negative value of the G-factor indicates that the protein chain has some residues in a low-probability conformation. From this analysis, the H21(Nε)–H78(Nδ) model appears to have the best structural parameters.

Table 4. Structural analysis of the models of *Sy*InrS binding nickel in square planar coordination geometry (see Table 3, see Tables S2 and S3 for the same analysis on all the models).

Ramachandran Plot Region	Holo-*Sy*InrS Ni(II) Bound Model	
	H21(Nδ)–H78(Nε)	H21(Nε)–H78(Nε)
Most favoured	94.2%	90.7%
Additionally allowed	5.8%	9.3%
Generously allowed	-	-
Disallowed	-	-
G-factor	0.09	0.00

Taken together, the analysis of the nickel coordination geometry and of the structural parameters suggests that the best model is that obtained starting from holo-*Sy*InrS and including H21(Nε) and H78(Nδ) as nickel ligands (Figure 5). This results suggest that, as observed for FrmR [12], a conformational transition is necessary to achieve the correct Ni(II)-induced crosslink between two *Sy*IrnS monomers at the interface between the C-terminal of α-helix α1 of one monomer and Cys53 located at the N-terminal of α-helix α3 of the interacting monomer.

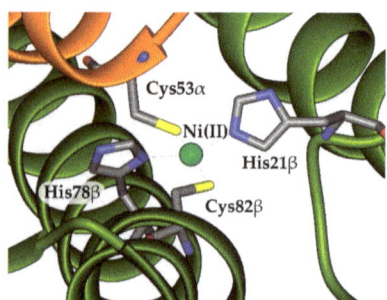

Figure 5. Results for the H21(Nε)–H78(Nδ) modelling of Ni(II) bound *Sy*InrS starting from the protein in the holo conformation. The *Sy*InrS backbone is reported as ribbons coloured by polypeptide chains, with chain α in orange and chain β in green. Putative metal binding residues are reported as sticks coloured according to atom types. The Ni(II) ion is shown as a green sphere.

3. Materials and Methods

The three missing residues (His19, Val20 and His21) at the N-terminal in chain α and α' of apo-SyInrS crystal structure (PDB id: 5FMN [17]) were modelled using Modeller v9.18 [32] and chains β and β' of the same structure as template. The model structure of tetrameric apo-SyInrS in the putative Ni(II)-bound conformation was generated using the same procedure and considering the available crystal structures of Cu(I)-bound CsoR from *Geobacillus thermodenitrificans* NG80-2 and *Mycobacterium tuberculosis* as templates (Figure S1, PDB id: 4M1P [8] and 2HH7 [7], respectively). See Figure 1 for the PROMALS3D [19] multiple sequence alignment used in the latter modelling. In both cases, 100 models were generated [32]. Symmetry restraints were also applied in the calculation in order to obtain four symmetrically identical monomers. The best model was selected on the basis of the lowest value of the DOPE score in Modeller v9.18 [33] and the analysis performed with PROCHECK [24] and QMEAN [25].

The Ni(II) bound form of both models was generated in a subsequent modelling stage by including four Ni(II) ions in the model (one per subunit), in agreement with the metal content analyses carried out in previous studies [6]. The van der Waals parameters for Ni(II) were derived from the Zn(II) parameters included in the CHARMM22 force field [34] implemented in the Modeller v9.18 package by applying a scale factor of 1.12 calculated on the basis of the Ni(II) ionic radius. In all modelling calculations that included Ni(II) ions, constraints were imposed using a Gaussian-shaped energy potential for distances, angles, and dihedrals in order to correctly position the Ni(II) ions with respect to the experimentally identified ligated residues (see Table 2). As in the previous modelling stage, symmetry restraints were applied, and the best model was selected on the basis of the lowest value of the DOPE score. A loop optimization routine was used to refine the regions that showed higher than average energy as calculated using the DOPE score. The molecular surfaces and graphics of SyInrS structures, together with analysis of the Ni(II) sites, were performed using the UCSF Chimera package [31] (see Supplementary Material for additional details). The study of the inter-chain interactions was performed by using PDBsum [27,28].

4. Conclusions

The results obtained in this study indicate that is it possible to successfully model a SyInrS structure able to bind Ni(II) ions in a square planar geometry, and that the structure of apo-SyInrS is not sufficient for such modelling. Thus, a quaternary structure rearrangement is necessary for metal binding upon shifting from the apo- to the holo-form of the protein. This result agrees with what has been proposed for other members of the RcnR/CsoR family [8,9,12,20,22,35]. The cross-link between two monomer chains occurring at the dimer interface appears to stabilize a specific protein conformation, interfering with the capability of SyInrS to bind the DNA. In particular, a rearrangement of surface charges upon metal binding appears to disfavour the DNA binding conformer in the case of CsoR [20,22] and FrmR [12]. It could be envisaged that also nickel sensors belonging to the RcnR/CsoR family are able to adopt an analogue strategy. The mechanism by which the Ni(II) ion interacts with SyInrS, which in principle can occur through an induced fit mechanism, a conformation selection in solution, or a combination of both, is far from being fully understood. To this aim, InrS and the other members of the RcnR/CsoR family can be interesting targets for further computational studies aimed to the exploration of the conformational space of these proteins through molecular dynamics techniques, as recently proposed by some of us [36]. On the other hand, the study of the Ni(II) bound form of SyInrS requires the development of specific force field parameters able to model a Ni(II) ion in the square planar coordination. In recent years the improvement of the computational capabilities and the development of accelerated sampling techniques for molecular simulations [37,38] have made available the tools for the exploration of large conformational changes at the atomistic level [39–41]. In the case of metalloproteins, the use of accelerated sampling techniques requires the use of ad hoc designed force field parameters for the metal ions, able to reproduce structural, thermodynamic, and kinetic observables [42].

Finally, the modelling of the apo-SyInrS/DNA complex and the characterization of the sensor-nucleic acid interface can also be useful for the development of new molecules able to interfere with the formation of the protein/DNA complex. This requires the identification of the *nrsD* operator region and can be done for instance through the use of hydroxyl radical foot-printing essays and opportune knowledge-based docking protocols [43–45]. Some of us already applied this technique in the case of two other metal-dependent transcriptional regulators from the human pathogen *H. pylori*: Fur, a sensor able to repress the transcription of both Fe(II)-repressible and Fe(II)-inducible promoters [46], and NikR, a Ni(II)-responsive sensor repressing or activating genes that code for Ni(II) enzymes and for proteins involved in nickel homeostasis [47].

Supplementary Materials: The following are available online at http://www.mdpi.com/2304-6740/7/6/76/s1. Figure S1: RcnR/CsoR family solved structures. Figure S2: Ribbon diagram and inertia ellipsoid of SyInrS in the apo and holo conformations. Figure S3: Detail of the α-helices α1 and α2 rotation occurring during the apo to holo conformational transition. Figures S4 and S5: Results for the modelling of Ni(II) bound SyInrS starting from the protein in the apo and holo conformation, respectively. Table S1: CsoR/RcnR family representative sequences. Tables S2 and S3: Structural analysis of the models of Ni(II) bound SyInrS generated starting from the protein in the apo and holo conformation, respectively.

Author Contributions: Conceptualization, F.M.; methodology, F.M.; investigation, E.B.; formal analysis, E.B. and F.M.; writing—original draft preparation, F.M.; data curation, F.M.; resources, F.M.; writing—review and editing, F.M.; visualization, E.B. and F.M.; supervision, F.M.; project administration, F.M.

Funding: This research received no external funding.

Acknowledgments: The Department of Pharmacy and Biotechnology of the University of Bologna is acknowledged for financial support.

Conflicts of Interest: The authors declare no conflict of interest.

Abbreviations

The following abbreviations are used in this manuscript:

NTA: Nitrilotriacetic acid (2,2',2''-Nitrilotriacetic acid)
EGTA: Ethylene glycol-bis(2-aminoethylether)-N,N,N',N'-tetraacetic acid
EDTA: Ethylenediaminetetraacetic acid (2,2',2'',2'''-(Ethane-1,2-diyldinitrilo)tetraacetic acid)

References

1. Zamble, D.; Rowińska-Żyrek, M.; Kozlowski, H. *The Biological Chemistry of Nickel*; The Royal Society of Chemistry: London, UK, 2017; Volume 10, pp. 1–380.
2. Organization, W.H. Global Priority List of Antibiotic-Resistant Bacteria to Guide Research, Discovery, and Development of New Antibiotics. Available online: https://www.who.int/medicines/publications/global-priority-list-antibiotic-resistant-bacteria/en/ (accessed on 27 January 2017).
3. Huertas, M.J.; Lopez-Maury, L.; Giner-Lamia, J.; Sanchez-Riego, A.M.; Florencio, F.J. Metals in cyanobacteria: Analysis of the copper, nickel, cobalt and arsenic homeostasis mechanisms. *Life (Basel)* **2014**, *4*, 865–886. [CrossRef] [PubMed]
4. Foster, A.W.; Patterson, C.J.; Pernil, R.; Hess, C.R.; Robinson, N.J. Cytosolic Ni(II) sensor in cyanobacterium: Nickel detection follows nickel affinity across four families of metal sensors. *J. Biol. Chem.* **2012**, *287*, 12142–12151. [CrossRef] [PubMed]
5. Zambelli, B.; Musiani, F.; Ciurli, S. Metal ion-mediated DNA-protein interactions. *Met. Ions Life Sci.* **2012**, *10*, 135–170. [PubMed]
6. Musiani, F.; Zambelli, B.; Bazzani, M.; Mazzei, L.; Ciurli, S. Nickel-responsive transcriptional regulators. *Metallom. Integr. Biometal Sci.* **2015**, *7*, 1305–1318. [CrossRef] [PubMed]
7. Liu, T.; Ramesh, A.; Ma, Z.; Ward, S.K.; Zhang, L.; George, G.N.; Talaat, A.M.; Sacchettini, J.C.; Giedroc, D.P. CsoR is a novel *Mycobacterium tuberculosis* copper-sensing transcriptional regulator. *Nat. Chem. Biol.* **2007**, *3*, 60–68. [CrossRef]

8. Chang, F.M.; Coyne, H.J.; Cubillas, C.; Vinuesa, P.; Fang, X.; Ma, Z.; Ma, D.; Helmann, J.D.; Garcia-de los Santos, A.; Wang, Y.X.; et al. Cu(I)-mediated allosteric switching in a copper-sensing operon repressor (CsoR). *J. Biol. Chem.* **2014**, *289*, 19204–19217. [CrossRef]
9. Higgins, K.A.; Giedroc, D. Insights into protein allostery in the CsoR/RcnR family of transcriptional repressors. *Chem. Lett.* **2014**, *43*, 20–25. [CrossRef]
10. Capdevila, D.A.; Edmonds, K.A.; Giedroc, D.P. Metallochaperones and metalloregulation in bacteria. *Essays Biochem.* **2017**, *61*, 177–200. [CrossRef]
11. Herring, C.D.; Blattner, F.R. Global transcriptional effects of a suppressor tRNA and the inactivation of the regulator *frmR*. *J. Bacteriol.* **2004**, *186*, 6714–6720. [CrossRef]
12. Denby, K.J.; Iwig, J.; Bisson, C.; Westwood, J.; Rolfe, M.D.; Sedelnikova, S.E.; Higgins, K.; Maroney, M.J.; Baker, P.J.; Chivers, P.T.; et al. The mechanism of a formaldehyde-sensing transcriptional regulator. *Sci. Rep.* **2016**, *6*, 38879. [CrossRef]
13. Grossoehme, N.; Kehl-Fie, T.E.; Ma, Z.; Adams, K.W.; Cowart, D.M.; Scott, R.A.; Skaar, E.P.; Giedroc, D.P. Control of copper resistance and inorganic sulfur metabolism by paralogous regulators in *Staphylococcus aureus*. *J. Biol. Chem.* **2011**, *286*, 13522–13531. [CrossRef]
14. Luebke, J.L.; Shen, J.; Bruce, K.E.; Kehl-Fie, T.E.; Peng, H.; Skaar, E.P.; Giedroc, D.P. The CsoR-like sulfurtransferase repressor (CstR) is a persulfide sensor in *Staphylococcus aureus*. *Mol. Microbiol.* **2014**, *94*, 1343–1360. [CrossRef] [PubMed]
15. Chivers, P.T. Nickel Regulation. In *The Biological Chemistry of Nickel*; The Royal Society of Chemistry: London, UK, 2017; pp. 259–283.
16. Foster, A.W.; Pernil, R.; Patterson, C.J.; Robinson, N.J. Metal specificity of cyanobacterial nickel-responsive repressor InrS: Cells maintain zinc and copper below the detection threshold for InrS. *Mol. Microbiol.* **2014**, *92*, 797–812. [CrossRef] [PubMed]
17. Foster, A.W.; Pernil, R.; Patterson, C.J.; Scott, A.J.P.; Pålsson, L.-O.; Pal, R.; Cummins, I.; Chivers, P.T.; Pohl, E.; Robinson, N.J. A tight tunable range for Ni(II) sensing and buffering in cells. *Nat. Chem. Biol.* **2017**, *13*, 409. [CrossRef] [PubMed]
18. Carr, C.E.; Foster, A.W.; Maroney, M.J. An XAS investigation of the nickel site structure in the transcriptional regulator InrS. *J. Inorg. Biochem.* **2017**, *177*, 352–358. [CrossRef] [PubMed]
19. Pei, J.; Kim, B.H.; Grishin, N.V. PROMALS3D: A tool for multiple protein sequence and structure alignments. *Nucl. Acids Res.* **2008**, *36*, 2295–2300. [CrossRef]
20. Tan, B.G.; Vijgenboom, E.; Worrall, J.A. Conformational and thermodynamic hallmarks of DNA operator site specificity in the copper sensitive operon repressor from *Streptomyces lividans*. *Nucl. Acids Res.* **2014**, *42*, 1326–1340. [CrossRef]
21. Chang, F.M.; Lauber, M.A.; Running, W.E.; Reilly, J.P.; Giedroc, D.P. Ratiometric pulse-chase amidination mass spectrometry as a probe of biomolecular complex formation. *Anal. Chem.* **2011**, *83*, 9092–9099. [CrossRef]
22. Porto, T.V.; Hough, M.A.; Worrall, J.A. Structural insights into conformational switching in the copper metalloregulator CsoR from *Streptomyces lividans*. *Acta Crystallogr. D* **2015**, *71*, 1872–1878. [CrossRef]
23. Chang, F.M.; Martin, J.E.; Giedroc, D.P. Electrostatic occlusion and quaternary structural ion pairing are key determinants of Cu(I)-mediated allostery in the copper-sensing operon repressor (CsoR). *Biochemistry* **2015**, *54*, 2463–2472. [CrossRef]
24. Laskowski, R.A.; MacArthur, M.W.; Moss, D.S.; Thornton, J.M. PROCHECK: A program to check the stereochemical quality of protein structures. *J. Appl. Crystallogr.* **1993**, *26*, 283–291. [CrossRef]
25. Benkert, P.; Biasini, M.; Schwede, T. Toward the estimation of the absolute quality of individual protein structure models. *Bioinformatics* **2011**, *27*, 343–350. [CrossRef] [PubMed]
26. Baker, N.A.; Sept, D.; Joseph, S.; Holst, M.J.; McCammon, J.A. Electrostatics of nanosystems: application to microtubules and the ribosome. *Proc. Natl. Acad. Sci. USA* **2001**, *98*, 10037–10041. [CrossRef] [PubMed]
27. Laskowski, R.A.; Hutchinson, E.G.; Michie, A.D.; Wallace, A.C.; Jones, M.L.; Thornton, J.M. PDBsum: A web-based database of summaries and analyses of all PDB structures. *Trends. Biochem. Sci.* **1997**, *22*, 488–490. [CrossRef]
28. Laskowski, R.A.; Jablonska, J.; Pravda, L.; Varekova, R.S.; Thornton, J.M. PDBsum: Structural summaries of PDB entries. *Protein Sci.* **2018**, *27*, 129–134. [CrossRef] [PubMed]

29. Carr, C.E.; Musiani, F.; Huang, H.T.; Chivers, P.T.; Ciurli, S.; Maroney, M.J. Glutamate ligation in the Ni(II)- and Co(II)-responsive *Escherichia coli* transcriptional regulator, RcnR. *Inorg. Chem.* **2017**, *56*, 6459–6476. [CrossRef]
30. Martin-Diaconescu, V.; Bellucci, M.; Musiani, F.; Ciurli, S.; Maroney, M.J. Unraveling the *Helicobacter pylori* UreG zinc binding site using X-ray absorption spectroscopy (XAS) and structural modeling. *J. Biol. Inorg. Chem.* **2012**, *17*, 353–361. [CrossRef]
31. Pettersen, E.F.; Goddard, T.D.; Huang, C.C.; Couch, G.S.; Greenblatt, D.M.; Meng, E.C.; Ferrin, T.E. UCSF Chimera—A visualization system for exploratory research and analysis. *J. Comput. Chem.* **2004**, *25*, 1605–1612. [CrossRef]
32. Marti-Renom, M.A.; Stuart, A.C.; Fiser, A.; Sanchez, R.; Melo, F.; Sali, A. Comparative protein structure modeling of genes and genomes. *Annu. Rev. Biophys. Biomol. Struct.* **2000**, *29*, 291–325. [CrossRef]
33. Shen, M.Y.; Sali, A. Statistical potential for assessment and prediction of protein structures. *Protein Sci.* **2006**, *15*, 2507–2524. [CrossRef]
34. MacKerell, A.D.; Bashford, D.; Bellott, M.; Dunbrack, R.L.; Evanseck, J.D.; Field, M.J.; Fischer, S.; Gao, J.; Guo, H.; Ha, S.; et al. All-atom empirical potential for molecular modeling and dynamics studies of proteins. *J. Phys. Chem. B* **1998**, *102*, 3586–3616. [CrossRef] [PubMed]
35. Huang, H.T.; Bobst, C.E.; Iwig, J.S.; Chivers, P.T.; Kaltashov, I.A.; Maroney, M.J. Co(II) and Ni(II) binding of the *Escherichia coli* transcriptional repressor RcnR orders its N terminus, alters helix dynamics, and reduces DNA affinity. *J. Biol. Chem.* **2018**, *293*, 324–332. [CrossRef] [PubMed]
36. Sala, D.; Musiani, F.; Rosato, A. Application of molecular dynamics to the investigation of metalloproteins involved in metal homeostasis. *Eur. J. Inorg. Chem.* **2018**, *2018*, 4661–4677. [CrossRef]
37. Bernardi, R.C.; Melo, M.C.R.; Schulten, K. Enhanced sampling techniques in molecular dynamics simulations of biological systems. *Biochim. Biophys. Acta* **2015**, *1850*, 872–877. [CrossRef] [PubMed]
38. Mori, T.; Miyashita, N.; Im, W.; Feig, M.; Sugita, Y. Molecular dynamics simulations of biological membranes and membrane proteins using enhanced conformational sampling algorithms. *Biochim. Biophys. Acta* **2016**, *1858*, 1635–1651. [CrossRef] [PubMed]
39. Orozco, M. A theoretical view of protein dynamics. *Chem. Soc. Rev.* **2014**, *43*, 5051–5066. [CrossRef] [PubMed]
40. Yang, L.Q.; Sang, P.; Tao, Y.; Fu, Y.X.; Zhang, K.Q.; Xie, Y.H.; Liu, S.Q. Protein dynamics and motions in relation to their functions: Several case studies and the underlying mechanisms. *J. Biomol. Struct. Dyn.* **2014**, *32*, 372–393. [CrossRef] [PubMed]
41. Chang, C.A.; Huang, Y.M.; Mueller, L.J.; You, W. Investigation of structural dynamics of enzymes and protonation states of substrates using computational tools. *Catalysts* **2016**, *6*, 82. [CrossRef] [PubMed]
42. Masetti, M.; Musiani, F.; Bernetti, M.; Falchi, F.; Cavalli, A.; Ciurli, S.; Recanatini, M. Development of a multisite model for Ni(II) ion in solution from thermodynamic and kinetic data. *J. Comput. Chem.* **2017**, *38*, 1834–1843. [CrossRef]
43. Van Dijk, M.; van Dijk, A.D.; Hsu, V.; Boelens, R.; Bonvin, A.M. Information-driven protein-DNA docking using HADDOCK: It is a matter of flexibility. *Nucl. Acids Res.* **2006**, *34*, 3317–3325. [CrossRef]
44. Van Dijk, M.; Bonvin, A.M. Pushing the limits of what is achievable in protein-DNA docking: Benchmarking HADDOCK's performance. *Nucl. Acids Res.* **2010**, *38*, 5634–6547. [CrossRef] [PubMed]
45. Musiani, F.; Ciurli, S. Evolution of macromolecular docking techniques: The case study of nickel and iron metabolism in pathogenic bacteria. *Molecules* **2015**, *20*, 14265–14292. [CrossRef] [PubMed]
46. Agriesti, F.; Roncarati, D.; Musiani, F.; Del Campo, C.; Iurlaro, M.; Sparla, F.; Ciurli, S.; Danielli, A.; Scarlato, V. FeON-FeOFF: The *Helicobacter pylori* Fur regulator commutates iron-responsive transcription by discriminative readout of opposed DNA grooves. *Nucl. Acids Res.* **2014**, *42*, 3138–3151. [CrossRef] [PubMed]
47. Mazzei, L.; Dobrovolska, O.; Musiani, F.; Zambelli, B.; Ciurli, S. On the interaction of *Helicobacter pylori* NikR, a Ni(II)-responsive transcription factor, with the urease operator: in solution and in silico studies. *J. Biol. Inorg. Chem.* **2015**, *20*, 1021–1037. [CrossRef] [PubMed]

© 2019 by the authors. Licensee MDPI, Basel, Switzerland. This article is an open access article distributed under the terms and conditions of the Creative Commons Attribution (CC BY) license (http://creativecommons.org/licenses/by/4.0/).

Communication

Nickel-Induced Oligomerization of the Histidine-Rich Metallochaperone CooJ from *Rhodospirillum Rubrum*

Marila Alfano, Julien Pérard and Christine Cavazza *

Univ. Grenoble Alpes, CEA, CNRS, CBM, F-38000 Grenoble, France
* Correspondence: christine.cavazza@cea.fr

Received: 27 May 2019; Accepted: 25 June 2019; Published: 1 July 2019

Abstract: [NiFe]-carbon monoxide dehydrogenase reversibly catalyzes the oxidation of CO to CO_2. Its active site is a unique $NiFe_4S_4$ cluster, known as C-cluster. In *Rhodospirillum rubrum*, three nickel-dependent proteins, CooC, CooT and CooJ are required for Ni insertion into the active site. Among them, CooJ is a histidine-rich protein, containing two distinct and spatially separated Ni(II)-binding sites: a strictly conserved N-terminal site and a variable histidine tail at the C-terminus. Here, using biophysical techniques, we study the behavior of the protein upon Ni(II) addition. Using circular dichroism and chemical denaturation, we show that the binding of Ni(II) to the protein increases its stability. Moreover, high-order oligomers are formed through nickel–histidine tail interactions, both in vitro and in cellulo, via a dynamical and reversible process.

Keywords: histidine-rich protein; carbon monoxide dehydrogenase; nickel chaperone; nickel-induced oligomerization

1. Introduction

Monofunctional nickel-dependent carbon monoxide dehydrogenases (CODH) reversibly catalyzes the oxidation of CO to CO_2 [1]. CODH's active site, called C-cluster, is a $NiFe_4S_4$ cluster, unique in biology, whose atypical nature was revealed by X-ray crystallography [2,3]. In the hydrogenogenic carboxydotroph *Rhodospirillum rubrum*, CODH plays an essential role in the energy metabolism when CO is the sole energy source [4]. In this bacterium, the structural gene of CODH, called *cooS*, is found in the *cooFSCTJ* operon [5]. Three of the proteins encoded by the *coo* operon, namely CooC, CooT and CooJ, are nickel chaperones dedicated to nickel insertion into CODH, while CooF is a ferredoxin [6]. In vivo studies have shown that the ATPase CooC and the Ni(II)-binding proteins CooT and CooJ play a significant role in the maturation pathway leading to a fully active enzyme. Moreover, Ni insertion is a key step in the enzyme activation process [5,7]. Among these chaperones, CooJ from *R. rubrum* (*Rr*CooJ) is a histidine-rich protein, with a histidine tail comprising 16 histidines and 2 cysteines at its C-terminus [8]. Initially thought to be only present in *R. rubrum*, our recent study revealed the existence of at least 46 CooJ homologues in bacteria [9]. In solution, *Rr*CooJ forms a 25-kDa homodimer with a central coiled coil and two independent C-terminal his-tails and possesses several Ni(II) binding sites. The first one is present in the N-terminal region and binds one Ni(II) per dimer via a "H-X3-H-X3-H" motif, strictly conserved in the CooJ family. In addition, the two Histidine-rich regions bind at least 2 Ni(II) ions each (Figure 1), via both histidine and cysteine residues, as shown by site-directed mutagenesis of cysteines 109 and 111 that led to decreased stoichiometry [9].

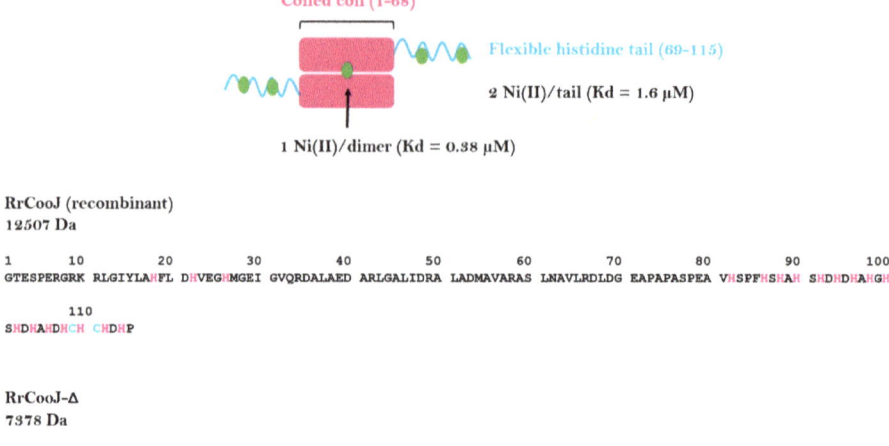

Figure 1. Schematic representation of *RrCooJ* and its amino acid sequence. Ni(II) ions are depicted as green spheres. Histidine residues are depicted in magenta and cysteines 109 and 111 are in blue.

Histidine-rich proteins, such as histatin 5 or aß peptides [10], are predicted to be intrinsically disordered with a tendency to form oligomers driven either by self-association or through interaction with multivalent ions. In nickel chaperones, the presence of histidine-rich regions is often proposed to be related to nickel storage, and/or detoxification, due to their ability to quickly bind and release nickel ions. Among them, one remarkable example is Hpn from *Helicobacter pylori*, a small protein of 7 kDa composed of histidines for about half of the total amino-acid sequence, which is expected to form high-order oligomers, with 20-mers as predominant species, even in the absence of metal [11].

Here, we report the behavior of *RrCooJ* towards nickel in vitro and in cellulo, using a combination of biophysical techniques. *RrCooJ* exists as an equilibrium of oligomeric states in solution in the presence of Ni(II) ions, while the apo-protein forms a stable dimer. We reveal that the His-rich region is responsible for the oligomeric process.

2. Results

Recently, we showed that the two histidine tails present in *RrCooJ* are highly flexible and predicted to be partly disordered [9]. Moreover, Ni addition to the apo-dimer led to its conformational change, as observed by size-exclusion chromatography coupled to multi-angle laser light scattering (SEC-MALLS) experiments. Here we used circular dichroism (CD) in the far UV region (190–250 nm) and chemical denaturation of *RrCooJ*, either in its apo- or holo-form (Ni-*RrCooJ*), to get more information about the impact of Ni(II) ions on *RrCooJ* conformation. The CD spectrum of apo-*RrCooJ* is characterized by a positive peak at 193 nm and two negative peaks at 220 nm and 208 nm, attributed to the presence of a high α-helix content in its secondary structures. Upon the addition of Ni(II) ions to the apo-protein solution, the negative peaks increased gradually, as an index of a conformational change (Figure 2a). Quantitative analysis, performed using the web server BeStSel [12], showed that the increase in ellipticity was correlated to a slight increase in the helical content (from 83% for apo-*RrCooJ* to 85% for Ni-*RrCooJ*) and a decrease in the turns content (from 6% for apo-*RrCooJ* to 0.4% for Ni-*RrCooJ*). Interestingly, the CD signal at 208 nm saturated when the Ni(II) concentration reached 20 µM, corresponding to 5 molar eq. of Ni(II) per dimer, the same stoichiometry as the one previously determined by ITC (Figure 2b). Considering the predominance of α-helices in the protein, the $\Theta_{222/208}$ ratio, which increases if the protein gains in stability [13], can be used to get information about *RrCooJ*'s behavior towards nickel. For a Ni(II) concentration greater than 12 µM (corresponding to 3 molar eq. of Ni(II)), the ratio is >1, reflecting an increase in the α-helix content and likely the stabilization of the protein. To further investigate the difference in stability between the apo- and the holo-forms, chemical denaturation using guanidine hydrochloride (GuCl) was performed. Their stability was compared by monitoring the variation of the 222 nm CD ellipticity signal after the addition of increasing concentrations of GuCl. Interestingly, at 1.2 M of GuCl, Ni-RrCooJ was still perfectly folded, whereas apo-RrCooJ started to denaturate, as shown by an ellipticity fractional change of 0.09 vs. 0.4, respectively. The midpoint GuCl concentration was determined to be 1.6 M for apo-RrCooJ and 2.0 M for Ni-RrCooJ. This result reveals a higher stability induced by Ni(II)-binding to the protein (Figure 2c).

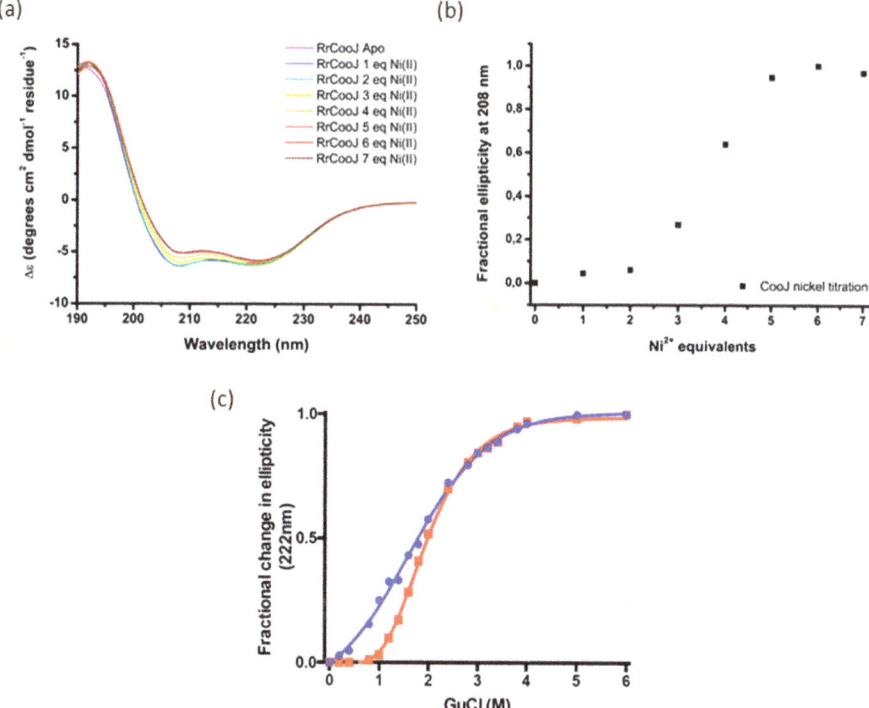

Figure 2. Far-UV circular dichroism (CD) and chemical denaturation of RrCooJ at 4 µM of dimer upon Ni(II) addition. (**a**) Evolution of far-UV signal upon the addition of increasing Ni(II) concentrations. (**b**) Fractional values of the ellipticity registered at 208 nm plotted versus Ni(II) molar equivalents. (**c**) Chemical denaturation of apo-RrCooJ (blue) and Ni-RrCooJ (red) using different concentrations of GuCl at room temperature.

SEC-MALLS experiments were conducted to investigate the oligomeric state of RrCooJ in the presence of Ni(II). We previously showed that a truncated version of RrCooJ (RrCooJ-Δ), lacking the histidine tail (Figure 1), is a stable dimer even in the presence of an excess of nickel [9]. Here, the column was equilibrated with a buffer supplemented with 10 µM of NiSO$_4$, corresponding to a concentration about six times higher than the affinity of the His tails for Ni(II) (Kd = 1.6 µM). The injection of 50 µM of RrCooJ dimer pre-incubated with 5 molar eq. of Ni(II) led to the appearance of a mixture of oligomeric forms ranging from dimers to high-order oligomers (Figure 3a).

To characterize these high-order oligomers, single particle analysis was performed using Transmission Electron Microscopy (TEM). In order to separate the different oligomeric states, size exclusion chromatography was used. Apo-RrCooJ dimer pre-incubated with 5 molar eq. of Ni(II) was injected onto the column and a series of fractions was selected, based on the SEC-MALLS results. Four samples were collected at different elution volumes (9, 10.5, 13 and 14.5 mL) and subsequently analyzed, revealing the existence of high-order oligomers with sizes ranging from 11 ± 2 nm to 28 ± 5 nm in the three first peaks. In the latter one, corresponding to the elution of RrCooJ tetramers as shown by SEC-MALLS, high-order oligomers could not be detected as expected. Neither aggregation nor fibrils were observed under any of the tested conditions (Figure 3b). The size of high-order oligomers decreased along the elution with an optimum distribution of particles at an elution peak of 13 mL corresponding to a homogenous particle size of about 11 nm.

The oligomerization process is both protein and Ni(II) concentration dependent. Indeed, for a protein concentration of 50 µM of dimer, RrCooJ started to form insoluble aggregates when the Ni(II) concentration reached 500 µM. Moreover, the incubation of 100 µM RrCooJ dimer with 10 molar eq. of NiSO$_4$ (1.0 mM) led to its complete precipitation. However, the addition of 10 mM EDTA can fully redissolve the precipitate within seconds (as shown by protein assay) revealing the easy reversibility of the metal-dependent oligomerization process.

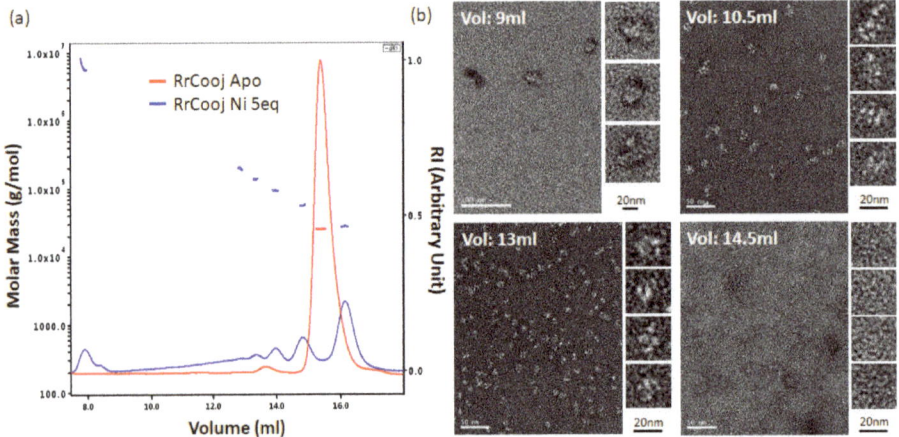

Figure 3. Nickel-dependent oligomerization of RrCooJ by Size Exclusion Chromatography and Multi Angle Laser Light Scattering (SEC-MALLS) and TEM. (**a**) 50 µM of apo-RrCooJ (red) were injected onto a column equilibrated with buffer A (96% of dimer); The injection of 50 µM of RrCooJ dimer preincubated with 250 µM of NiSO$_4$ (blue) in buffer A supplement with 10 µM of NiSO$_4$ shows multiple peaks corresponding to a mixture of oligomers: dimer (35%), tetramer (16%), hexamer (10%), octamer (6.2%) and high-order oligomers (32.8%). (**b**) Nickel-dependent oligomerization analysis of RrCooJ by negative stain after size exclusion chromatography. Image of negatively stained particles (23,000× and 30,000× of magnification) of high-order oligomers of Ni-RrCooJ with highlights of particles measured for the following elution volumes: 9 mL (28 ± 3 nm), 10.5 mL (15 ± 2 nm), 13 mL (11 ± 2 nm) and 14.5 mL (below 5 nm). Oligomer size was determined after merging 20 independents particles by "ImageJ v 1.51" software. Vol: Elution volume.

In order to study their in-cell Ni(II)-binding capability and behavior, RrCooJ and its truncated version RrCooJ-Δ were overproduced in *E. coli* in growth medium supplemented with a range of Ni(II) concentrations from 0 to 2.5 mM NiSO$_4$. As a control, the impact of nickel was evaluated on a strain of *E. coli* transformed with an empty vector. The growth of the control culture was strongly affected by nickel, decreasing remarkably between 0.5 mM and 1.5 mM, and stopping for concentrations higher than 1.5 mM. The presence of RrCooJ or RrCooJ-Δ increases the nickel resistance of *E. coli*, resulting in a constant growth even at high concentration of metal (Figure 4a). Although RrCooJ and RrCooJ-Δ are both able to bind Ni(II) in-cell, we observed that RrCooJ tends to form insoluble oligomers upon addition of increasing concentrations of Ni(II) in the growth medium. At 1.0 mM of NiSO$_4$, despite a bacterial growth rate similar to the one in the absence of nickel, the overproduced *Rr*CooJ was mainly found in the insoluble fraction, as shown by SDS gel (Figure 4b). This behavior was not observed with the overproduced *Rr*CooJ-Δ, mainly found in the soluble fractions independently on the nickel concentration (Figure 4b). This is a strong indication that the His-tail triggers the oligomerization process of *Rr*CooJ in the presence of nickel, in agreement with the in vitro studies, without drastically affecting the bacterial growth.

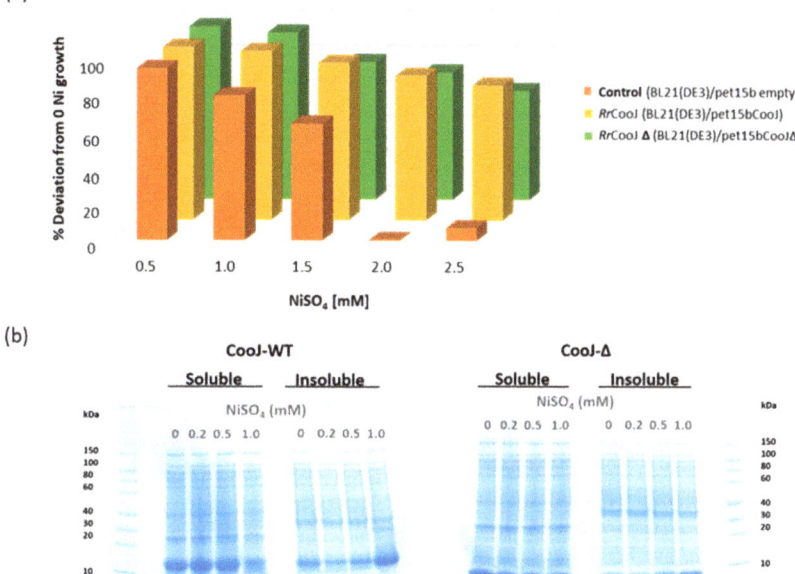

Figure 4. Growth response to nickel content in *E. coli* strains overproducing *Rr*CooJ or *Rr*CooJ-Δ. (a) The resistance towards nickel for the control (orange), *Rr*CooJ (yellow) and *Rr*CooJ-Δ (green) cultures is plotted for each different nickel concentration after 16 h from the induction. The resistance is reported as percentile deviation of the OD_{600} measured at 16 h considering as maximum value the OD_{600^-16h} measured for the cultures at 0 mM NiSO$_4$ versus the ones measured for the cultures at different nickel concentrations. The deviation is calculated using the formula $\sum_{n=1}^{3}\left(\frac{OD_{600}\ at\ x\ mM\ 16\ h - OD_{600}\ before\ induction}{OD_{600}\ at\ 16\ h\ for\ 0\ mM\ Ni}\right) * 100$, using the OD_{600^-16h} measured for each different nickel concentration (x mM = 0.5, 1.0, 1.5, 2.0, 2.5 mM NiSO$_4$). (b) SDS-PAGE of soluble and insoluble extracts of *E. coli* cultures overproducing *Rr*CooJ or *Rr*CooJ-Δ in the presence of Ni(II) in the growth media (0 to 1.0 mM NiSO$_4$).

3. Materials and Methods

Recombinant *Rr*CooJ and *Rr*CooJ-Δ were overproduced in *E. Coli* and purified as previously described [9]. Protein concentration was determined either by SEC-MALLS or by Rose Bengal protein assay to determine the protein concentration for the EDTA solubilization experiment.

3.1. Circular Dichroism (CD) Spectroscopy

CD spectra were recorded using a J-1500 circular dichroism spectrometer (JASCO Analytical Instruments, Easton, PA, USA). A stock solution of 10 mM NiSO$_4$ was used to monitor metal-dependent secondary structure content upon Ni(II) titration (from 0 to 7 molar equivalents of NiSO$_4$). Spectra were recorded from 190 to 250 nm using a 1 mm cuvette, with ten accumulations to increase the signal-to-noise ratio. Proteins were thawed and diluted to 4 µM dimer in CD buffer (5 mM potassium phosphate pH

7.0, containing 0.5 mM TCEP). For the determination of the secondary structure elements, the spectra were analyzed using the BeStSel online software [12]. The guanidine hydrochloride denaturation experiments were performed via additions of different concentrations of GuCl on 4 µM dimer samples (apo- and Ni-RrCooJ), followed by 2 h of incubation. The Ni-RrCooJ sample was prepared by adding 5 molar equivalents of Ni(II) per protein dimer (20 µM final concentration) and incubated for 30 min prior to its denaturation. The fractional change in ellipticity at 222 nm was used to determine the variation of the secondary structure elements due to the denaturation of the protein. Empiric fit was done via a sigmoidal equation.

3.2. Size Exclusion Chromatography and Multi Angle Laser Light Scattering (SEC-MALLS)

Purified and frozen RrCooJ was thawed and diluted to 50 µM of dimer in 50 mM HEPES pH 7.5, 150 mM NaCl, 1 mM TCEP (buffer A). Ni-RrCooJ was incubated in the presence of 250 µM $NiSO_4$ for 30 min at room temperature prior to injection onto the SEC-MALLS system (Wyatt Dawn HELEOS-II 18-angle light scattering detector and Wyatt Optilab rEX refractive index monitor linked to a Shimadzu HPLC system comprising a LC-20AD pump, a SPD20A UV/Vis detector, and a Superdex 200 10/300 increase column (GE Healthcare Life Sciences, Pittsburg, KS, USA). Injections were carried out using a 20 µL loop. The size exclusion column was equilibrated using buffer A ± 10 µM $NiSO_4$. Protein concentration in all samples was determined by integration of the differential refractive index (dRI) peak (dn/dc = 0.185). It is important to note that all used samples came from the same stock protein. The data were analyzed using the ASTRA software (version 6) (WYATT Technology Corporation, Santa Barbara, CA, USA).

3.3. Electron Microscopy

100 µL of Ni-RrCooJ dimer at 50 µM were injected onto a Superdex 200 10/300 column (GE Healthcare) equilibrated with buffer A + 10 µM $NiSO_4$. 200 µL-fractions were collected and 4 fractions corresponding to elution volumes of 9, 10.5, 13 and 14.5 mL were selected. The negative stain Mica-carbon Flotation Technique (MFT) was used to prepare the samples. Briefly, 3–5 µL of protein sample (0.1 to 0.01 mg/mL) were applied to a clean side of a carbon layer (between a carbon and a mica layer). The carbon was then floated on stain (2% w/v uranyl acetate (AcU) pH 4.5) and covered by a copper grid. The images were captured under low-dose conditions (<10 e– $Å^{-2}$) at a magnification of 23K× and 30K× with defocus values between 1.2 and 2.5 µm on a Tecnai 12 LaB6 electron microscope at 120 kV accelerating voltage using a CCD Camera Gatan Orius 1000 (Gatan, Inc., Pleasanton, CA, USA).

3.4. Growth Response to Nickel in E. coli Strains Overproducing RrCooJ or RrCooJ-Δ

E. coli BL21 (DE3) strain was transformed with pET15b, pET15b-RrCooJ or pET15b-RrCooJ-Δ. The three different cultures (100 mL each) were grown in LB medium, with the appropriate antibiotic, at 25 °C and 180 rpm to an OD_{600} of ~0.6 and induced with 0.25 mM IPTG. At this point, each culture was divided into five 50 mL-Falcon tubes (10 mL-culture each) and a range of $NiSO_4$ concentrations was added to the culture medium (0, 0.5, 1.0, 1.5, 2.0 and 2.5 mM). The cells were grown overnight at 25 °C and 180 rpm, and the OD_{600} was measured after 16 h for all the cultures. The cultures were done in triplicate for each different condition. The effect of the overproduction of RrCooJ or RrCooJ-Δ according to the amount of Ni(II) was evaluated by normalizing the culture optical densities, measured at the different nickel concentrations, with the values obtained from the control. To study the solubility of RrCooJ in the presence of Ni(II) in-cell, E. coli BL21 (DE3) strain transformed with either pET15b-RrCooJ or pET15b-RrCooJ-Δ were grown as described above, except that 25 mL-cultures were grown in 50 mL-flasks with 0, 0.2, 0.5 and 1.0 mM $NiSO_4$ concentration. After overnight growth, the OD_{600} was measured (5.5 ± 0.2 for BL21(DE3)/pET15b-RrCooJ-Δ cultures and 5.8 ± 0.3 for BL21(DE3)/pET15b-RrCooJ cultures). The cell pellets were collected by centrifugation and then resuspended in 4 mL of 50 mM Tris-HCl, pH 8.0 containing a complete Protease Inhibitor cocktail tablet (Roche Life Science, Indianapolis, IN, USA). After three freeze-thaw cycles, the cultures were sonicated and centrifuged at 14,000 rpm for 20 min. The 16 pellets were resuspended in 50 mM Tris-HCl pH 8.0, containing 4 M urea and 0.1% SDS. Ten microliters of each sample were examined using SDS-PAGE (Bio-Rad Corporate, Hercules, CA, USA).

4. Conclusions

Recently, structural studies of apo-RrCooJ revealed its atypical topology with two independent histidine tails flanking a central coiled coil region. Here, we focused our attention on the histidine-rich region behavior. Two phenomena were observed upon nickel addition. Firstly, Ni(II)-binding directly impacts the secondary structure content of the protein, with a gain in α-helices related to higher stability. Secondly, nickel induces a dynamic and reversible oligomerization process in vitro, leading to the protein precipitation observed both in vitro and in cellulo, in the presence of elevated Ni(II) levels. Further studies are necessary to investigate the functional role of RrCooJ oligomerisation in nickel homeostasis under physiological conditions.

Author Contributions: Conceptualization, C.C. and M.A.; formal analysis, C.C., M.A., J.P.; investigation, C.C., M.A.; writing—original draft preparation, C.C.; writing—review and editing, M.A.

Funding: This work was supported by "the ITERLIS PhD program, CEA Life sciences" for MA's PhD funding, the "FUNBIOCO" project (IDEX-UGA, Initiatives de Recherche stratégiques) and the "COSYNBIO" project (Projets exploratoires, Cellule energie-CNRS). This work has been partially supported by Labex ARCANE and CBH-EUR-GS (ANR-17-EURE-0003). The research leading to these results has received funding from the networking support from the COST Action FeSBioNet (Contract CA15133).

Acknowledgments: This work used the platforms of the Grenoble Instruct Centre (ISBG; UMS 3518 CNRS-CEA-UJF-EMBL) with support from FRISBI (ANR-10-INSB-05-02) and GRAL (ANR-10-LABX-49-01) within the Grenoble Partnership for Structural Biology (PSB). The electron microscope facility is supported by the Rhône-Alpes Region, the Fondation Recherche Medicale (FRM), the fonds FEDER, the Centre National de la Recherche Scientifique (CNRS), the CEA, the University of Grenoble, EMBL, and the GIS-Infrastrutures en Biologie Santé et Agronomie (IBISA). We thank Daphna FENNEL and Dr Guy SCHOEHN, from the Electron Microscopy platform of the Integrated Structural Biology of Grenoble (ISBG, UMI3265).

Conflicts of Interest: The authors declare no conflicts of interest.

References

1. Kung, Y.; Drennan, C.L. A role for nickel–iron cofactors in biological carbon monoxide and carbon dioxide utilization. *Curr. Opin. Chem. Biol.* **2011**, *15*, 276–283. [CrossRef] [PubMed]
2. Dobbek, H.; Svetlitchnyi, V.; Gremer, L.; Huber, R.; Meyer, O. Crystal structure of a carbon monoxide dehydrogenase reveals a [Ni–4Fe–5S] cluster. *Science* **2001**, *293*, 1281–1285. [CrossRef] [PubMed]
3. Drennan, C.L.; Heo, J.; Sintchak, M.D.; Schreiter, E.; Ludden, P.W. Life on carbon monoxide: X-ray structure of Rhodospirillum rubrum Ni–Fe–S carbon monoxide dehydrogenase. *Proc. Natl. Acad. Sci. USA* **2001**, *98*, 11973–11978. [CrossRef] [PubMed]
4. Alfano, M.; Cavazza, C. The biologically mediated water–gas shift reaction: Structure, function and biosynthesis of monofunctional [NiFe]-carbon monoxide dehydrogenases. *Sustain. Energy Fuels* **2018**, *2*, 1653–1670. [CrossRef]
5. Kerby, R.L.; Ludden, P.W. In vivo nickel insertion into the carbon monoxide dehydrogenase of Rhodospirillum rubrum: Molecular and physiological characterization of cooCTJ. *J. Bacteriol.* **1997**, *179*, 2259–2266. [CrossRef] [PubMed]
6. Singer, S.W.; Hirst, M.B.; Ludden, P.W. CO-dependent H2 evolution by Rhodospirillum rubrum: Role of CODH: CooF complex. *Biochim. Biophys. Acta Bioenerg.* **2006**, *1757*, 1582–1591. [CrossRef] [PubMed]
7. Jeon, W.B.; Cheng, J.; Ludden, P.W. Purification and Characterization of Membrane-associated CooC Protein and Its Functional Role in the Insertion of Nickel into Carbon Monoxide Dehydrogenase from Rhodospirillum rubrum. *J. Biol. Chem.* **2001**, *276*, 38602–38609. [CrossRef] [PubMed]
8. Watt, R.K.; Ludden, P.W. The Identification, Purification, and Characterization of CooJ. *J. Biol. Chem.* **1998**, *273*, 10019–10025. [CrossRef] [PubMed]

9. Alfano, M.; Pérard, J.; Carpentier, P.; Basset, C.; Zambelli, B.; Timm, J.; Crouzy, S.; Ciurli, S.; Cavazza, C. The carbon monoxide dehydrogenase accessory protein CooJ is a histidine-rich multidomain dimer containing an unexpected Ni(II)-binding site. *J. Biol. Chem.* **2019**, *294*, 7601–7614. [CrossRef] [PubMed]
10. Iadanza, M.G.; Jackson, M.P.; Hewitt, E.W.; Ranson, N.A.; Radford, S.E. A new era for understanding amyloid structures and disease. *Nat. Rev. Mol. Cell Biol.* **2018**, *19*, 755–773. [CrossRef] [PubMed]
11. Ge, R.; Zhang, Y.; Sun, X.; Watt, R.M.; He, Q.-Y.; Huang, J.-D.; Wilcox, D.E.; Sun, H. Thermodynamic and Kinetic Aspects of Metal Binding to the Histidine-rich Protein, Hpn. *J. Am. Chem. Soc.* **2006**, *128*, 11330–11331. [CrossRef] [PubMed]
12. Micsonai, A.; Wien, F.; Bulyáki, É.; Kun, J.; Moussong, É.; Lee, Y.-H.; Goto, Y.; Réfrégiers, M.; Kardos, J. BeStSel: A web server for accurate protein secondary structure prediction and fold recognition from the circular dichroism spectra. *Nucleic Acids Res.* **2018**, *46*, W315–W322. [CrossRef] [PubMed]
13. Kwok, S.C.; Hodges, R.S. Stabilizing and destabilizing clusters in the hydrophobic core of long two-stranded α-helical coiled-coils. *J. Biol. Chem.* **2004**, *279*, 21576–21588. [CrossRef] [PubMed]

© 2019 by the authors. Licensee MDPI, Basel, Switzerland. This article is an open access article distributed under the terms and conditions of the Creative Commons Attribution (CC BY) license (http://creativecommons.org/licenses/by/4.0/).

MDPI
St. Alban-Anlage 66
4052 Basel
Switzerland
Tel. +41 61 683 77 34
Fax +41 61 302 89 18
www.mdpi.com

Inorganics Editorial Office
E-mail: inorganics@mdpi.com
www.mdpi.com/journal/inorganics

www.ingramcontent.com/pod-product-compliance
Lightning Source LLC
LaVergne TN
LVHW071943080526
838202LV00064B/6662